TECHNOLOGY AND SOCIETY

TECHNOLOGY AND SOCIETY

Advisory Editor
DANIEL J. BOORSTIN, author of
The Americans and Director of
The National Museum of History
and Technology, Smithsonian Institution

THE

MANUFACTURE OF PAPER:

BEING A

DESCRIPTION OF THE VARIOUS PROCESSES FOR THE FABRICATION,
COLORING, AND FINISHING OF EVERY KIND OF PAPER

BY

CHARLES THOMAS DAVIS

ARNO PRESS
A NEW YORK TIMES COMPANY
New York • 1972

Reprint Edition 1972 by Arno Press Inc.

Reprinted from a copy in Princeton University
Library

Technology and Society
ISBN for complete set: 0-405-04680-4
See last pages of this volume for titles.

Manufactured in the United States of America

———————

Library of Congress Cataloging in Publication Data

Davis, Charles Thomas.
    The manufacture of paper.

    (Technology and society)
    Reprint of the 1886 ed.
    1. Paper making and trade. I. Title. II. Series.
TS1105.D26 1972   676'.23   72-5042
ISBN 0-405-04694-4

THE

MANUFACTURE OF PAPER:

BEING A

DESCRIPTION OF THE VARIOUS PROCESSES FOR THE FABRICATION, COLORING, AND FINISHING OF EVERY KIND OF PAPER;

INCLUDING THE

DIFFERENT RAW MATERIALS AND THE METHODS FOR DETERMINING
THEIR VALUES; THE TOOLS, MACHINES, AND PRACTICAL DETAILS
CONNECTED WITH AN INTELLIGENT AND A PROFITABLE
PROSECUTION OF THE ART, WITH SPECIAL REFERENCE
TO THE BEST AMERICAN PRACTICE.

TO WHICH ARE ADDED

A HISTORY OF PAPER, COMPLETE LISTS OF PAPER-MAKING MATERIALS,
LISTS OF AMERICAN MACHINES, TOOLS, AND PROCESSES USED IN
TREATING THE RAW MATERIALS, AND IN MAKING,
COLORING, AND FINISHING PAPER.

BY

CHARLES THOMAS DAVIS,

AUTHOR OF THE "MANUFACTURE OF LEATHER," "A PRACTICAL TREATISE ON THE
MANUFACTURE OF BRICKS, TILES, AND TERRA-COTTA," ETC. ETC.

ILLUSTRATED BY ONE HUNDRED AND EIGHTY ENGRAVINGS.

PHILADELPHIA:
HENRY CAREY BAIRD & CO.,
INDUSTRIAL PUBLISHERS, BOOKSELLERS AND IMPORTERS,
No. 810 WALNUT STREET.

LONDON:
SAMPSON LOW, MARSTON, SEARLE & RIVINGTON,
CROWN BUILDINGS, 188 FLEET STREET.

1886.

COLLINS PRINTING HOUSE,
705 Jayne Street.

PREFACE.

To tell the story of the manufacture of paper and describe concisely the processes employed in the various stages of papermaking, and give an account of all the known substances that have ever been used for the purposes for which we employ paper is by no means an easy task. At the present time, when books, newspapers, and other periodicals are issued to the world in numbers almost beyond calculation, certainly no apology is due for a dissertation on the manufacture of paper, on the score of the lack of importance of that article.

The remarkable advance which has been made in the United States in all classes of manufactures during the past twenty years has excited alike the admiration and the envy of the civilized world, but in no industry has greater progress been made than in the fabrication of paper. The great ingenuity and intelligence of our people, coupled with an abundance of all the raw materials used in paper-making, have aided in developing that industry in this country to proportions entirely unknown in any other country on the globe.

It was the writer's first intention to have devoted the book entirely to a description of the "practical" processes employed in paper-making in various parts of the world; and to this end he collected in the United States, Germany, France, Belgium, and Great Britain a large amount of so-called "practical information." But upon a reconsideration of the subject, and by the advice of some of the leading paper-makers in the United States and in Europe, he decided that the publication of such information would prove of little value to the paper trade, for the reason that paper-makers in the United States have little to learn from the paper manufacturers of Europe. The machinery and processes in use in Great Britain and on the continent of Europe are far behind those employed in the United States, consequently the author determined that the book should be confined almost exclusively to a description of the machines and processes employed in the leading mills of this country. In order that such descriptions should not be too meagre the size of the volume has been materially enlarged over that at first contemplated. If American readers of the present volume complain that the author has described at length processes and machines which are already familiar to them through trade publications or otherwise, they should remember that such inventions are probably not so familiar to those engaged in paper-making in foreign countries whence a considerable proportion of the present edition of this book will go.

On account of the importance of the acid or bisulphite processes for making pulp from wood, etc., the author has given considerable space to a description of such processes, and while he thinks the apparatus employed in working these processes is much too cumbersome and expensive for general use, he could not on that account omit a description of them; nor should he be held accountable for the defects in such processes, the aim in view being to present a description of the state of the art as it exists at the present time.

It would, of course, be impossible, in a book the size of the present one, to describe all the processes and machines which have been invented and used in paper-making in the United States; but the lists of patents which this volume contains give reference to the number and date of nearly all the inventions which have been patented in the United States since the year 1790 to the end of the year 1885, and copies of all patents issued subsequent to the year 1866 can be obtained, upon application to the Commissioner of Patents, at a cost of twenty-five cents each. Patents issued prior to the year 1866 require to be copied in manuscript, and the cost of such copies depends upon the number of words contained in the patent.

CHARLES T. DAVIS.

Washington, D. C.,
1114 Pennsylvania Ave., July 1, 1886.

CONTENTS.

CHAPTER I.

THE HISTORY OF THE MANUFACTURE OF PAPER.

CHAPTER II.

MATERIALS USED FOR PAPER—MICROGRAPHICAL STUDY OF THE MANU-
FACTURE OF PAPER—CELLULOSE — DETERMINATION OF CELLULOSE —
RECOGNITION OF VEGETABLE FIBRES.

CHAPTER III.

COMMERCIAL CLASSIFICATIONS OF PAPER—SIZES OF PAPER—COMMERCIAL
CLASSIFICATIONS OF PAPER-MAKING MATERIALS.

CHAPTER IV.

MANUFACTURE OF PAPER BY HAND.

CHAPTER V.

DISINFECTING RAGS—PURCHASING RAGS.

CHAPTER VI.

SORTING RAGS—SORTING WASTE PAPER—SORTING OR "DRY-PICKING" ESPARTO—MACHINE FOR FACILITATING THE SORTING OF PAPER STOCK.

CHAPTER VII.

CUTTING RAGS BY HAND—CUTTING RAGS BY MACHINERY—LIST OF PAT-
ENTS FOR RAG CUTTERS AND DUSTERS—CUTTING WOOD FOR CHEMICAL
FIBRE—TREATING WOOD BEFORE GRINDING—VOELTER'S MACHINE FOR
CUTTING OR GRINDING WOOD—LIST OF PATENTS FOR WOOD-GRINDERS
—CORN-HUSK CUTTER.

CHAPTER VIII.

DUSTING RAGS—WET DUSTING—WASTE-PAPER DUSTER AND WASHER.

CHAPTER IX.

BOILING RAGS—STATIONARY BOILERS—REVOLVING BOILERS—TREATING COLORED RAGS—BOILING WASTE PAPER—BOILING STRAW—BOILING ESPARTO—BOILING MANILLA AND JUTE—BOILING WOOD—SODA RECOVERY—ACID OR BISULPHITE PROCESSES OF TREATING WOOD—LIST OF PATENTS FOR PREPARING CELLULOSE FROM WOOD BY THE ACID OR BISULPHITE PROCESSES—LIST OF PATENTS FOR DIGESTERS WITH LEAD LININGS—LIST OF ALL AMERICAN PATENTS FOR DIGESTERS FOR PAPER PULP—METHODS OTHER THAN THE MECHANICAL, SODA, AND BISULPHITE PROCESSES FOR THE TREATMENT OF WOOD.

CHAPTER X.

Washing Rags—Washing Waste Paper or "Imperfections"—Washing Straw—Washing Wood-Pulp—Washing and Pouching Esparto—Wash-Water—List of Patents for Pulp-Washing and Straining.

CHAPTER XI.

BLEACHING POWDER—ESTIMATION OF CHLORINE IN BLEACHING POWDER
—PREPARING AND USING THE BLEACHING SOLUTION—ZINC BLEACH
LIQUOR—ALUMINA BLEACH LIQUOR—DRAINING—SOUR BLEACHING—
BLEACHING WITH GAS—BLEACHING PULP MADE FROM OLD PAPERS OR
IMPERFECTIONS — BLEACHING STRAW — BLEACHING WOOD FIBRE —
METHOD FOR BLEACHING WOOD, STRAW, ETC.—BLEACHING JUTE—
BLEACHING OF MATERIALS COMPOSED OF HEMP, FLAX, ETC.—BLEACHING
VEGETABLE TISSUES WITH PERMANGANATE OF POTASH—BLEACHING
PAPER PULP BY APPLYING THE BLEACHING AGENT IN A SPRAYED CON-
DITION—BLEACHING IN ROTARIES—LIST OF PATENTS FOR BLEACHING
PULP.

B

CHAPTER XIV.

Coloring.

CHAPTER XV.

Making and Finishing.

CHAPTER XVI.

The Preparation of Various Kinds of Paper.

THE

MANUFACTURE OF PAPER.

CHAPTER I.

THE HISTORY OF THE MANUFACTURE OF PAPER.

THE origins of nearly all those arts which have been slowly developed and which have so largely contributed to the progress and civilization of man are surrounded with so much obscurity that it is not possible to treat their early history in a satisfactory manner, for the reason that all such explanations must be more or less hypothetical.

But even in these times of rapid development in all branches of mechanics and the arts, we cannot entirely free ourselves from that irresistible law of our nature which impels us to seek acquaintance with primitive past events in connection with matters under discussion, not so much with a view to gathering practical ideas as from interest.

Men have in all ages been proud of their own achievements, and it is to this that we owe our present state of civilization. Labor and a certain amount of self-conceit are the basis of progress.

If all men were willing that their experiences and observations should pass unrecorded, there could be no progression in the human understanding beyond a certain point.

2

Men who have recorded their own thoughts and actions or those of others are the ones who have exerted the greatest influence for good or for evil in all ages.

To have at one's disposal a plane surface upon which it is possible to delineate in more or less indelible and conventional signs the conceptions of the brain, has been a necessity which man has felt from the moment he emerged from the first stages of the savage state; but before arriving at the present ingenious methods by means of which paper is manufactured, various primitive efforts to solve the problem have been made.

The mineral, vegetable, and animal kingdoms, each, in turn, have been utilized to supply man with a convenient substance upon which he could record the results of his studies or mark out his plans.

The ancients used stone, clay, palm leaves, tablets of wax, of ivory, and of lead, linen and cotton tissues, guts, or skins; also the interior barks of various plants. The skins of fishes and of snakes, and the shells of the tortoise and of the oyster have been in their turn used for the same purposes, and from them we derive the expressions of biblos, cordex, charta, etc., indicating the various substances used for writing upon.

Stone has been largely employed, but clay is the most prominent mineral used in very ancient times for many purposes for which we now employ paper, and this is particularly true in regard to Assyria and Chaldea, in which countries almost every transaction of a public or private character was first written upon thin tablets of clay, or tiles, and then baked.

Clay was probably used for writing upon, more than 2000 years before Christ, but the prophet Ezekiel, who was among the captives near the River Chebar, in the land of the Chaldeans, is among the first to describe the use to which the clay tile was sometimes put for receiving drawings or portraying plans. In 596 B. C., Ezekiel was commanded to make use of this common Assyrian practice at the time when the siege of Jerusalem was prefigured, the commandment being in the following language: "Thou also, son of man, take thee a tile, and lay it before thee, and portray upon it the city, even Jerusalem."—Ezekiel iv. 1.

Bank notes, notes of hand, deeds of property, private transactions, public records, transcripts of astronomical observations, and many things of this character have been and can still be found in a good state of preservation among the ruins of ancient Nineveh and Babylon; but they are not traced upon papyrus or parchment, but are in the indestructible terra-cotta.

The best histories of Chaldea, Babylonia, and Assyria come to us in this shape. There is something in these tablets of clay that forbids any desire on our part to discredit them. They seem to appeal to our practical understanding, and the tendency to doubt them is not so strong as with some modern written histories.

In the British Museum, in the Kouyunyik gallery, which contains the collection of bas-reliefs procured by Mr. Layard in 1849 and 1850 from the remains of a very extensive Assyrian edifice at Kouyunyik, possibly the palace of Sennacherib, who commenced his reign 705 B. C., there are arranged in six table-cases along the middle of the room, a large number of Assyrian antiquities. One table-case contains terra-cotta tablets referring to the language, legends, and mythology of the Assyrians, together with a selection of despatch or report tablets and letters. One series of these clay tablets is supposed to record the creation of the world. The first of the series gives an account of the first three days of the creation, in which it is stated that the Water-deep was the begetter of all the creatures then existing, for there was not even a seed in the earth, and none of the gods had come forth. The remainder of the texts, which are extremely difficult to translate, refer to the creating and placing of the heavenly bodies, the creation of creeping things, and of mankind instead of certain rebellious gods or angels, the war between the gods and Bisbistiamtu (the Water-chaos) and her servants, in which the latter were overthrown. Another tablet refers to the misfortunes of certain men who went forth and returned not, and mentions a flood. Three tablets, copies of the eleventh of the series entitled "The Record of Gistubar," are also of interest. This text contains the account of the flood, which is told to the hero by Umnapistim—the Babylonian Noah— who states that the gods within Suripak, a city on the Euphrates, determined to make a flood, and Umnapistim was commanded to build a ship, and to put within it all his property, the members of his family, and the beasts and cattle of the field. The coming of the flood, its abatement, the resting of the ship on the mountain of Nizir, and the sending forth of a dove, a swallow, and a raven on the seventh day, are also told, together with the coming forth from the ship. The god Bêl, however, was angry that all the race of mankind had not been destroyed; but the god Hea appeased his wrath, the patriarch and his family were allowed to live, and the gods took him and his wife to a "remote place at the mouth of the rivers," supposed to be the region of the Persian Gulf.

The Egyptian papyri in the British Museum are arranged in glass cases on the northwestern staircase, and they show the three forms of writing in use among the Egyptians: 1. The Hieroglyphic, in which all the characters or figures are separately and distinctly defined. 2. The Hieratic, in which the same characters are represented in what may be termed a running hand. 3. The Demotic, or Enchorial, a still more cursive form, in which the language of the common people was written; it was principally employed in civil transactions during the Ptolemaic period, and continued in use to the third or fourth century of our era.

The hieroglyphic character was in use in Egypt as early as the third dynasty, the date of which is placed about 4000 B. C. by some chronologists; but no hieroglyphic papyri of that remote age are extant, and the oldest examples known appear to be of the eighteenth dynasty—about 1700 B. C. Hieroglyphic writing seems to have been employed almost exclusively for religious purposes, and the papyri written in it are Rituals, or the Book of the Dead, as it is called, a copy of which has been published by Professor Lepsius, under the title of 'Das Todtenbuch der Agypten,' 4to, Leipzig,

1842. The chapters of this book contained in the work of Lepsius are as old as the eleventh dynasty—about 2000 B. C.—and continued in use till the thirty-first dynasty— about 340 B. C.

The hieratic or written form of the hieroglyphics appears first about the age of the fifth dynasty, and continued in existence till the first century of our era, when it became superseded for all purposes by the demotic. The entire ritual is rarely found in the hieratic character at an early period, portions only having been rarely transcribed into that character till the twentieth dynasty. Other religious works, however, appear in it as early as the eleventh dynasty, when the linen wraps of mummies were inscribed with ritualistic formulas. Other works occur in hieratic. A few papyri of later date contain the Shai en Sinsin, or the Book of the Respirations, *i. e.*, the sighs or lamentations of Isis, containing extracts of portions of chapters in the Ritual, or expressions similar to them. The affairs of official and private life were written hieratic, and amongst the papyri exhibited are found literary compositions, scientific treatises, law documents, criminal police reports, registers or inventories of valuable or other objects.

The demotic papyri consist of rituals, literary compositions, deeds of sale, contracts of marriage, all indorsed by witnesses. At an early period these witnesses were few in number, but as many as sixteen are found in later times. These deeds, which are dated in the regnal years of the monarchs at the time of execution, commence in the age of

Tirhakah, nearly 800 B. C., and run on till the end of the first century, A. D. The religious books continue, however, until apparently about the end of 300 B. C. Letters, memoranda, and registers were also written in the demotic.

The width of the Egyptian papyri rarely exceeded 15 inches, but their length sometimes, though rarely, extended to 150 feet. Papyrus, both before use and afterwards, was rolled up into a cylindrical roll, and, when opened for the purpose of reading, unrolled from the ends. Besides these methods, the papyri were occasionally placed in wooden figures, always colored black, of the god Osiris standing on a pedestal, either in the hollowed body of the god, or else in a place in the pedestal, covered by a small slip, the whole so carefully painted over as not to give any indication of the papyrus within.

The Roman scholar, Varro, is indorsed by Pliny in the statement that the discovery of the use of papyrus was an incident in the victorious expedition of Alexander of Macedon. But when we consider that Egyptian tombs plainly demonstrate that papyrus was used not only long prior to the time of Alexander, but also previous to any authentic historical account of Greece, it becomes manifest that Varro's statement is erroneous; doubtless, however, the expedition of Alexander materially aided in introducing the papyrus among the western nations.

Papyrus is an aquatic plant common in many warm countries and especially in Egypt, and from the layer between the flesh and thick bark of this reed or flag the Greeks and the Romans obtained the paper which they used for a long time. The strips or ribbons of different lengths obtained by peeling the interior of the bark were bleached in the sun, then spread open upon a table and covered crosswise by other strips; then moistened with water and pressed, thus causing the adhesion of the strips by means of the vegetable mucilage naturally present in the bark. In this manner sheets were obtained which the Romans sized with a starch of fecula or flour.

It is difficult to determine at exactly what time the use of Egyptian papyrus was entirely superseded by cotton paper, as the latter material could have been introduced only by degrees, and papyrus was also employed for special purposes long after the general introduction of cotton paper.

Leather was used by the Israelites as a material to write upon. Parchment for writing upon with ink was probably invented by Eumenes, king of Pergamos, whence the name is derived.

The parchment used by the Ionians in the time of Herodotus was coarser than that invented by Eumenes, and was probably painted upon with especially prepared pigments.

The skins of sheep, lambs, and calves are principally employed for the manufacture of parchment, and although the use of this material for common purposes has greatly diminished of late years, still it continues to be extensively produced for special purposes. For diplomas, parchment is even now almost exclusively employed.

In addition to the materials which have been named, thin boards of wood covered with wax or some similar composition, and plates of ivory and of metal, have also been used.

Most convenient materials were also afforded by the bark and the leaves of some species of trees. In the time of Confucius the Chinese wrote with a style or bodkin, on the inner bark of the bamboo.

It is probably not desirable that we should further enlarge upon the mineral, vegetable, and animal substances previously named, and which have in times past been so extensively employed in lieu of the material which is now commonly known as paper.

Our word *paper* is derived from the Latin *papyrus*, which is the Greek πάπυρος, and the Egyptian *papu*, meaning a reed. Paper is now commonly made by machinery from cotton and linen rags, and also from wood, straw, esparto grass, and numerous other vegetable fibres, the material being reduced to a pulp and afterwards formed into a thin sheet which is subjected to pressure, and finally dried. The sheets of paper may be of greater or less thickness, width, and length, or the paper may be produced in indefinite lengths and formed into rolls. The finishing of the paper, with or without vegetable or animal size, of course depends upon the purpose for which it is to be employed.

In addition to being the common material for printing and writing upon, and for bags, boxes, and wrapping, paper also finds numerous secondary employments, such as for toilet purposes, and is also manufactured into barrels, berry, and grape baskets, and pails, buckets, carpets, mattings, carwheels, collars, cuffs, curtains, dishes, elevator seats, and panelling, napkins, observatory domes, picture frames, roofing-felt, roofing-tiles, racing-shells, and twine, and in mechanical construction. Paper has also been used for journal bearings, packing for steam engines, for belting, etc. A manufacturer in Breslau, Germany, is said to have built a chimney over fifty feet in height out of compressed paper blocks, used in place of bricks. Paper has also been used in place of wood in the manufacture of lead pencils.

It is a source of much regret in tracing the origin of so valuable an art as that of the fabrication of modern paper that no accurate estimate can be formed as to the precise time of its adoption.

In 1755 and 1763, in order to stimulate researches in this direction, the Royal Society of Sciences at Göttingen offered valuable premiums for that especial object, but the result sought to be accomplished was altogether unattained, and all such investigations, however directed, proved fruitless. In 1762, M. Miserman offered a prize for the oldest manuscript written on rag paper.

The different minutes of the proceedings of this competition, printed at The Hague, in 1767, unite in admitting that paper of this kind was used prior to the commencement of the fourteenth century.

In the tract of Peter, Abbot of Cluny (A. D. 1122–50), *adversus Judæcos*, cap. 5, among the various kinds of books, we find the first mention of rag paper when he refers to such books as are written on material made " *ex rasuris veterum pannorum.*" It has been thought probable that at this early period woollen cloth is intended; but of this point we shall have more to say later on in discussing the invention of linen paper. The process of making writing paper from fibrous materials, and, among other substances, from the wool of the cotton plant, reduced to a pulp, appears, according to the works of Mr. Stanislas Julien, to have been practised by the Chinese as early as the year 152 of our era.[1] But it was not until the commencement of the eighth century of the Christian era that the art of manufacturing cotton paper became known to the western world, and this was accomplished through the Arabs, who, in 704, captured Samarcand, and there learned the method of using and making paper.

The Arabs at once took up the manufacture of paper in

---

[1] The raw materials which were used at first and which have continued in use, are the barks of trees (*Morus, Broussonetia papyrifera*), hemp, bamboo, straw, and old linen.

The paper most used is obtained from bamboo stems chipped in small fragments. The fibres become disintegrated after a few weeks by means of lime-water, and are bruised energetically and the pulp is treated with an alkaline wash. The washed pulp is then made in sheets by means of moulds similar to those used for hand-made paper. After dessication in the air the sheets are dried upon heated plates. These sheets are used for letter paper, commercial books, and wrapping paper. Rolled up in cylindrical shape, and, provided it offers to the spark a part already carbonized, this paper will ignite and burn like tinder; in this form it is used as lighting matches.

In the north of China, where bamboo does not grow, they use the bark of broussonetia, care being observed to disintegrate the fibres very little, so that they will felt. By this method there are obtained large sheets, transparent enough, of a variable thickness, which are used for wrapping, and in the manufacture of umbrellas and window-panes. The paper, improperly called rice paper, is obtained from the marrow of the *Olrelia papyrifera*. The operator rolls with the left hand the roll of marrow upon a plane surface, while with his right he engages, in an almost tangential direction, a thin and sharp blade in the marrow. By this equal and continuous rotatory movement he cuts a sheet more or less thin, more or less long. This paper, white and soft to the eye, is used in the manufacture of those colored designs which are produced in the shops of Canton. (Champion, L'Orient, Archives de l'industrie au xix. Siecle, t. v. p. 297.)

Samarcand, and a thorough knowledge of the curious art rapidly spread through all their empire.

Charta[1] *Damascena* is one of the titles which was applied in the Middle Ages to the cotton paper which was manufactured in large quantities at Damascus.

The statement that cotton paper was early extensively adopted by the Arabs for literary employments is corroborated by the numerous Arabic manuscripts written on paper during the ninth and tenth centuries.[2]

Bombacinum was used at Rome in the tenth century.

[1] In addition to being termed *charta* and *papyrus*, cotton paper was also known in the Middle Ages as *charta bombycina, gossypina, cuttunea, xylina, Damascena,* and *serica;* the latter title being probably suggested by its glossy and silken appearance.

[2] The following, compiled from the 'Encyclopædia Britannica,' may be instanced as a few of the earliest dated examples. The 'Gharibu l'Hadith,' a treatise on the rare and curious words in the sayings of Mohammed and his companions, written in the year 866, is probably one of the oldest paper MSS. in existence (Pal. Soc., Orient Ser., pl. 6). It is preserved in the University Library of Leyden. A treatise by an Arabian physician on the nourishment of the different members of the body, of the year 960, is the oldest dated Arabic MS. on paper in the British Museum (Or. MS. 2600, Pal. Soc., pl. 96). The Bodleian Library possesses a MS. of the 'Diwann l'Adab,' a grammatical work of 974 A. D., of particular interest as having been written at Samarcand on paper, presumably made at that seat of the first Arab manufacture (Pal. Soc., pl. 60). Other early examples are a volume of poems written at Bagdad, 990 A. D., now at Leipsic, and the Gospel of St. Luke, 993 A. D., in the Vatican Library (Pal. Soc., pls. 7, 21). In the great collection of Syriac MSS., which were obtained from the Nitrian desert in Egypt, and are now in the British Museum, there are many volumes written on cotton paper of the tenth century. The oldest two dated examples, however, are not earlier than 1075 and 1084 A. D.

It may not be amiss to include in this note a few words regarding the extant samples of cotton paper MSS. written in European countries.

Several which have been quoted by former writers as early instances have proved, on more recent examination, to be nothing but vellum. The ancient fragments of the Gospel of St. Mark, preserved at Venice, which were stated by Moffer to be of cotton paper, by Montfaucon of papyrus, and by the Benedictines

The Empress Irene, wife of Alexes Commene, at about the close of the eleventh century or the commencement of the twelfth, in the statutes for regulating some religious houses at Constantinople, states that she had left three copies of these statutes, two on parchment and one on cotton paper (de Martin, ' Essais chimiques sur les arts et manufactures,' t. iii. p. 161).

"*Tolle pergamenam Græcam, quæ fit ex lana lingi*," are the words used by Theophilus, presbyter, who wrote in the twelfth century.[1]

But notwithstanding this early reference to cotton paper by Theophilus, under the name of Greek parchment, paper probably was not used to any great extent in Greece, much prior to the second half of the thirteenth century, for there are no reliable Greek MSS. on paper which bear an earlier date than about the middle of that century.

After the capture of Samarcand in 704, the Arabians transplanted the art of fabricating paper to Spain, and to

---

of bark, are in fact written on skin. The oldest European document on cotton paper is a diploma of King Roger of Sicily, of the year 1102. The oldest known imperial deed on the same material is a charter of Frederick II., to the nuns of Goess in Styria, of the year 1228, now at Vienna. In 1231, however, the same emperor, on account of the liability of paper made from cotton to be affected by the damp atmosphere, forbade further use of paper for official documents, which were in future to be inscribed on vellum. In France the *Liber plegiorum*, the entries of which began with the year 1223, is made of rough cotton paper; and similarly the registers of the Council of Ten, beginning in 1325, and of the Emperor Henry VII. (1308–13), preserved at Tunis, are also written on a like substance. The letters addressed from Castile to the English king, Edward I., in 1279 and following years (Pauli in Bericht. Berl. Akad., 1854), are instances of Spanish-made paper; and other specimens in existence prove that in this latter country a rough kind of *charta bonbycina* was manufactured to a comparatively late date. ' Encyclopædia Brit.'

[1] Schedula diversarum artium, 1, 23.

the Moors, the credit of first manufacturing paper in Europe is undoubtedly due.

Valencia and its neighbor Xativa, as also Toledo, the latter city being about forty miles distant from Madrid, were the primitive seats of the cotton-paper industry in Europe.

When the Crusaders visited Byzantium, Palestine, and Syria, they became acquainted with the great convenience and the value of paper, and they carried back with them some knowledge of its manufacture. But it was not until after the fourth crusade that the first paper-mills were established in France, the art of making cotton paper being introduced in the year 1189, in the district of Hérault.

The French were a very energetic and cultured race, and as the Norman buildings of the twelfth century plainly show, they took excessive delight in construction; their princes and nobles seem to have taken their greatest pleasure in dwelling in and constantly beautifying their magnificent castles.

They did not care so much for feasting and high living as their English neighbors, but devoted their time and talents to the development of those arts and manufactures which contribute so greatly to the refinement of society.

In the twelfth century new ideas everywhere appeared at once in France, and the people prosecuted their acquired knowledge and their own inventions with so much energy and skill that we are not surprised at the very rapid development of the paper trade in France during the fourteenth century, for she was not only soon in a position to provide for her own wants, but also to supply all her neighbors. Troyes and Essonnes are the oldest centres of the paper industry in France.

The great progress of France in paper manufacture stimulated the fabrication of paper in the Netherlands, and for a long period the French and Dutch papers were the best produced in Europe.

That England was far less progressive in the manufacture of paper than either of the two countries which have been last mentioned, is incontestibly shown by the language of the English patent of John Briscoe, granted July 4, 1685, and which is for: "The true art and way of making English paper for writing, printing, and for other uses, *both as good and as serviceable in all respects and as white as any French or Dutch paper.*"

It seems almost incredible that no paper was made in England prior to the time of the Tudors, but such indeed seems to be the fact.

A manufacturer by the name of Tait is stated to have operated a paper manufactory in Hertfort early in the sixteenth century, and a German named Spielman is said to have had a paper-mill at Dartford in 1588; but if paper was produced at these works it was undoubtedly of the common sort.

In corroboration of the last statement we have the language of the first English patent for making paper granted as late as February 16, 1665, to Charles Hildegerd for "the way and art of making blew paper used by sugar bakers and others," and also the second English patent for paper granted in January, 1675, to Eustace Barneby, for "The art and skill of making all sorts of white paper for the use of writing and printing, being a *new manufacture and never practised in any way in any of our kingdoms or dominions*," thus showing the incipiency in England of the manufacture of writing and printing paper.[1]

After the fall of the Moorish power and the decline of the paper-making industry in Spain, the little town of Fabriano in the province of Ancona, in Central Italy, rose into prominence as a centre for the fabrication of fine paper. In 1340 a paper-mill was established at Padua, this was followed by one at Treviso, and then a little later other paper manufactories were established in the territories of Florence, Bologna, Milan, Venice, etc.

Southern Germany, even as late as the fifteenth century, imported most of its paper from the line of factories in Northern Italy.

Italian workmen were, however, induced to aid in establishing paper-mills in Germany, and the manufacture of paper was commenced in the latter country at an early date, and numerous mills were operated during the fourteenth century near Cologne and in Mainz, in the latter even as early as about 1320.

Nuremberg did not possess a paper-mill until 1390; but Augsburgh and Ratisbon were places of early manufacture.

---

[1] In regard to the early use of cotton paper in England for writing upon there is evidence that it was employed for registers and accounts even as early as 1309. The register of the Hustings Court of Lyme Regis, now in the British Museum, contains entries which commenced in the last-named year. But the appearance of the paper shows that it was without doubt imported either from Spain or from France.

The Arabians in Spain were the first to mix rags with the cotton pulp in the fabrication of paper, and in that imported into England from Spain at the commencement of the fourteenth century the threads of rags are plainly visible imbedded in the pulp.

Linen paper was first made in Europe in the fourteenth century, but until about the middle of that century woollen fabrics probably formed a large percentage of the material from which the pulp was produced, but in individual instances this fact requires to be established by the assistance of the microscope.

The period and manner of the invention of linen paper are thus described in 'Trextinum Antiquorum,' by James Yates, M.A., Part I. pp. 383–388 : " No part of the *res diplomatica* has been more frequently discussed than the question respecting the origin of paper made from linen rags. The inquiry is interesting on account of the unspeakable importance of this material in connection with the progress of knowledge and all the means of civilization, and it also claims attention from the philologist as an aid in determining the age of manuscripts.

" Wehrs refers to a document written A. D. 1308, as the oldest known specimen of linen paper; and, as the invention must have been at least a little previous to the preparation of this document, he fixes upon 1300 as its probable date.[1] Various writers on the subject, as Von Murr, Breitkopf, Schönemann, etc., concur in this opinion.

---

[1] 'Vom Papier,' pp. 309, 343.

" Gotthelf Fischer, in his essay on paper marks,[1] cites an extract from an account written in 1301 on linen paper. In this specimen the mark is a circle surmounted by a sprig, at the end of which is a star. The paper is thick, firm, and well grained; and its water-lines and water-marks (*vergures et pontuseaux*) may readily be distinguished.

" The date was carried considerably higher by Schwandner, principal keeper of the Imperial Library at Vienna, who found among the charters of the Monastery of Göss in Upper Styria one in a state of decay, only seven inches long and three wide. So highly did he estimate the value of this curious relic as to publish in 1788 a full account of his discovery in a thin quarto volume, which bears the following title: ' Chartam linteam antiquissiman, omnia hactenus producta specimina ætate suâ superantem, ex cimeliis Bibliothecæ Augustæ Vindobonensis, exponit Jo. Ge. Schwandner,' etc. The document is a mandate of Frederick II., Emperor of the Romans, entrusting to the Archbishop of Saltzburg and the Duke of Austria the determination of a dispute between the Duke of Carinthia and the Monastery of Göss, respecting the property of the latter in Carinthia. Schwandner proves the date of it to be 1243. He does not say whether it has any lines or water-marks, but is quite satisfied from its flexibility and other qualities that it is linen. Although on the first discovery of this document some doubt was expressed as to its genuineness, it appears to have risen in estimation with

---

[1] This essay translated into French is published by Jansen, in his ' Essai sur l'origine de la gravure en bois et en taille-douce.' Paris, 1808, tome i. pp. 357–385.

succeeding writers; and it is probable rather from inadvertence than from any deficiency in the evidence, that it is not noticed at all by Schönemann, Ebert, Delandine, or by Horne. Due attention is, however, bestowed upon it by August Friedrich Pfeiffer, 'Uber Bücher-Handschriften, Erlangen,' 1810, pp. 39, 40.

" With regard to the circumstances which led to the invention of the paper now in common use, or the country in which it took place, we find in the writers on the subject from Polydore Virgil to the present day nothing but conjectures or confessions of ignorance. Wehrs supposes, and others follow him, that in making paper linen rags were either by accident or through design at first mixed with cotton rags, so as to produce a paper which was partly linen and partly cotton, and that this led by degrees to the manufacture of paper from linen only.[1] Wehrs also endeavors to claim the honor of the invention for Germany, his own country; but Schönemann gives that distinction to Italy, because there, in the district of Ancona, a considerable manufacture of cotton paper was carried on before the fourteenth century.[2] All, however, admit that they have no satisfactory evidence on the subject.

" A clear light is thrown upon these questions by a remark[3] of the Arabian physician Abdollatiph, who visited Egypt

---

[1] 'Vom Papier,' p. 183.   [2] 'Diplomatik,' vol. i. p. 494.
[3] Chapter iv. p. 188 of Silvestre de Sacy's French translation, p. 221 of Wahl's German translation. This interesting passage was translated as follows by Edward Pococke, the younger. "Et qui ex Arabibus, incolisve Rifae aliieve has arcas indagant, hæc integumenta diripiunt, quodque in iis rapiendum invenitur; et conficiunt sibi vestes, aut ea chartariis vendunt ad conficiendam

A. D. 1200. He informs us '*that the cloth found in the cata-combs, and used to envelop the mummies, was made into gar-ments or sold to the scribes to make paper for shopkeepers.*' This cloth was linen, and the passage of Abdollatiph is proof, which, however, has never been produced as such, of the manufacture of linen paper as early as the year 1200.

" This account coincides remarkably with what we know from various other sources. Professor Tychsen, in his learned and curious dissertation on the use of paper from papyrus (published in the 'Commentationes Reg. Soc. Gottingensis Recentiores,' vol. iv. A. D. 1820), has brought abundant testimony to prove that Egypt supplied all Europe with this kind of paper until towards the end of the eleventh century. The use of it was then abandoned, cotton paper being employed instead. The Arabs in consequence of their conquests in Bucharia had learned the art of making cotton paper about the year 704, and through them or the Saracens it was introduced into Europe in the eleventh century.[1] We may therefore consider it as in the highest degree probable, that the mode of making cotton paper was known to the paper-makers of Egypt. At the same time endless quantities of linen cloth, the best of all materials for the manufacture of paper, were to be obtained from the catacombs.

" If we put together these circumstances we cannot but perceive how they concur in illustrating and justifying the

chartam emporeaticam." Silvestre de Sacy (Notice, etc.), animadverting on White's version, which is entirely different, expresses his approbation of Pococke's, from which Wahl's does not materially differ.

[1] Wehrs, ' vom Papier,' pp. 131, 144, note. Breitkopf, p. 81.

statement of Abdollatiph. We perceive the interest which the great Egyptian paper-manufacturers had in the improvement of their article, and the unrivalled facilities which they possessed for this purpose; and thus the direct testimony of an eye-witness of the highest reputation for veracity and intelligence, supported as it is by collateral probabilities, tends to clear up in a great measure the long-agitated question respecting the origin of paper such as we now commonly use for writing.

" The evidence being carried thus far, we may now take in connection with it the following passage from Petrus Cluniacensis :—

" ' *Sed cujusmodi librum ?   Si talem quales quotidie in usu legendi habemus, utique ex pellibus arietum, hircorum, vel vitulorum, sive ex biblis, vel juncis orientalium paludum, aut ex rasuris veterum pannorum, seu ex qualibet alia forte viliore materia compactos, et pennis avium vel calamis palustrium locorum, qualibet tinctura infectis descriptos.*' *Tractatus adv. Judaeos, c. v. in Max. Bibl. vet. Patrum,* tom. xxii. p. 1014.

" All the writers upon this subject, except Trombelli, suppose the Abbot of Cluny to allude in the phrase '*ex rasuris veterum pannorum*' to the use of woollen and cotton cloth only, and not of linen. But, as we are now authorized to carry up the invention of linen paper higher than before, and as the mention of it by Abdollatiph justifies the conclusion that it was manufactured in Egypt some time before his visit to that country in 1200, we may reasonably conjecture that Petrus Cluniacensis alluded to the same fact. The treatise above quoted is supposed to have been written A. D. 1120. The account of the materials used for making books appears to be full and accurate. The expression '*scrapings of old cloths*' agrees exactly with the mode of making paper from linen rags, but is not in accordance with any facts known to us respecting the use of woollen or cotton cloth. The only objection I can suppose to arise to this view of the subject is, that, as Peter of Cluny had not when he wrote this passage travelled eastward of France, we can scarcely suppose him to have been sufficiently acquainted with the manners and productions of Egypt to introduce any allusion to their newly invented mode of making paper. But we know that the Abbey of Cluny had more than three hundred churches, colleges, and monasteries dependent on it, and that at least two of these were in Palestine and one at Constantinople. The intercourse which must have subsisted in this way between the Abbey of Cluny and the Levant may account for the Abbot Peter's acquaintance with the fact, and I therefore think it probable that he alludes to the manufacture of paper in Egypt from the cloth of mummies, which on this supposition had been invented early in the twelfth century.[1]

" Another fact, which not only coincides with all the evidence now produced, but carries the date of the invention still a little higher, is the description of the manuscript No.

---

[1] Gibbon says (vol. v. p. 295, 4to. edition), " The inestimable art of transforming linen into paper has been diffused from the manufacture of *Samarcand* over the western world." This assertion seems to me entirely destitute of foundation.

787, containing an Arabic version of the 'Aphorisms of Hippocrates,' in Casiri's 'Bibliotheca Arabico-Hispana Escurialensis,' tom. i. p. 235. This MS. was probably brought from Egypt, or the East. It has a date corresponding to A. D. 1100, and is of linen paper, according to Casiri, who calls it ' Chartaceus.'

" ' *Cordices chartacei*,' *i. e.*, MSS. on linen paper, as old as the thirteenth century, are mentioned not unfrequently in the catalogues of the Escurial, the Nani, and other libraries.

" The preceding facts coincide with the opinion long ago expressed by Prindeaux, who concluded that linen paper was an Eastern invention, because 'most of the old MSS. in Arabic and other Oriental languages are written on this sort of paper,' and that it was first introduced into Europe by the Saracens of Spain."[1]

In the first years of the fourteenth century the art of fabricating paper had become a truly European industry, and it, therefore, becomes manifest that it is of much practical importance to seek to define the line of demarcation between the two classes of linen paper then made.

The papers into which linen first entered as a constituent of the pulp may be described as water-marked and non-water-marked, the first-named variety of paper making its appearance in the early years of the fourteenth century.

The Oriental fashion was to make the paper without water-marks, and while instances of cotton paper of the Oriental pattern made in the first half of the fourteenth

[1] Old and New Testament connected, part i., chapter vii., p. 393, 3d edition, folio.

century are extant, still they occur but seldom, if ever, in the north of Europe.

The water-marks on paper have been partially investigated with a view to discovering the various channels in which the trade in paper of different nations travelled; but up to this time there has been no thorough and systematic collection of water-marks and consequently no classification can now be made.

The student will be greatly aided in fixing very approximate periods to undated documents by acquiring a knowledge of the different varieties of paper and of water-marks.

" Rag paper of the fourteenth century may generally be recognized by its firm texture, its stoutness, and the large size of its wires. The water-marks are usually simple in design; and being the result of the impress of thick wires, they are, therefore, strongly marked. In the course of the fifteenth century the texture gradually becomes finer and the water-marks more elaborate. While the old subjects of the latter are still continued in use, they are more neatly outlined, and, particularly in Italian paper, they are frequently inclosed in circles. The practice of inserting the full name of the maker in the water-marks came into fashion in the sixteenth century. The variety of subjects of water-marks is quite extensive. Animals, birds, fishes, heads, flowers, domestic and warlike implements, armorial bearings, etc., are found from the earliest times. Some of these, such as armorial bearings, and national, provincial or personal cognizances, as the imperial crown, the crossed keys, or the cardinal's hat, can be attributed to particular countries or districts, and the wide dissemination of the paper bearing these marks in different countries seems to prove how large and international was the paper trade in the fourteenth and fifteenth centuries." (*'Encyc. Brit.'*)

Some idea of the rapidity with which paper-making has been developed in Europe may be judged from the fact that at this writing (Jan. 1886) there are probably not less than three thousand five hundred paper-mills in Europe.

Germany possesses by far the largest number of mills of any country in Europe, after which comes France, next Great Britain, then Austro-Hungary, Italy, Russia, Spain, Sweden, Holland, Norway, Switzerland, Belgium, Portugal, and Denmark, all following in regular order according to their present relative importance in paper manufacture.

The paper-mills of Asia are few, and the aggregate of all of them would probably not be equal to the number of mills in the smallest paper-making country of Europe, which is Denmark, with about fifteen mills.

In China there are probably no modern paper-mills. But Japan has several in operation, and, judging from the progress shown in the machine-made paper on which are printed the first, second, and third statistical reports on agriculture published by the department of agriculture and commerce, Tokio, Japan, and which now lie before the author, he hazards the opinion that Japan will in the near future become a paper-making country of considerable importance.

India is also rapidly developing her paper manufacture, and is probably abreast of Japan in this department.

The Australasian Continent contains but few mills, Australia of course leading.

On the Western Continent we find the Dominion of
Canada producing paper in about the same quantity as
Norway, which latter country occupies but an insignificant
position among the paper-making countries of Europe.

All the paper-mills in South America do not probably
exceed in number those of Japan, and in Cuba and Mexico
there are not more than two or three mills.

But in the United States we find the greatest paper-
manufacturing country in the world, and the great and re-
lentless energy with which our country is developing her
productions in excess of all other nations is shown in one
instance in the strides which she has made in the fabrication
of paper.

We have already shown that the manufacture of paper
for writing and printing upon was probably not under way in
England prior to the year 1685, but it is probable that the
fabrication of paper did not flourish in that country until
after the year 1688, for in the ' British Merchant' of the
latter year we find the statement that hardly any sort of
paper except brown was made in England previous to the
Revolution, and in 1689 Bohun, in his autobiography, says
" paper became so dear that all printing stopped almost,
and the stationers did not care to undertake anything."

If England was not entirely dependent upon other nations
her publishers would not have been reduced in 1689 to such
dire straits as stated by Bohun to have existed.

The first paper-mill was established in America in 1690,
the mill being near Philadelphia, Pa., and thus we find the
manufacture of paper to have commenced almost simultaneously in America and in England.

But now, after the lapse of almost two centuries, we discover that the United States possesses mills for the manufacture of paper which exceed in number the aggregate of those in England, Ireland, Scotland, and Wales, with the addition of France, Belgium, Portugal, Sweden, Norway, and Holland.

When we realize that this remarkable development has been made in spite of almost insurmountable obstacles and serious foreign competition, added to the drain upon our energies caused by the two wars with Great Britain, our own civil war, disastrous industrial depressions, and numerous financial panics, and complications in currency, it forcibly reminds us how surely and rapidly the great advance which our country is making in mechanics and the arts is unsettling the commerce of the world.

There are in this industry, as in all others, times of depression, but the natural facilities for obtaining the raw material, the great ingenuity of our people, added to the large home consumption of paper, with an increasing export demand, are certain to keep the United States in the position of the leading paper-producing country in the world.

The art of printing was, of course, the immediate cause stimulating the manufacture of paper, and it was through the efforts of William Bradford, one of the earliest printers in the American colonies, that the first paper-mill was established on our soil.

Bradford realized the advantage of having a constant supply of paper near at hand, and he, therefore, readily joined William Rittenhüysen, who had emigrated from

Broich, in Holland, in the project of starting a paper-mill in Roxborough, near Philadelphia.   The location for the mill was selected on a small tributary of the Wissahickon, the pure water of the little stream, which is still called Paper Mill Run, making it very desirable for paper-making ; the abundant supply of cotton and linen rags in the neighborhood furnishing ample raw material for continuing the mill for a long time.   The name of Rittenhüysen in time became anglicized into that of Rittenhouse, which is now in common use.

In 1710 and 1728 other paper-mills were established in Pennsylvania by relations and apprentices of Rittenhouse.

In 1724, William Bradford made an effort to induce the council of New York to grant him an exclusive privilege for manufacturing paper in the province for the space of fifteen years, but was unsuccessful.   But, in 1728, Bradford succeeded in establishing a mill in Elizabethtown, N. J., which was the first paper-mill in the State.

In 1727, Thomas Willcox, a native of England, built the Ivy Mill, on Chester Creek, in what is now Delaware County, Pa., on land purchased from Wm. Penn.   This property has remained continuously in this family, who are still manufacturers of paper under the firm name of James M. Willcox & Co., the oldest existing commercial house in America, and who now have other mills on the same tract of land.   Ivy Mill still stands, although it has not been used for several years.

In the records of Massachusetts for 1730, there is an Act for the encouragement of the first paper-mill built in New England, passed September 13th, 1728, granting a patent to Daniel Henchman, Gillam Phillips, Benjamin Faneuil,

Thomas Hancock, and Henry Dering, for the sole manufacture of paper in the province for ten years.

The granting of the patent was conditional upon the production of a stated quantity of paper yearly, until at the end of the third year and each year thereafter the total annual produce of the various qualities of paper described in the patent was not to be less than five hundred reams. The proprietors mentioned in the above Act soon erected a small paper-mill in Milton, afterward in the county of Norfolk, on a site adjoining the Neponset River. The master-workman at the mill was an Englishman, by the name of Henry Woodman, under whose management very satisfactory qualities of paper were produced, samples of which were exhibited to the Court in General Term, in 1731, by one of the proprietors of the mill, Daniel Henchman, an enterprising bookseller of Boston.

It is not probable that the mill at Milton suffered for the want of raw material, for in 1732 we find Richard Fry, of Cornhill, Boston, who was an agent for the mill, thanking the public for saving and selling him their rags, of which he had already received upwards of seven thousand weight. But on account of the scarcity of good workmen the mill was finally compelled to stop, and was afterward sold to Mr. Jeremiah Smith, who also for the lack of suitable workmen was unable to utilize his purchase.

In 1760 the mill was again started by James Boies, of Boston, who became acquainted with a soldier by the name of Hazleton who was a practical paper-maker and a member of a British regiment then stationed in Boston.

It was not a difficult matter to obtain a furlough for Hazleton, who at once put the mill in proper condition and started it to work, his main assistant being Abijah Smith then living in Milton. But Hazleton was compelled to leave the mill and join the regiment when the latter was ordered to Quebec, and, like Wolfe, the private soldier received a mortal wound on the Plains of Abraham.

A short interval then took place before another Englishman by the name of Richard Clarke arrived from New York and again set the mill in operation. In a few years Clarke, who was an excellent workman, was joined by his son George, a young man of about twenty years of age, who also proved himself to be a good paper-maker. Hazleton's assistant, Abijah Smith, developed into a good workman, and continued at the business until of an advanced age.

In 1768 a paper-mill was established in Norwich, Conn., by Christopher Leffingwell, who was promised a bounty by the Legislature of 2d. per quire on all good writing paper and 1d. per quire on all printing and common paper; but this subsidy was withdrawn in 1770.

When the American Revolution commenced, there were probably not more than an aggregate of fifty paper-mills in all the American colonies; the produce of these mills was inadequate in quantity to supply the home demand, which made the price of paper very high. For a long time there was a great scarcity of rags, and as but little labor could be bestowed upon paper then made, the quality was very inferior, and these conditions continued until after the adoption of the Constitution.

While the manufacture of paper was being prosecuted under such difficulties in the United States, the important announcement was made in France that N. L. Robert, a machinist connected with Didot's paper-mill at Essonnes, had invented a machine for making sheets of paper of very large size, even twelve feet wide and fifty feet long.

On account of the importance of Robert's invention, and the great influence which it afterwards exerted in building up our paper manufactures, we will here divert from an account of the history of paper-making in the United States, and follow the history of the Robert's machine.

On the 18th of January, 1799, the government of France granted Mr. N. L. Robert a patent (No. 329) for fifteen years, for his invention for making paper by machinery, and during the same year a working model of the machine was constructed, but its work was not fully satisfactory. But the French government, appreciating the usefulness of Robert's invention, and realizing that its imperfections would be overcome by experience, granted him a bounty of 8000 francs. The experiments necessary for perfecting the machine were both troublesome and expensive, and on account of the difficulties in which France was then involved it became manifest that the machine could be better perfected in England.

M. Leger Didot, of Essonnes, agreed to purchase the patent and model from Robert, and accompanied by John Gamble, an Englishman, Didot sailed for England. Some improvements were made on Robert's model by Didot before leaving France; but on reaching England, Didot was fortunate in

securing the aid of a Mr. Bryan Donkin, a man of considerable mechanical skill, and by means of the experiments and
observations of the trio—Didot, Gamble, and Donkin—the
invention of Robert was perfected. On April 2, 1801, a
patent (No. 2487) was granted in England to John Gamble,
for the improved invention of Robert, the title of the patent
being "An invention for making paper in single sheets
without seam or joining, from one to twelve feet and upwards wide, and from one to forty-five feet and upwards in
length."

On June 7, 1803, a patent (No. 2708) was granted by
the English government to John Gamble for "Improvements and additions to a machine for making paper in single
sheets without seam or joinings, from one to twelve feet and
upwards wide, and from one to fifty feet and upwards in
length," and during the same year the first machine for
making paper by machinery was successfully put in operation at Frogmore, England.

During the next year, 1804, the second paper-making
machine was put in operation at Two Waters, England.

In 1804 Messrs. Henry and Sealy Fourdrinier purchased
the interest of Didot and Gamble in the improved Robert
machine. Henry Fourdrinier on July 24, 1806, was granted
a patent (No. 2951) for "The method of making a machine
for manufacturing paper of an indefinite length, laid and
wove ' with separate moulds.' " On August 14, 1807, an
Act of Parliament was obtained for prolonging the term of
certain letters patent assigned to Henry Fourdrinier and
Sealy Fourdrinier for the invention of making paper by

means of machinery, and the machine described by John Gamble in the specifications of his patents (Nos. 2487 and 2708), with any other improvements added to it, was also fully described by diagrams.

During the next year, 1808, John Gamble assigned to the Messrs. Fourdrinier all his right in the patents as extended by the Act of Parliament, thus making the latter gentlemen the sole proprietors of the patent for the only satisfactory paper-making machine in England, and in this manner the machine invented by Robert, and improved by Gamble, assisted by the skill of Didot and Donkin, came to be and continues to be known as the Fourdrinier machine.

But let us now return to the history of paper-making in our own country.

After the adoption of the Constitution a stimulus was given to manufactures, and, although rags continued to be scarce for many years, we learn from estimates of Isaiah Thomas that there were in 1810 about one hundred and eighty-five paper-mills in the United States, distributed and producing as follows:—

	No. of mills.	Value of products.
Pennsylvania	60	$626,749
Massachusetts	38	290,951
New York	28	233,368
Connecticut	17	82,188
Vermont	9	70,050
New Hampshire	7	42,450
Kentucky	6	18,600
Rhode Island	4	52,297
Delaware	4	75,000
Virginia	4	22,400
Tennessee	4	15,500
Maryland	3	77,515
North Carolina	1	6,000

4

But as New Jersey is credited by Tench Coxe with producing paper in 1810 to the value of $49,750, Maine with $16,000, and Ohio with $10,000, it is probable that Mr. Thomas did not gather all the paper-mills into his estimate. The total value of all the paper produced in the United States in 1810 was therefore about $1,689,718.

On account of the increased consumption of rags used for the manufacture of paper, the United States commenced in 1810 to import rags from Europe.

It does not appear of record that our American inventors gave any great amount of attention prior to the war of 1812 to the matter of making paper by machinery.

By Act of Congress of April 10, 1790, the first American patent system was founded; but during the years from 1790 to 1812, our inventors confined themselves almost wholly to agricultural and commercial objects. Implements for tilling the soil and converting its products and machinery for navigation attracted most attention.

The war of 1812, however, forced our people to attempt production in many branches of manufacture and industry heretofore almost wholly uncultivated, and the result was the most remarkable development of human ingenuity ever known to any age or country. It is a source of great regret that no well-preserved history of American inventions dating from this time is in existence, and that no classified list of models which were in the Patent Office at the time of the fire in 1836 can be obtained. The earliest date that can be reached is January 21, 1823, and that is only partially complete.

After the fire in 1836, the United States government advertised for the patents which had been issued prior to the conflagration, and in this way some copies of the earlier patents were received.

The records of the United States Patent Office show that patents for manufacturing paper were issued to J. Biddis, May 31, 1794; C. Austin, December 14, 1798; and R. R. Livingston, October 28, 1799.

T. Langstroth was granted a patent for a paper-mill May 1, 1804; J. Tatterson, Southampton, Long Island, N. Y., was granted a patent on December 7, 1805, for a machine for preparing and hacking tow for paper; and a patent for a paper-making machine was issued May 8, 1807, to C. Kinsey, of Essex, N. J. But there are no specifications or other descriptions of any of these early patents now extant.

In 1809, Mr. John Dickinson invented and patented in England a new system of making paper in a continuous sheet by machinery; the apparatus consisted of a cylinder the periphery of which was covered with a metallic cloth properly supported; this cylinder revolved in a vat kept filled with pulp, in which the cylinder was half immersed. By a special system of suction a partial vacuum was created in the cylinder, thus causing the pulp to adhere to the metallic cloth and thereby forming the sheet, which being immediately detached passed upon a cylinder covered with felting.

Later, in 1826, Mr. Canson applied suction pumps to the Fourdrinier machine, causing a suction underneath the metallic cloth upon which the sheet is formed, uniting thereby to that machine the only advantage of the Dickinson invention.

While the Fourdrinier and Dickinson machines were being perfected in England, American inventors were also working in the same line, and in corroboration of this statement we have the paper-making machine, which, on December 24, 1816, was patented by Thomas Gilpin, of Philadelphia, and, which, in 1817, was put in operation in the paper manufactory of Messrs. Thomas Gilpin & Co., their mill being located on the Brandywine. This machine was doubtless suggested by Dickinson's invention, and was what is known as a cylinder machine, and it is stated that it would do the work of ten paper vats, and deliver a sheet of greater width than any other made in America, and of an indefinite length.

The war with England gave a great impulse to all branches of manufactures, and Thomas Gilpin covered the water-power on the Brandywine with large structures for the manufacture of wool and cotton, in addition to those for the manufacture of paper, which continued in their previous perfection.

The great reverses which, in a few years, befell the manufacturing establishments of the United States, produced disastrous effects on these large works, and under the circumstances it seemed expedient to suspend them until better times should come. Thomas Gilpin determined in this emergency to augment the paper works. So far, in the United States, all such works had been conducted upon the ancient system; but, in England, considerable advances had been made by the introduction of machinery, which produced paper in an endless sheet. Every publication on this subject had been carefully noted by Mr. Gilpin, and availing himself of all the published drawings explaining the parts of the new machinery, he became convinced, by careful study, that he could construct a machine which, if not exactly similar to, or as perfect as those of England, would enable him to produce paper of an indefinite length, and of a merchantable quality.

This effort on the part of Mr. Gilpin was attended with almost infinite trouble, but success crowned his efforts, and in February, 1817, he sent to Philadelphia paper cut from a continuous sheet.

Poulson's 'Daily Advertiser,' a leading gazette of the city, was the first publication printed upon this paper.

The enterprising firm of Mathew Carey & Sons, then the largest publishing house in the United States, were preparing an edition of the 'Historical Atlas of Lavoisne,' and the work appeared in 1821, printed on paper made by Thomas Gilpin's machine.

Mr. Gilpin was greatly encouraged by the success of his experiment, and he continued for several years after the machine was set to work in 1817 to make successive improvements, until it altogether superseded his other machinery, and promised a result not less valuable to the arts than remunerative to Mr. Gilpin for the years of anxious labor that the work had cost, and the large expense which he had incurred in perfecting the machine.

But Mr. Gilpin, like many other pioneers, was not destined to enjoy his well-earned reward, for, in the spring of 1822, the paper-mill and the valuable machinery were destroyed during a flood of unprecedented violence and magnitude, which occurred on the Brandywine.

The thousands of dollars which had been expended by Mr. Gilpin in perfecting his machine was a small loss compared with the study of many years, the numerous experiments, and the many mechanical improvements which had in a few hours been rendered abortive by the fury of the flood.

The failure of Congress to impose a suitable tariff on paper was a great drawback to the development of that industry in the United States after the close of the war of 1812.

The heavy importations of paper from Europe, which commenced soon after the war was ended, and the greatly depressed financial condition of the country, caused the almost total destruction of paper manufacture in the States of Pennsylvania and Delaware.

By Act of Congress of April 26, 1816, the rate of duty on paper imported into the United States after June 30, 1816, was to be fixed at thirty per cent.; but in the year 1820 we find the paper-makers of the last named States with those of Maryland earnestly petitioning Congress for an increased tariff on paper, the paper-makers of Pennsylvania and Delaware stating that in their district there were seventy paper-mills with ninety-five vats in operation until the importations after the war, since which they had been reduced to seventeen vats.

Congress itself was at this time using and continued for

some years afterwards to use paper imported from England and from France, but, under the continued criticism of the newspapers, paper made in the United States came finally to be used by Congress.

We have seen on page 50 that the value of the paper produced in the United States in 1810 was $1,689,718. In 1820 the value was estimated at $3,000,000, thus showing a great increase in spite of the large importations of paper from Europe, and the widespread and deeply seated period of bankruptcy which had intervened.

The period from 1820 to 1830 was not remarkable for progress in paper-making, either in the United States or in England.

The following list of patents will convey an idea of what was accomplished in this line in the United States from 1820 to 1830:—

Date.	Name.	Residence.	Nature of Invention.
May 14, 1822.	J. Ames,	Springfield, Mass.	Paper-making machine.
Sept. 1, 1822.	J. Ames,	Springfield, Mass.	Paper sizing.
Sept. 8, 1824.	I. Burbank,	Worcester, Mass.	Manufacturing paper.
April 12, 1826.	G. Burbank,	Worcester, Mass.	Paper-making machinery.
Feb. 28, 1827.	J. White and L. Gale,	Newburg, Vermont.	Finishing paper.
April 15, 1828.	E. H. Collier,	Plymouth Co., Mass.	Making paper from "Ulvamarina."
May 22, 1828.	W. Magaw,	Meadville, Pa.	Preparing hay, straw and other substances for making paper.
July 17, 1828.	M. Haddock,	New York, N. Y.	Machine for making paper in the sheet.
Sept. 11, 1828.	M. T. Beach,	Springfield, Mass.	Machine for cutting rags for paper.
Oct. 30, 1828.	A. & N. A. Sprague,	Fredonia, N. Y.	Manufacturing paper from corn husks.
Oct. 20, 1828.	M. Hunting,	Watertown, Mass.	Top press roller for making paper.
Jan. 13, 1829.	W. Debit,	East Hartford, Conn.	Machinery for cleaning rags for paper-mills.

Date.	Name.	Residence.	Nature of Invention.
Feb. 7, 1829.	J. W. Cooper,	Washingtown Twp., Pa.	White paper from rags, straw, and corn husks.
April 18, 1829.	I. Sanderson,	Milton, Mass.	Cylinder paper machine.
May 4, 1829.	R. Fairchild,	Trumbull, Conn.	Machine for manufacturing paper.
Sept. 10, 1829.	L. Bomeisler,	Philadelphia, Pa.	Manufacture of white paper from straw.

Efforts were made during the decade from 1820 to 1830 to introduce paper-making machinery from England; but on account of the high price few orders were given for it.

Cylinder machines of American invention met with some encouragement in the States of Massachusetts, Connecticut, and Pennsylvania after the year 1822, but for a long time they were very crude, and were used mostly for the lower grades of paper.

In 1831 it was estimated that the quantity of paper manufactured in the United States during the year 1830 amounted to more than seven millions of dollars.

From 1830 to 1840 there were no remarkable advances made in the manufacture of paper in the United States. Machinery was more commonly employed, the use of bleaching and other chemicals came to be better understood, and with the advances in its manufacture the price of paper declined, but its quality was better and the importations of paper gradually diminished while the exports increased.

In 1842 a convention of paper-makers was held in New York City, and an estimate then made, placed the value of the machinery and paper-mill property in the United States at $16,000,000, and the value of the paper manufactured at $15,000,000 per annum.

But there is great difference between the above estimate

and the census report for 1840, which places the value of all the paper manufactured in the United States during the year 1840 at \$5,641,495 and the capital invested at \$4,745,239.

During the whole of the decade from 1840 to 1850 the importations of paper constantly increased while the exports did not average more than one-fourth of the imports of that material.

In 1844 there was patented in Germany a machine for grinding wood for the manufacture of pulp. The inventor, Keller, sold the patent to the firm of Henry Voelter's Sons, who afterwards used the pulp in the manufacture of news paper.

The Voelters made numerous improvements in Keller's invention, and a quarter of a century after it was patented in Germany by Keller this wood-pulp machine was destined to play an important part in the United States, when in response to the demand for the rapid printing of daily news-papers the web press was to come into use. The Voelters, Christian and Henry, made numerous improvements in the machine, Christian Voelter obtaining patents in various European countries, in France even as early as April 11, 1847. Henry Voelter patented his improvement on the pulp machine in Wurtemburg, Germany, August 29, 1856, and in the United States, August 10, 1858.

Pearson C. Chenney, ex-governor of New Hampshire, has described the difficulty of introducing paper made from wood. In his testimony before the Senate Committee on Education and Labor, Mr. Chenney said: " When Mr. Russell built his mill at Franklin, those of us who were

engaged in the manufacture of paper and had no knowledge of what could be done with wood supposed that his
enterprise would ruin him. We supposed that his material
would be more like sawdust or clay. Mr. Russell completed his mills at Franklin, but after manufacturing the
pulp, he could not find a paper manufacturer who would
buy a pound of his wood pulp, because they did not believe
in it—they had no faith in it, and he was compelled to buy
a paper-mill in order to make a good test of it, which he
did in Franklin, right beside his pulp-mill, and made the
test, and a successful test, and showed a very good paper.
After the paper was made he found great difficulty in selling
it. The printers felt that they could not use it; they were
afraid to use paper made from raw wood; they were afraid
it would injure their type or ruin it, and they declined to
use it. His selling agents were the firm of Rice & Kendall,
of Boston. They resorted to all sorts of devices to get this
paper used, but they were finally obliged to resort to something that did not appear on the surface, but seemed to be
necessary in order to secure the introduction of the paper
into use. They had an order from, I think the ' Boston
Herald' for about 500 reams of paper. They were supplying
that journal regularly from month to month, and, without
saying anything as to the nature of the paper, they sent
paper made from this wood; the paper passed, and was used,
and when the next order came and they delivered the regular paper which they had been in the habit of sending before,
the ' Herald' people came to Mr. Rice in some displeasure,
and asked him why he could not send such paper as he had

sent the month previous. He told them that he could do so if they preferred it, and they said they did. They said that it worked very well—very much better than the other. So he told them that the next order they gave him he would send some of that paper. The next month he again delivered 500 reams of the wood paper, and that was used and gave great satisfaction. But I think they were using it for 'six months before they knew that it was wood paper. That established the use of that class of paper, and there was no trouble after that in selling it. The fact is that it absorbs the ink better and works much better for printing than other paper does, and works particularly well in rapid presses."

If we compare the methods of manufacturing paper in the United States during the decade from 1840 to 1850 with those in use at the present time, the result will, of course, in many respects be greatly in favor of the present methods. But as the plants and products were small, wages low, and the margins of profits large, it was easier in those days to do business with a small capital than it now is with a large one.

Forty years ago the present method of boiling rags in rotary boilers under pressure was not employed, the boiling was then done in open kettles or tubs. Instead of working different grades of stock separately and uniting them in the beating engines as required, hard and soft stock was in those days commonly boiled and worked together; the proportions varying according to thickness, strength, and quality desired. The washing and beating engines then employed were small, averaging from about one-fourth to one-eighth the size of those now used. Light, narrow, slow-running cylinder machines were almost exclusively employed. There was still considerable waste in the use of chemicals and loss in labor. Elevators from drainers to engine-room and from the machine-room to the loft were unknown, the paper being carried to the loft on men's shoulders.

The almost general employment of the Fourdrinier machine in Europe and the numerous improvements which were made in paper-making in England from 1840 to 1850, added to the enforcement in 1843 of a new duty on rags, operated to check our exports and increase our imports of paper, and it is questionable whether any material advance was made in the quantity of paper manufactured in the United States during the last-named decade over that from 1830 to 1840.

The year 1850 opened with a bright prospect for all branches of trade in the United States; the new empire which was arising and following the discovery of gold on the Pacific coast, and the new markets abroad which were being opened for our grain and cotton acted as powerful stimulants to many branches of manufacture.

It was not long before many of the paper-makers, especially in Massachusetts, New York, and Connecticut, were compelled to enlarge their capacities, and the Fourdrinier machine came into more common use. Many old and narrow cylinder machines were removed and superseded by new and wider ones having steam dryers and other improvements. Larger beaters and washers were also introduced;

but the demand for paper was greatly in excess of the supply, and notwithstanding the continued introduction of improved machinery the price of paper constantly advanced until about 1855.

But on account of the large amount of capital which had been drawn into the business of paper-making the effects of competition were severely felt from 1855 to 1860, and the profits, especially on fine papers, were reduced below the average profits of other branches of manufacture; an overstocked market soon played sad havoc with prices, and the final result was that many mills were compelled to shut down.

The census of 1860 showed that the State of Massachusetts alone in that year produced paper to the value of $5,968,469; New York returned paper to the value of $3,516,276; Connecticut, $2,528,759; and Pennsylvania, $1,785,900.

The decade from 1860 to 1870 was a memorable one in the annals of paper-making in the United States. The enormous rise in the price of cotton which immediately followed the outbreak of the civil war caused paper to be used for many purposes for which cotton had formerly been employed. Paper twine, paper collars, cuffs, and shirt fronts, are some of the new applications of paper which at once consumed very large quantities of the material.

Early in 1862 the price of ordinary news paper ruled at 8 cents per pound, but before ten months of the year had passed the price was increased to 17 cents, and No. 1 printing was 30 cents, while all writing papers were 40 cents.

The prices of paper continued to increase, until in 1864 news paper sold at 28 cents per pound, and fine book paper at 45 cents. But in January, 1865, the price of common news paper, straw, etc., in New York City ranged from 20 to 22 cents per pound, and the price of good news paper, rag, was from 22 to 25 cents; fair white book was 25 to 28 cents, and extra book from 28 to 32 cents. The average price for first class writing, folded, during 1865, was $52\frac{1}{2}$ cents per pound; superfine writing, folded, 50 cents; superfine flats, 45 cents. Manilla wrapping paper ranged from 18 to 21 cents per pound during 1865.

From 1865 to 1870 the price of all kinds of paper continued rapidly to decline, and from 1870 to 1885 there remains but little to notice, excepting the introduction of wood pulp into paper-making, and the enormous consumption of straw, etc., for news paper.

For Manilla paper, jute butts and threads, coming principally from India, but partly from Dundee, are largely used in the United States; old cordage, which is chiefly supplied by the shipping yards of England, is also largely employed in the manufacture of Manilla paper.

During the twenty years from 1865 to 1885 the largest number of patents relating to paper making were issued by the government of the United States that has ever been known in the history of any country.

The invention and employment of a large number of mechanical appliances and chemical processes have done much to stimulate and cheapen paper production, especially during the fifteen years just past (1870–1885), and many manufacturers who have failed to keep fully abreast of the times have found their business absorbed by more enterprising firms. Such results are only natural, but it is even more important in the future than in the past that small manufacturers should be fully informed regarding all the improvements in the art, for, under the enormous output of paper from many of the large mills in this country, it is becoming a very serious question whether even mills of moderate capacity can afford to manufacture paper on the small margin of profit which produces a meagre but satisfactory dividend to the wealthy owners of the capital stock in the larger mills.

However, such questions as the latter are matters which do not come within the province of this work, and we shall, therefore, not further discuss them in connection with the history of paper-making, but proceed in the following chapters to describe the various materials, processes, and appliances employed in the modern methods of manufacturing paper.

CHAPTER II.

MATERIALS USED FOR PAPER—MICROGRAPHICAL STUDY OF THE
MANUFACTURE OF PAPER — CELLULOSE — DETERMINATION OF
CELLULOSE—RECOGNITION OF VEGETABLE FIBRES.

DURING the past century the consumption of paper has
increased in a much greater ratio than the supply of rags,
and on account of the consequent advance in the price of
the latter, numerous other raw materials have been employed in the manufacture of the lower grades of paper.

Cotton and linen rags continue to be employed for the
higher class of writing and record papers; but the pulps
for news and cheap book papers are almost wholly produced from wood, straw, esparto, corn stock, old papers, etc.

It would be almost an impossibility to enumerate all the
materials which have been used for the manufacture of pulp
for paper; but in the following list the author has arranged
in alphabetical order all those paper-making substances
concerning which he has acquired any information, through
diligent research :—

MATERIALS USED FOR PAPER.

Abaca, same as Manilla hemp.

Abelmoschus esculentus, { Ochra,
Okra,
Okro.

Abutilon avicennæ.

Abutilon Bedfordianum, Hollyhock-
tree.

Abutilon Indicum, Indian mallow.

Abutilon mollis, Hollyhock-tree.
Soft-leaved abutilon.

Abutilon strictum, Hollyhock-tree.
    Vined lantern flower.
Abrasive cloth and paper, waste.
Acacia.
Acacia, *Robinia Pseud-Acacia,* wood of.
Adam's Needle, *Hibiscus heterophyllus.*
    Sparmannia Africana.
    Yucca gloriosa.
Adamsonia digitata, Baobab.
Æsculus hippocastanum, Horse-chestnut, wood of.
Agave Americana, Maguey.
Agave, *Agave Americana.*
Agave Americana, leaves of.
Agave, *Fourcroya gigantea.*
Agave Mexicana, Maguey.
Agrostis spica venti, Bent-grass.
Ailanthus, bark of.
Alder, *Alnus glutinosa,* wood of.
Alder-buckthorn. Same as Black alder.
Alfa fibre.
Alga marina.
Algæ, fresh-water.
Alnus glutinosa, Alder, wood of.
Aloe fibre.
Aloes.
Alpina magnifica, Mulberry.
Alsimastrum.
Althea.
Althea frutex, Cockle-burr.
Ambaree.
Amianthus.
Amomum.
Anacharsis.
Ananassa.
Ananassa Sativa, Pineapple.
Animal excrement.
Animal substances.
Anona reticulator, Nona.
Anonaceæ.
Apocineæ.
Aporentype.
Aralia papyrifera.
Arrache.

Arrowroot, refuse stems and leaves of.
Artemisia bark.
Artemisia wood.
Artichoke.
Artiplex.
Artocarpeæ.
Arunda conspicua, Plume-grass.
Arundinaria macrosperma.
Asclepiadiæ.
Asclepias.
Ash, *Fraxinus excelsior,* wood of.
Asparagus stalks, *Asparagus officinalis.*
Aspen-tree, *Populus tremula,* wood of.
Asthoder.
Avena Sativa, Oats.

Bagasse.
Bagasse refuse.
Bagging, old.
Baldengera Arundinacia, Ftomenteau.
Bamboo, *Bambusa thonarsu,* wood of.
Bambusa arundinacea.
Bambusa thornarsu, Bamboo.
Bambusa vulgaris, Bamboo.
    Inner bark of.
    Leaves of.
    Young shoots of.
Banana fibre and leaves.
Banhinia racemosa.
Baobab, *Adamsonia digitata.*
    Heliconia gigantea.
    Strelitzia regina.
Bark of various kinds of woods, including resinous, etc.
Bark of coniferous trees after extracting the resin.
Barley straw.
Barriala, *Sida rhomboida.*
Basswood.
Bastard Cedar or Guazuma.
Bean leaves and vines.
Beans.
Beech, *Fagus sylvatica,* wood of.

5

Beet and mangel-wurzel root.

Beets.

Begonaceæ.

Bent-grass, *Agrostis spica-venti.*

Berries.

Betula alba, Birch, wood of.

Betula Bhojpattra, Birch.

Bhenda, *Hibiscus esculentus.*

Bhurja (Birch), *Betula Bhojpattra.*

Birch, *Betula alba,* wood of.

    Betula Bhojpattra.

Black alder, *Rhamnus Fragula,* wood of.

Blackberries.

Black-moss, *Tillandsia.*

Black reed (cutting-grass), *Cladium radula.*

Blue cabbage stalks.

Blue-flag, *Enodium cœruleum.*

Blue-grass, *Agrostis spica-venti.*

Bœhmeria.

Bœhmeria vivea, China-grass (Rhea).

Bombax.

Bon-dhenras, *Hibiscus ficulneus.*

Bottle-tree, *Sterculia diversifolia.*

    Sterculia fœtida.

    Sterculia lucida.

    Sterculia repestris.

Bowstring hemp, *Sanseviera Zeylanica.*

Brachychiton acerifolium, Flame-tree.

Bracken.

Brake.

Bran.

Brank. Same as Buckwheat.

Brazil-wood.

Brazilian-grass.

Brewery refuse.

Bromeliaceæ.

Bromelia Pinguin, Pineapple.

Bromelia sylvestris, Pineapple.

Broom.

Broomcorn.

Broom-leaved tea-tree, *Melaleuca genistifolia.*

Broom swamp.

Broussonetia.

Broussonetia papyrifera, Paper mulberry.

Brown-grass.

Brown-hemp (see Sunn).

Bryon.

Buckwheat, *Fagopyrum esculentum.*

Bulrush, *Typha angustifolia.*

Burdock.

Burlap bagging.

Button-tree, } Same as Plane-tree.
Buttonwood. }

Cabbage.

Cabbage stumps.

Cactus.

Calamus verus, Rattan, wood of.

Calotropis gigantea, Yucca. Mudar.

Camelina, *Camelina sativa.*

Canadian poplar, *Populus Canadensis,* wood of.

Canary-grass, *Phalaris canariensis.*

Canes.

Cannabis sativa, Kangra hemp.

Canvas, old.

Cardamom, refuse stems and leaves of.

Carex, Sedge.

Carex appressa, { Sedge-grass.
{ Stems of.

Carex pseudo-cypherus, { Gallingall
{ rush.
{ Stems of.

Carludovica palmata, Panama hat straw.

Caryota urens, { Jaggery palm.
{ Stems and leaves of.

Catkins of black poplar.

Cat-tail, *Typha latifolia.*

Cecropia.

Cedar-wood.

Cenodruli.

Cereal and leguminous plants, straws of.

Chaff.

China-grass (Rhea fibre), *Bœhmeria nivea.*

Chinese sugar-cane.

Cissus family.

Claudium radula, Black reed.

Clematis bark and wood.

Clematite.

Cloth, refuse.

Clover.

Club rush, *Scirpus fluviatilis.*

Coast sword-grass, *Lepidosperma elatius.*

Cockle-burr, *Althea frutex.*

Cocoons, silk, refuse of.

Coir.

Colocasia antiquorum, Sago-palm.

Coltsfoot.

Commersonia Fraseri, Lye-plant.

Common rush, *Juncus pauciflorus.*

Common tea-tree, *Melalenca ericifolia.*

Compositæ.

Conferva.

Conferva sp., Swamp moss, stems and leaves of.

Coniferæ, leaves of.

Coniferous trees, bark of, after extracting the resin.

Convolvulaceæ.

Coral moss.

Corchorus olitorius, Jute.

Cordage.

Cord-grass, *Spartina cynosuroides.*

Cordia Myxa, Mulberry.

Cordyline indivisa, Tall palm-lily, stems and leaves of.

Cork.

Corn cobs, husks, leaves, and stalks.

Corylus Avellana, Filbert-tree, wood of.

Cow-itch-tree, *Lagunaria Patersonii.*

Coton du peuplier.

Cotton.

Cotton plant, fibre of the.

Bark of the root and bark of the stalk, pith, seed, and stalks of the different species.

Cotton rags.

Cotton-seed waste.

Cotton waste.

Couch-grass. Same as Dog-grass.

Crotalaria juncea, Sunn.

Crotalaria tenuifolia, Jubbulpore hemp.

Cruciferæ.

Cryptogams.

Cucumbers.

Cucurbitaceæ.

Cudweed. Same as Everlasting.

Cupheæ.

Currijong, *Dais cotinifolia.*

Pimellea axiflora.

Pittosporum cornifolium.

Cyperus dives, Diss.

Cyperus lucidus, Galingale rush, stems and leaves of.

Cyperus tegetum.

Cyperus sp., stems and leaves of.

Cyprian asbestos.

Dais continifolia, Currijong.

Danchi, *Sesbanio aculeata.*

Daphne.

Daphne cannabina.

Daphne oleoides.

Datura stramonium, Stramonium.

Decayed wood.

Dhoncha, *Sesbania aculeata.*

Diannella latifolia, Sea-coast rush, stems and leaves of.

Diannella longifolia.

Dismodium argenteum.

Diss, *Cyperus dives.*

Dog-grass, *Triticum repens.*

Dog-wheat. Same as Dog-grass.

Dolbega Natalensis, Lye-plant.

Doryanthes excelsa, Spear lily.

Dracæna Draco, Dragon-tree.

Dungar.

Durst.

Dust.
Dwarf-palm.
Dyer's wood, stalks of.
Dye-woods, spent.

Earth-moss.
Edgeworthia Gardneri.
Ehrharta tenacissima, Wire-grass, stems of.
Ejoo.
Elder.
Elm, *Ulmus campestris,* wood of.
Enodium cœruleum, Blue-flag.
Emery cloth, } waste.
Emery paper, }
Ericaceœ.
Erica vulgaris, Heath, wood of.
Eriphorum cannabinum.
Erigeron.
Erythroxylon guttafereœ.
Esclepias, down of.
Esparto, all varieties of.
Eucalyptus fissilis, Messmate.
Eucalyptus obliqua, Stringy bark.
Euphorbiaceœ.
Everlasting, *Gnaphalium.*
Excrement of animals.

Fagopyrum esculentum, Buckwheat.
Fenequen.
Ferns.
Few-flowered rush, *Juncus pauciflorus.*
Fibres, various.
Fibrila.
Ficus speciosa, Fig-tree.
Filbert-tree, *Corylus Arellana,* wood of.
Fir-cones.
Fir, *Pinus sylvestris,* wood of.
Fishing nets, old.
Flame-tree, *Brachychiton acerifolium.*
              Sterculia acerifolia.
Flax, hemp, etc.

Floss silk.
Fourcroya gigantea, Agave.
              Giant lily (Agave), stems and leaves of.
Fourdrini.
Fraxinus excelsior, Ash, wood of.
Frog-spittle.
Ftomenteau, *Baldenegra Arundinacea.*
Fucus vesiculosus, Varec.

Gahnia psittacorum, Sword-grass.
      Var. Erythrocarpum, stems and leaves of.
Galega officinalis.
Galega orientalis.
Galingale.
Galingale rush, *Carex pseudo-cyperus.*
              Cyperus lucidus.
Geld, *Marsdenia tenacissima.*
Genista.
              After extracting dye.
Giant nettle, *Urticati divaricata.*
Ginger, refuse stems and leaves of.
Glyheria aquatica, Marsh-grass. Called also Reed-grass.
Gnaphalium.
Gombo, *Hibiscus syriacus.*
Gossypium.
Grains.
Grape-vine, inner and outer bark.
Grape-vines.
Grass-cloth plant (Chinese), *Bœhmeria nivea.*
Grasses.
Grass, Spanish.
      Tule.
Grewia oppositifolia.
Guazuma or Bastard cedar.
Guazuma ulmifolia, Urania.
Gunny.
Gunny bags, old.
Gun-cotton.
Gutta-percha.

Gynerium argenteum, Pampas-grass, stems and leaves of.

Hair.

Halfa, *Lignum Spatium*.

Hawthorne, bark of.

Hay.

Heath, *Erica vulgaris*, wood of.

Heather.

Heliconia gigantea, Baobab.

Hemp.

Hemp-fibres.

Hemp, flax, etc.

Hemp, jute, dressed.

Herbaceous plants—

	Yield per cent. of fibre.
Asparagus stalks, *Asparagus officinalis*	32.56
Banana, *Musa ensete*	31.81
Barley, *Hordeum vulgare*	36.21
Bent-grass, *Agrostis spica-venti*	45.82
Blue-flag, *Enodium cæruleum*	40.07
Buckwheat, *Fagopyrum vsculentum*	30.60
Camelina, *Camelina sativa*	29.16
Canary-grass, *Phalaris Canariensis*	44.16
Canna, *Canna*	20.29
Dog-grass, *Triticum repens*	28.38
Ftomenteau, *Baldengera Arundinacia*	46.17
Giant nettle, *Urtica divaricata*	21.66
Hop, *Humulus Lupulus*	34.84
Maize, *Zea Mays*	40.24
Marsh-grass, *Glyceria aquatica*	38.80
Marsh rush, *Scirpus palustris*	41.70
Mateva, *Hyphœne Thebaica*	26.08
New Zealand flax, *Phormium tenax*	32.71
Oats, *Avena sativa*	35.08
Reed, *Phragmites vulgaris*	41.57
Rye, *Secale cereale*	44.12
Sedge, *Carex*	33.86
Sugar-cane, *Saccharum officinarum*	29.15
Wheat, *Triticum sativum*	43.14
Wild broom, *Spartium scoparium*	32.43

Hibiscus arboreus, Mohant-tree.

Hibiscus cannabinus, Meshta.

Hibiscus esculentus, Bhendi.

Hibiscus fibre.

Hibiscus ficulneus, Bon-dheuras.

Hibiscus heterophyllus, Adam's Needle.

Hibiscus Moscheutos.

Hibiscus mutabilis, Stolpoddo.

Hibiscus Palustris.

Hibiscus Rosa-sinensis, Joba.

Hibiscus Sabdariffa, Meshta.
        Reselle.

Hibiscus splendens, Hollyhock-tree.

Hides.

Hollyhock-tree, *Abutilon Bedfordianum*.

 Abutilon mollis.

 Abutilon striatum.

 Abutilon venosum.

 Hibiscus splendens.

Hop bark.

Hop-bind.

Hop-plant, *Humulus Lupulus*.

Hops.

Hop vines.

Hordeum murinum, Rye-grass.

Hordeum vulgare, Barley.

Hornets' nests.

Horse-chestnut, *Æsculus Hippocastanum*, wood of.

Horse-chestnut leaves.

Horse-radish.

Humulus Lupulus, Hop-plant.

Hydrangea sp.

Hyphœne Thebaica, Mateva.

Ife-tree, *Sanseviera Zeylanica*.

Immortelle, same as Everlasting.

Indian-corn husks.

Indian mallow, *Abutilon Indicum*.

India-rubber fibre.

Indigo.

Imperfections (waste papers).

Iris.

Isolepus nodosa, { River rush.
                   { Stems of.
Italian poplar, *Populus Italica,* wood of.
Ivory.
Ivory shavings.
Ixora cuneifolia, Mulberry.

Jaggery palm, *Caryota urens.*
Joba, *Hibiscus Rosa-sinensis.*
Jubbulpore hemp, *Crotalaria tenuifolia.*
Jucca, Yucca.
Juncaceæ.
Juncus Gesneri.
Juncus maritimus, Sea-coast rush.
        Stems and leaves of.
Juncus pauciflorus, Common rush.
        Stems of.
Juncus vaginatus, Small sheathed rush.
        Stems and leaves of.
Juniper.
Jute butts.
Jute canvas.
Jute, *Corchorus olitorius.*
Jute rags.
Jute rejections.
Jute rope.
Jute sackcloth.

Kangra hemp.
Killed paper-stock.
Knot-grass.   Same as Dog-grass.

Lace from aloe fibre.
Lagunaria Patersonii, Cow-itch-tree.
Laportia gigus, Tree nettle.
Leather.
Leather cuttings and skivings.
Leaves.
Leaves of trees.
        Withered.
Leguminous plants.
Leguminous and cereal plants, stems of.
Lentils.

Lepidosperma elatius, { Coast sword-
                        {   grass
                        { Sword-grass.
                        { Tall sword-
                        {   grass, leaves
                        {   and stems of.
Lepidosperma flexuosa, Slender sword-
        grass.
Lepidosperma gladiata, Coast sword-
        rush, stems and leaves of.
Libitaæ.
Libertia formosa.
Lichens.
Ligneous meal (Wood-flour pulp).
Lignum Spartium, Halfa.
Liliaceæ.
Lily of the valley leaves.
Lily roots.
Lily stalks.
Lime-tree, { *Tilia Europæa,* wood of.
           { Bark of the.
Linden.
Linden leaves.
Linen.
Linen rags.
Liquorice root.
Liquorice wood.
Livistonia mauritiana, Sago-palm.
Lolium perenne, Rye-grass.
Long moss, *Tillandsia.*
Lucerne.
Lucitodium equisetum.
Lychnophora.
Lye-plant, *Commersonia Fraseri.*
        Dombega Natalensis.

Macrochola Tenacissima.
Madar.   (See Mudar.)
Madder.
Madgascariensis, Urania.
Madras hemp.   (See Sunn.)
Maguey, *Agave Americana.*
        Agave Mexicana.
Maize, *Zea mays.*

Maize, cobs, husks, leaves, and stalks.
Mallow.
Malpighiceæ.
Malva.
Malvaceæ.
Mandioc, *Jatropha manihot.*
Mangel-wurzel.
Manilla.
Manilla hemp, *Musa textilis.*
Manispernum.
Manure.
Maple.
Marica Northiana, stems and leaves of.
Marsdenia tenacissima, Geld.
Marsh-grass, *Glyceria aquatica.*
Marsh rush, *Scirpus palustris.*
Marzi.
Masse-d'eau.
Mateva, *Hyphœne Thebaica.*
Medichey.
Melaleuca ericifolia, { Common tea-tree. Swamp tea-tree
Melalenca genistifolia, Broom-leaved tea-tree.
Melalanca squarrosa, Victorian yellow wood.
Melochia liliacefolia, Urania.
Melastomaceæ.
Melic-grass, *Molinea cœrulea,*
Meshta, *Hibiecus cannbinus.*
    Hibiscus Sabdariffa.
Messmate, *Eucalyptus fissilis.*
Milkweed, Asclepias.
Mineral fibre.
Mohant-tree, *Hibiscus arboreus.*
Molinea cœrulea, Melic-grass.
Moorva.
Moova.
Morus bark.
Morus tartarica, Mulberry.
Mosses.
Mothwort.
Mudar, *Calotropis gigantea.*

Mugwort, stalks of.
Mulberry (Kuwa, Japan).
    Alpina magnifica.
    Cordia myxa.
    Ixona cuneifolia.
    Morus tartarica.
    Inner bark.
Mulberry trees.
Mulberry wood.
Mummies and cloth.
Mummy cloth.
Musaceæ.
Musa ensete, Banana.
Musa paradisiaca, Plantain.
Musa sapientum, Plantain.
Musa textilis, Manilla hemp.
Muscovy match.
Mustard.
Mya-grass.
Myrtaceæ.

Native tussock-grass, *Xerotes longifolia.*
Navy stores, old.
Neilgherry nettle, *Urtica heterophylla.*
Nepal paper plant, *Daphne cannabina.*
Nets, old fishing.
Nettle, *Urtica incisa.*
Nettle bark.
Nettle wood.
Nettles, stinging, *Urtica.*
Nettles, stingless, *Bœhmeria.*
New Zealand flax. *Phormium tenax.*
Nona, *Anona reticulata.*

Oak, *Quercus robur,* wood of.
Oak leaves.
Oakum.
Oats, *Avena sativa.*
Okra, { *Abelmoschus esculentus.*
Okro,
Onocarpus batava.
Orach (written also Orache, Arrache, and Orrach).

Oryza, Rice plant.
Osier, *Salix alba*, wood of.
Oryza, Rice plant.

Paddy straw.
Palm, dwarf.
Palm fibre.
Palm, palmetto.
Palmetto and chamoerops (Palmetto cabbage).
Palmetto fibre.
Palmyra, leaves of.
Palygaleæ.
Pampas-grass, *Gynerium argenteum*.
Panama-hat straw, *Carludovica palmata*.
Pandanus.
Pandanus utilis, Screw pine.
Pandamus utilis, stems and leaves of.
Paper cuttings.
Paper mulberry, *Broussonetia papyrifero*.
Paper mulberry, bark of.
Paper, old.
Pappus.
Papyrus.
Papyrus, raw fibre of.
Parkinsonia aculeata.
Pasteboard scraps.
Pat, same as Jute.
Peas.
Pea-stalks.
Peat.
Pederic fœtida.
Phalaria Canariensis, Canary-grass.
Phormium tenax, New Zealand flax.
    Stems and leaves of.
Phragmites vulgaris, Reed.
Pimelea axiflora, Currijong.
Pineapple, *Ananassa Sativa*.
    Bromelia Pinguin.
    Bromelia sylvestris.
Pineapple leaves.
Pine-cones.

Pine-leaves.
Pine-shavings.
Pine-tree, inner bark of.
Pinus Australis, Pitch-pine, wood of.
Pinus sylvestris, Fir, wood of.
Pinus sylvestris rubra, Red pine, wood of.
Pipturus propinquus, Queensland grass-cloth plant.
Pisang.
Pita.
Pitch pine, *Pinus Australis*, wood of.
Pittosporum cornifolium, Currijong.
Pittosporum crassifolium, Thick-leaved pittosporum.
Plagianthus betulinus, Ribbon-tree.
Plane-tree, *Platanus occidentalis*.
Plantain, *Musa paradisiaca*.
    Musa sapientum.
Platanus occidentalis, Plane-tree.
Plume-grass, *Arunda conspicua*.
Poa Australis, Wire-grass.
Poke-weed, *Phytolacca decandra*.
Pollen of plants.
Poplar.
Poplar down.
Poppy.
Popylus Italica, Italian poplar.
Populus ciliata.
Populus tremula, Aspen-tree, wood of.
Potari, *Abutilon Indicum*.
Potatoes.
Potato skins.
Potato vines.
Pourretia plantanifolia.
Printed waste.
Pteris.
Pterospermum acerifolium, Urania.
Pulps, wood, straw, and other.
Pulu.
Puyba Bœhmeria.

Queensland grass-cloth plant, *Pipturus propinquus*.

Queensland hemp, *Sida retusa.*
Quercus robur, Red oak, wood of.
Quickens, same as Dog-grass.

Rag-bagging.
Rags.
Ramie.
Ram turai of India.
Raspberry.
Rattan, *Calamus verus,* wood of,
Red pine, *Pinus sylvestris rubra,* wood of.
Reed, *Phragmites vulgaris.*
Reed-grass, meadow, *Glyceria aquatica.*
Reeds.
Rhamneæ.
Rhamnus Fragula, Black alder, wood of.
Rhea-fibre.
Rhubarb.
Ribbon-tree, *Plagianthus betulinus.*
Rice-plant, *Oryza.*
Rice, stalks of the wild.
Rice-straw.
Ricinus.
River rush, *Isolepis nodosa.*
Robinia Pseudo-Acacia, Acacia, wood of.
Roots.
Roots of grasses.
Rope.
Rosaceæ.
Roselle, *Hibiscus Sabdariffa.*
Rose-mallow.
Rubiaceæ.
Rushes.
Rutaceæ.
Rye, *Secale cereale.*
Rye-grass, *Hordem murinum.*
       Lolium perenne.

Saccharum munja.
Saccharum officinarum, Sugar-cane.

Sacks, old.
Sago.
Sago-palm, *Colocasia antiquorum.*
       Livistonia mauritiana.
       Sagus ruffia.
       Sanseviera cylindrica.
       Sanseviera latifolia.
       Sanseviera Zebrina.
Sagus ruffia, Sago-palm.
Sagus saccherifera, Sago-palm.
Sails, old.
Salix alba, Osier, wood of.
       Willow, wood of.
Salt hay, *Spartina juncea.*
Salvia Canariensis.
Sandpaper waste.
Sanseviera cylindrica, ⎫
Sanseviera latifolia,   ⎬ Sago-palm.
Sanseviera Zebrina,   ⎭
                  ⎧ Bowstring
Sanseviera Zelanica, ⎨ hemp.
                  ⎩ Ife-tree.
Satin.
Sawdust.
Scirpus fluviatilis, Club rush.
       Stems and leaves of.
Scirpus lœnstris.
Scirpus palustris, Marsh rush.
Scotch ferns.
Screw pine, *Pandanus utilis,* stems and leaves of.
Seacoast rush, *Dianella latifolia.*
       Dianella longifolia.
       Juncus maritimus.
       Juncus vaginatus.
Sea-grass.
Sea-mallow.
Sea-weeds.
Sea-wrack.   (See Varec.)
Secole cereale, Rye.
Secrate.
Sedge, stalks and roots of.
Sedge-grass, *Carex appressa.*
Seed down of thistles.

Seines, old.
Seratula ervansis.
Sesbania aculeata, Danchi.
      Dhonchi.
Shavings.
      Ground.
      Paper.
      Wood.
Shingles, old.
Sida.
Sida pulchella, Victorian hemp.
Sida rhomboida, Barriala.
Sida tiliæfolia.
Silk.
Silk cocoon, refuse of.
Silk plant, *Asclepias.*
Silk, refuse.
Sisal-grass, } *Agave Americana.*
Sisal-hemp, }
Skins, pieces of.
Slender sword-grass (Mat-grass), *Lepidosperma flexuosa.*
Small-sheathed rush, *Juncus vaginatus,* stems and leaves of.
Soft-leaved abutilon, *Abutilon mollis.*
Solaneæ.
Solonaceæ.
Sorghum.
Sorghum, refuse.
Sorgo sucre, Chinese sugar-cane.
Sotal tree.
Spanish Bayonet, *Yucca aloifolia.*
Spanish Broom, *Macrochola Tenacissima.*
Spanish-grass.
Sparganium family.
Sparmannia Africana, Adam's Needle.
Spartina cynosuroides, Cord-grass.
Spartina juncea, Salt hay.
Spartium junceum, Spanish-broom.
Spartium scoparium, Wheat.
Spear lily, *Doryanthes excelsa.*
Spindle-tree.
Spruce, firewood of.

Stems and leaves of—
    Coast rush, *Juncus maritimus.*
    Coast sword rush, *Lepidosperma gladiatum.*
    Cypherus lucidus.
    Cypherus sp.
    Dianella latifolia.
    Gahnia psittacorum, Var. erythrocarpum.
    Giant lily, Agave, *Fourcroya gigantea.*
    Jaggery palm, *Caryota urens.*
    Marica Northiana.
    Native bulrush.
    Native tussock-grass, *Zerotes longifolia.*
    New Zealand flax, *Phormium tenax.*
    Pampas-grass, *Arundo conspicuo.*
    Scirpus fluviatilis.
    Screw pine, *Pandanus utilis.*
    Small-sheathed rush, *Juncus vaginatus.*
    Swamp moss, *Confeva sp.*
    Tall palm lily, *Cordyline indivisa.*
    Tall sword rush, *Lepidosperma clatius.*
Stems of—
    Carex appressa.
    Carex pseudo-cyperus.
    Ehrharta tenacissima.
    Few-flowered rush, *Juncus pauciflorus.*
    Isonepeis nodosa.
    Victorian nettle, *Urtica incisa.*
    Sterculia acerifolia, Flame-tree.
    Sterculia diversifolia, Victorian bottle-tree.
    Sterculia fœtida, ⎫
    Sterculia lucida, ⎪
    Sterculia represtris, ⎬ Bottle-tree.
    Sterculia villosa. ⎪
    Sterculia urens. ⎭
    Stinging nettle.
    Stipa tenacissima.

Stolpoddo, *Hibiscus mutabilis.*
Stone.
Stramonium, *Datura Stramonium.*
Straws of cereal and leguminous plants.
Straw paper, old.
Strelitza regina, Baobab.
Stringy bark, *Eucalyptus obliqua.*
Stypa spartum.
Sugar-cane, *Saccharum officinarum.*
Sugar-cane leaves.
Sultana bark.
Sunflower.
Sun-hemp. (See Sunn.)
Sunn, *Crotalaria juncea.*
Swamp moss.
Swamp tea-tree, *Melaleuca ericifolia.*
   Melaleuca genistifolia.
   Melaleuca squarrosa.
Sweet broom.
Sword-grass, *Gahnia psittacorum.*
   Lepidosperma elatius.

Tall palm lily, *Cordyline indivisa,* stems
 and leaves of.
Tall sword rush, *Lepidosperma elatius,*
 stems and leaves of.
Tan.
Tan bark, etc.
Tan spent.
Tarred rope.
Terebinthenaceæ.
Thalipot, leaves of.
Thick-leaved pittosporum, *Pittosporum*
 crassifolium.
Thistle-down.
Thistle-stalks.
Thistles.
Threads.
Tikkur, refuse stems and leaves of.
Tilia Europæa, Lime-tree, wood of.
Tillandsia.
Timothy.
Tobacco.
Tousles Mois, annual stems and leaves.

Tow.
Tracena endivisa.
Tree moss.
Tree nettle, *Laportia gigus.*
Traphis.
Triticum repens, Dog-grass (also called
 Couch-grass, Dog-wheat, Knot-grass,
 Twitch-grass, Quitch and Quickens).
Triticum vulgare, Wheat.
Tulip leaves.
Turmeric, refuse stems and leaves of.
Turf.
Turnips.
Tussock-grass, *Xerotes longifolia.*
Twine.
Twitch-grass, same as Dog-grass.
Typha angustifolia, Bulrush.
  Stems and leaves of.
Typha latifolia, Cat-tail.

Ulmus.
Ulmus campestris, Elm, wood of.
Ulva marina.
Urania, *Guazama ulmifolia.*
 (Ravenda) *Madgascariensis.*
 Melochia liliacefolia.
 Pterospermum acerifolium.
Urtica divaricata, Giant nettle.
Urtica heterophylla, Neilgherry nettle.
Urtica incisa, Victorian nettle.
Urticeæ.
Usnea (Lichens).

Varec, } *Fucus vesiculosus.*
Vareek, }
Vegetable fibres, raw.
Velloziæ.
Victorian bottle-tree, *Sterculia diversi-*
 folia.
Victorian hemp, *Sida pulchella.*
Victorian nettle, *Urtica incisa,* stems of.
Victorian yellow-wood, *Melaleuca*
 squarrosa.
Vined lantern-flower, *Abutilon venosum.*

MICROGRAPHIC STUDY OF THE MANUFACTURE OF PAPER.[1]

From a microscopic study of the various vegetable fibres used in paper-making, A. Girard has determined the qualities such fibres ought to possess. Absolute length is not of much importance, but the fibre should be slender and elastic, and possess the property of turning upon itself with facility. Tenacity is of but secondary importance, for, when paper is torn, the fibres scarcely ever break. The principal substances employed in paper-making may be divided into five different classes:—

1. *Round, ribbed fibres*, as hemp and flax.
2. *Smooth or feebly ribbed fibres*, as esparto, jute, phormium, dwarf palm, hop, and sugar-cane.
3. *Fibro-cellular substances*, as the pulp obtained from the straw of rye and wheat, by the action of caustic lye.
4. *Flat fibres*, as cotton, and those obtained by the action of caustic lye upon wood.
5. *Imperfect substances*, as the pulp obtained from sawdust.

CELLULOSE.

Cellulose, $C_6H_{10}O_5$, constitutes the essential part of the solid framework or cellular tissue of plants, and hence is an especially characteristic product of the vegetable kingdom. The outer coating of ascidian animals is, however, apparently identical with cellulose.

[1] 'Comp. Rend.,' lxxx. 629-631 ; 'Chem. Soc. J.,' xxviii. 675.

Cellulose occurs nearly pure in cotton, linen, and the pith of certain plants. Swedish filter paper, linen rags, and cotton-wool are still purer forms of cellulose.

Cellulose is a white, tasteless, odorless, non-volatile body of about 1.45 specific gravity. It is insoluble in water and all ordinary menstrua, but dissolves, as first observed by Schweitzer, in a strong solution of cupric oxide in ammonia.

Hydrocellulose, $C_{12}H_{22}O_{11}$, is the product of the action of mineral acids (*e. g.*, sulphuric acid of 1.42 sp. gr., or fuming hydrochloric acid), and many other reagents on cellulose. It always retains the form of the cellulose from which it is derived, but differs therefrom in being extremely friable, more readily affected by reagents, and in the readiness with which it combines with coloring matters.

Cellulose undergoes gradual change by prolonged boiling with dilute acids, being converted into hydrocellulose, and is even affected by boiling water alone, especially if heated under pressure.

Oxycellulose appears to vary somewhat in composition according to the mode of preparation, but an apparently definite substance of the formula, $C_{18}H_{26}O_{16}$, was obtained by Cross and Bevan ('Journ. Soc. Chem. Ind.,' iii. 206) from several different sources. The cellulose was boiled with nitric acid containing 50 per cent. of HNO_3, whereby it was largely oxidized to oxalic acid, but yielded 30 per cent. of oxycellulose in the form of a fine white powder, readily soluble in dilute alkalies, and reprecipitable from the solution in a pectous form on addition of acids, salts, or alcohol. Oxycellulose dissolves in concentrated sulphuric acid with pink coloration, and yields a gummy dextro-rotatory substance resembling ordinary dextrin. By the action of concentrated nitric acid mixed with sulphuric acid oxycellulose yields a nitro-compound of the formula $C_{18}H_{23}(NO_2)_3O_{16}$.

The oxidation of cellulose by hypochlorites seems to depend on the presence of free acid,[1] even the atmospheric carbonic acid having a notable influence. When once converted into oxycellulose, no reducing agent (e. g., thiosulphate) will restore the fibre to its original condition. By immersing dyed oxycellulose tissue in a bleaching liquid, the dye can be made to disappear, and the fibre can be re-dyed of any color by immersion in the solution of a suitable coloring matter.

DETERMINATION OF CELLULOSE.

For the determination of cellulose in wood, vegetable fibres, and substances to be used for the manufacture of paper, Müller recommends the following processes: 5 grams weight of the finely divided substance is boiled four or five times with water, using 100 c. c. each time. The residue is dried at 100° C., weighed and exhausted with a mixture of equal measures of benzine and strong alcohol, to remove fat, wax, resin, etc. The residue is again dried, and boiled several times with water, to every 100 c. c. of which 1 c. c. of strong ammonia has been added. This treatment removes coloring matter and pectous substances.

---

[1] If paper be written on with a solution of potassium chlorate acidulated with hydrochloric acid, oxycellulose is formed, and on immersing the paper in a solution of a basic coal-tar dye the writing will appear in color.

The residue is further bruised in a mortar if necessary, and is then treated in a closed bottle with 250 c. c. of water, and 20 c. c. of bromine water containing 4 c. c. of bromine to the litre. In the case of the purer bark-fibres, such as flax and hemp, the yellow color of the liquid only slowly disappears, but with straw and woods decolorization occurs in a few minutes. When this takes place, more bromine water is added, and this is repeated till the yellow color remains and bromine can be detected in the liquid after twelve hours. The liquid is then filtered, and the residue washed with water and heated to boiling with a litre of water containing 5 c. c. of strong ammonia. The liquid and tissue are usually colored brown by this treatment. The undissolved matter is filtered off, washed, and again treated with bromine water. When the action seems complete, the residue is again heated with ammoniacal water. This second treatment is sufficient with the purer fibres, but the operation must be repeated as often as the residue imparts a brownish tint to the alkaline liquid. The cellulose is thus obtained as a pure white body. It is washed with water and then with boiling alcohol, after which treatment it may be dried at 100° C. and weighed.

Bevan and Cross, 'Chem. News,' xlii. 77, substitute a treatment with chlorine gas for the repeated digestion with dilute bromine water, prescribed in the foregoing process. A single repetition of the treatment is then always sufficient, and the results obtained are concordant with those given by the bromine process. Bevan and Cross also find that by boiling the chlorinated fibre for a few minutes in a 5 per cent. solution of sodium sulphite, and then in a 1 per cent.

solution of caustic potash, pure cellulose is at once obtained, the results by this method being 5 per cent. higher than those yielded by Müller's process.

RECOGNITION OF VEGETABLE FIBRES.

As vegetable fibres, when thoroughly bleached, all consist of nearly pure cellulose, chemical tests are not available for distinguishing one kind from another; but, owing to the impossibility of wholly removing the incrusting matter on the manufacturing scale, it is possible to distinguish between certain fibres, such as cotton and linen.

By far the best and most reliable means of differentiating vegetable fibres is to examine their structure with a microscopic power of 120 to 150 diameters.

Filaments of cotton appear under the microscope as transparent tubes about .04 millimetre in diameter, flattened and twisted round their axes, and tapering off to a closed point at each end. A section of the filament resembles somewhat a figure-of-eight, the tube, originally cylindrical, having collapsed most in the middle, forming semi-tubes on each side, which give to the fibre, when viewed in certain lights, the appearance of a flat ribbon with the hem or border at each edge. The uniform transparency of the filament is impaired by small irregular figures, in all probability wrinkles or creases arising from the desiccation of the tube. The twisted and corkscrew form of the dried filament of cotton distinguishes it from all other vegetable fibres, and is characteristic of the fully ripe and mature pod, M. Bauer having ascertained that the fibres of the unripe seed are simply untwisted cylindrical tubes, which never twist afterwards if separated from the plant; but when the seeds ripen, even before the capsule bursts, the cylindrical tubes collapse in the middle and assume the form already described. This form and character the fibres always retain, undergoing no change through the various operations of spinning, weaving, bleaching, printing, and dyeing, nor in all the subsequent domestic processes of washing, etc., and even the reduction of the rags to pulp for the manufacture of paper effects no change in the structure of the fibres.

Linen or flax fibre appears under the microscope as hollow cylindrical tubes open at both ends, and having a diameter of about .02 of a millimetre. The fibres are smooth, the inner tube very narrow, and joints or septa appear at intervals, but they are not furnished with hairy appendages, as is the case with hemp. The jointed structure of flax is only perceptible under a very excellent instrument, and with judicious management of the light.

When flax fibre (linen) is immersed in a boiling solution of equal parts of caustic potash and water for about a minute, and then removed and pressed between folds of filter paper, it assumes a dark-yellow color, whilst cotton, when similarly treated, either remains white or becomes a very bright yellow. The same solution of potash, employed cold, colors raw flax orange-yellow, whilst raw cotton becomes gray.

When flax or a tissue made from it is immersed in oil, and then strongly pressed to remove the excess of the liquid, it remains transparent, while cotton similarly treated becomes opaque.

Phormium tenax, or New Zealand flax, may be distinguished from ordinary flax or hemp by the red color produced on immersing it in nitric acid of 1.32 sp. gravity, containing lower oxides of nitrogen. A reddish color is also developed if New Zealand flax be immersed first in strong chlorine water and then in ammonia.

In machine-dressed New Zealand flax the bundles are translucent and irregularly covered with tissue.   Spiral fibres can be detected in the bundles, but less numerous than with sizal.   The bundles are flat, and numerous ultimate fibres project from them.   In Maori-prepared phormium the bundles are almost wholly free from tissue, and there are no spiral fibres.

Hemp fibre resembles flax, but has a mean diameter of about .04 mm., and exhibits small hairy appendages at the joints.

With Manilla hemp the fibrous bundles are oval, nearly opaque, and surrounded by a considerable quantity of dried-up cellular tissue composed of rectangular cells.   The bundles are smooth, very few partly detached ultimate fibres are seen, and no spiral tissue.

Sizal forms oval fibrous bundles surrounded by cellular tissue ; a few smooth ultimate fibres projecting from the bundles.   Sizal is more translucent than Manilla, and is characterized by the large quantity of spiral fibres mixed up in the bundles.

Jute fibre appears under the microscope as bundles of tendrils, each of which is a cylinder with irregularly thickened walls, the thickening often amounting to a partial interruption of the central lumen. The bundles offer a smooth cylindrical surface, to which fact the silky lustre of jute is due, and which is much increased by bleaching. By the action of sodium hypochlorite, the bundles of fibres can be disintegrated so that the single fibres can be more readily distinguished under the microscope. Jute is colored a deeper yellow by aniline sulphate than is any other fibre, and responds strongly to the bromine and sulphite test.

In examining fibres under the microscope the tissue should be cut up with sharp scissors, placed on a glass slide, moistened with water, and covered with a piece of thin glass.

NOTE.—For the portions of this chapter contained under the heads of Determination of Cellulose, and Recognition of Vegetable Fibres, the author desires to acknowledge his indebtedness to Allen's 'Commercial Organic Analysis,' vol. i. p. 316 *et seq.*

CHAPTER III.

COMMERCIAL CLASSIFICATIONS OF PAPER — SIZES OF PAPER — COMMERCIAL CLASSIFICATIONS OF PAPER-MAKING MATERIALS.

THE following list shows the manner in which new papers and boards are usually classified in the markets of the United States :—

Binders' boards No. 1.
Binders' boards No. 2.
Blotting, American.
Blotting, English.
Book, extra machine finish.
Book, fine white and tinted.
Book, machine finish, low grade.
Book, No. 1, shavings and imperfections.
Book, No. 2.
Book, superfine.
Book, super-sized and calendered.
Book, super-sized and tinted.
Card middles, ground wood.
Card middles, long fibre wood.
Card middles, rag and wood.
Cigarette straw tissue.
Colored papers, double mediums.
Colored papers, glazed mediums.
Colored papers, tobacco.
Colored papers, tissues.
Filter paper.
Flat caps, engine-sized.
Flat caps, fines.
Hanging, brown.
Hanging, buff.
Hanging, curtain.

Hanging, machine satin.
Hanging, superfine, No. 1.
Hanging, superfine, No. 2.
Hanging, white blank, No. 1.
Hardware, light-colored, No. 1.
Hardware, No. 1, glazed.
Hardware, No. 1, glazed, tarred.
Hardware, red.
Leather board, common.
Leather board, counter.
Leather board, extra.
Ledger.
Manilla, bleached, No. 1.
Manilla, bleached, No. 2.
Manilla, cream, rope.
Manilla, extra jute.
Manilla, jute and gunny.
Manilla, No. 1, heavy weight.
Manilla, No. 1, light weight.
Manilla, No. 1, rope.
Manilla, No. 2.
Manilla, ordinary.
Manillas, bogus.
Manillas, flour-sack, cream.
Manillas, flour-sack, drab.
Manillas, rope, unbleached, No. 1.
Manillas, rope, unbleached, No. 2.

News, No 1.

News, rag and wood.

News, No. 1, rag.

News, ordinary rag.

News, straw.

News, straw and wood.

Plate.

Record.

Straw boards, air-dried.

Straw boards, air-dried, No. 1.

Straw boards, air-dried, New York.

Straw boards, air-dried, Penn.

Straw boards, steam-dried.

Straw boards, steam-dried, No. 1.

Straw boards, steam-dried, No. 2.

Straw boards, steam-dried, No. 3.

Straw wrapping.

Straw wrapping, heavy weight.

Straw wrapping, light weight.

Super-calendered, white and tinted.

Tar boards.

Tea papers.

Test papers.

Tissue, Manillas.

Tissues, black.

Tissues, white.

Tracing paper.

SIZES OF PAPERS.

There is in the United States no one set of standard sizes for book papers and for news papers. The lists of sizes vary in Philadelphia, New York, Chicago, and Boston. Old names of sizes for writings and bank ledger papers have been retained; but names of sizes for printing papers have virtually disappeared. The sizes and weights of news, book, and other papers carried regularly in stock by paper dealers are as follows:—

NEWS.

Sizes, inches.	Weights (lbs. per ream).
24 × 38	25, 27, 30.
26 × 40	35.
28 × 42	40, 45.
31 × 44	40, 50, 60.
33 × 46	60.

MACHINE FINISHED BOOK, WHITE AND TONED.

Sizes, inches.	Weights (lbs. per ream).
24 × 38	30, 35, 40, 50, 60.
26 × 40	50, 60.
28 × 42	40, 45, 50.
31 × 44	50, 60.

SIZED SUPER CALENDERED BOOK, WHITE AND TONED.

Sizes, inches.	Weights (lbs. per ream).
22 × 28	30, 35, 40, 50, 60, 70, 80.
24 × 38	30, 35, 40, 45, 50, 60, 70, 80, 100.
26 × 40	50, 60, 70.
28 × 42	50, 60, 70.
31 × 44	50, 60, 70.

COLORED COVER PAPERS.

Sizes, inches.	Weights (lbs. per ream).
20 × 25	22, 32, 40, 48.
22 × 28	32, 50.
24 × 38	40.

MANILLAS.

Sizes, inches.	Weights (lbs. per ream).
24 × 36	20, 25, 30, 35, 40, 50, 60, 70, 80.
30 × 40	28, 30, 40, 50, 60, 70, 80.
40 × 48	100, 125, 150.

FLAT WRITINGS.

Names.	Sizes, inches.	Weights (lbs. per ream).
Letter	10 × 16	7, 8, 9, 10, 11, 12.
Small cap	13 × 16	12.
Cap	14 × 17	12, 14, 16, 18, 20.
Crown	15 × 19	18, 20, 22, 24.
Demy	16 × 21	16, 18, 20, 22, 24, 26, 28.
Folio	17 × 22	14, 16, 18, 20, 22, 24, 26, 28.
Medium	18 × 23	24, 26, 28, 30, 32, 36.
Royal	19 × 24	20, 24, 26, 28, 30, 32, 36.
Double cap	17 × 28	24, 28, 32, 36, 40.

"LINEN" BANK–LEDGER PAPERS.

Names.	Sizes, inches.	Weights (lbs. per ream).
Crown	15 × 19	22.
Demy	16 × 21	28, 30.
Medium	18 × 23	36, 40.
Royal	19 × 24	44.
Super royal	20 × 28	54.
Elephant	23 × 28	65.
Imperial	23 × 31	72.
Double cap	17 × 28	32, 36, 40.
Double demy	21 × 32	60.
"	16 × 42	60.
Double medium	23 × 36	80.
"	18 × 46	80.
Double royal	24 × 38	88.
Colombier	23 × 34	80.
Atlas	26 × 33	100.
Double elephant	27 × 40	125.
Antiquarian	31 × 53	200.

COMMERCIAL CLASSIFICATIONS OF PAPER-MAKING MATERIALS.

Rags.

Rags are imported into the United States from almost every civilized country in the world; but the alphabetical classifications (not classifications according to value) of paper-making materials given in this chapter embrace only those which are regularly quoted in the markets of the United States and of Great Britain.

ALEXANDRIA RAGS.

Blues.	Whites.
Colors.	

BELGIAN RAGS.

Dirty fines.	Linens, No. 4.
Fustians, dark.	Linens, white, No. 1.
Fustians, light.	Outshots.
Housecloths.	Prints, light.
Linens, gray.	Prints, tender, for blottings.
Linens, No. 1.	White cottons.
Linens, No. 2.	White cottons, superfine.
Linens, No. 3.	

BRITISH RAGS.

Black calicoes.	Fustians, light.
Burlaps, bagging, No. 1.	Light prints.
Canvas linen, first.	London fines, cotton.
Canvas linen, second.	New cuttings, cotton.
Checks and blues.	New print tabs.
Essex fines.	Outshots, cotton.
Flax tow.	Seconds.
Fustians, dark.	Thirds.

CONSTANTINOPLE RAGS.

Blues.	Whites, No. 2.
Reds.	Whites, No. 3.
Whites, No. 1.	

DOMESTIC RAGS.

Canvas, cotton.
Canvas, linen.
City whites, No. 1.
City whites, No. 2.
Colors.
Country, mixed.
Country, mixed, free of woollens.

Country, seconds.
Country, white.
Mill assorted, whites.
New seconds, dark.
New seconds, light.
Unbleached muslins.
White shirt cuttings.

DUTCH RAGS.

Blues.
Cottons, dark.
Fustians, light and brown.
Linens, fine whites, No. 1.

Linens, fine whites, No. 2.
Linens, fine whites, No. 3.
Prints, light.

HAMBURG RAGS.

N S C, new shirt cuttings.
S P F F F, No. 1, linens.
S P F F, No. 2, linens.
S P F, No. 3, linens.
F G, No. 4, linens.
F F, No. 5, linens.
Extra fine blue linen, light color.

L F B, blue linens.
C S P F F F, No. 1, cottons.
C S P F F, No. 2, cottons.
C S P F, No. 3, cottons.
C C C, colored cottons.
C F X, low-grade cottons.
Extra fine blue cottons.

JAPANESE RAGS.

Blues, ordinary.
Blues, selected.

Whites, ordinary.

KÖNIGSBERG RAGS.

S P F F F, No. 1, linens.
S P F F, No. 2, linens.
S P F, No. 3, linens.

F G, No. 4, linens.
F F, No. 5, linens.
L F X, low-grade linens.

LEGHORN RAGS.

P P, No. 1, white linens.
S S, No. 2, white linens.
T T, No. 3, white linens.
R R, linen stripes.
P C, No. 1, white cottons.

S C, No. 2, white cottons.
T C, No. 3, white cottons.
R C, cotton stripes.
C C, colored cottons.

LIBAN RAGS, LINENS.

S P F F.

S P F.

F G.

B G.

Sacking.

A.

L F B.

MEUREL RAGS, LINENS.

S P F F, No. 1, linens.

S P F, No. 2, linens.

F G, No. 3, linens.

F F, No. 4, linens.

L F B, blue linens.

RUSSIAN RAGS.

S P F F, No. 1, linens.

S P F, No. 2, linens.

F G, No. 3, linens.

F F, No. 4, linens.

L F X, No. 5, linens.

L F B, blue linens.

SMYRNA RAGS.

Blues.

Mixed.

Reds.

Whites.

TRIEST RAGS.

S P F.

S F F.

S F X.

S F B.

Rope, Bagging, and Threads.

Bagging, mixed.

Gunny bagging, No. 1.

Gunny bagging, No. 2.

Jute threads, best.

Jute threads, clean.

Rope, jute.

Rope, Manilla.

Rope, tarred Manilla.

Rope, white Manilla.

Rope, mixed.

Shavings and Old Papers.

Binders' board cuttings.

Bogus Manillas.

Book stock, No. 1, light.

Books, new, solid folios.

Books, old blank.

Books, old printed.

Briefs and letters.

Broken news and letters.

Commons.

Hardware, No. 1.

Ledger and writing.

Manillas, No. 1.

Newspapers and pamphlets, extra.

Newspapers and pamphlets, old.

Paper-collar cuttings.

Railway sheets, white and buff.

Railway tickets.

Shavings, cream post.

Shavings, mixed.
Shavings, mixed, part white.
Shavings, No. 1, hard.
Shavings, No. 1, soft.
Shavings, No. 1, white and colored.
Shavings, No. 2, white and colored.

Solid stock.
Straw board cuttings.
White collar cuttings.
White envelope cuttings.
White shavings, hard, No. 1.
White shavings, soft, No. 1.

Various Fibres.

Adansonia fibre.
Bamboo, crushed.

Palm fibre.
Palm leaves.

COIR GOODS.

Fibre, bales.
Fibre, ballots.

Fibre, rope.

CURLED FIBRE.

Good black.

Good green.

GUTTA-PERCHA.

Gutta-percha, good to fine.

Gutta-percha, low to medium.

INDIA-RUBBER.

Assam.
Borneo.
Central American.
Madagascar.
Mozambique.

Negrohead.
Para, fine.
Pegu.

Kitool fibre, brown.

Wastes.

Cotton waste.
Flax waste.
Jute bagging, special.

Jute bagging, good, clean.
Jute bagging, second quality.
Jute cuttings.

Wood Pulps.

Aspen, dry, in sheets.
Aspen, 50 per cent. moisture.
Brown pine, dry.
Brown pine, 50 per cent. moisture.
Brown pine (half chemical), "Heosfos" brand, 50 per cent. moisture.

Chemically prepared (acid), 50 per cent. moisture.
Chemically prepared, bleached.
Chemically prepared, unbleached.
Ligneous meal (wood flour), selected.
Ligneous meal (wood flour), extra fine.

Ligneous meal (wood flour), fine.
Pine, dry, in sheets.
Pine (long fibre), 50 per cent. moisture.

Pine, 50 per cent. moisture.
Pine, 50 per cent. moisture (single sorted).

Straw Pulp.

Straw pulp (bleached), 50 per cent. moisture.
Straw pulp (bleached), 50 per cent. moisture, extra quality.

These pulps are sold by the ton of dry weight.

Esparto-Grass.

Algerian (oran, etc.), first quality.
Algerian (oran, etc.), second quality.
Algerian (oran, etc.), third quality.
Gabes, sfax, or skira, good average.

Spanish, fine to best.
Tripoli, hand picked.
Tripoli, fair average.

Chemicals, Clays, Coloring Materials, Rosins, etc., employed in Paper-making.

Alkali (quoted according to per cent.).
Alum, ground.
Alum, lump.
Alum, pearl.
Alum, porous.
Aluminous, cake.
Anti-chlorine.
Bichromate of potash, American.
Bleaching powders.
Brazil wood.
Catechu.
Cochineal.
Caustic soda (quoted according to the per cent.).
Clay, China, English.
Clay, China, "Star."
Clay, South Carolina.
Clay, Terra alba, American.
Clay, Terra alba, French.
Corn-starch.
Copperas, American.
Extract of logwood.
Mineral, fibrous pulp.

Orange mineral.
Potato starch.
Prussian blue.
Prussiate potash.
Rosins, common to good, strained.
Rosins, good, No. 1.
Rosins, good, No. 2.
Rosins, low.
Rosins, No. 1.
Rosins, extra pale.
Rosins, pale.
Sal soda, caustic.
Sal soda, English.
Soluble blue.
Spanish brown.
Sugar of lead, brown.
Sugar of lead, white.
Sulphuric acid (quoted according to the per cent.).
Venetian red.
Vitriol, blue.
Yellow ochre, Rochelle.

Aniline Dyes, etc.

Blue, paper, 1.
Blue, paper, 2.
Blue, ultramarine.
Brown, Bismarck.
Diamond, magneta.
Eosine, pure.
Green, fast, No. 1.
Green, fast, No. 2.
Lac à la cochennille.
Magenta crystals, pure.
Magenta crystals, No. 2.

Methyl blue.
Methyl green.
Methyl violet.
Orange B.
Paris blue.
Red lake.
Rocceline, pure (fast red).
Silk green.
Violet, 2 B crystals.
Violet, 2 B powder.
Yellow, No. 6 P.

CHAPTER IV.

MANUFACTURE OF PAPER BY HAND.

WE who live in the present day with our newspapers issued every twenty-four hours by the millions, can form but an indistinct idea of what was the state of the art of paper manufacture even at so late a period as the commencement of the present century. Instead of paper being reeled off in webs many feet in width, and at the rate of nearly a mile in length in the hour, each sheet had at that time to be separately made on a mould by hand, and had then afterward to be subjected to various processes before it was in a state suitable for use.

To obtain an uniform and continuous supply of paper for any purpose was quite impossible, and the necessity of applying machinery to this manufacture was beginning to be urgently felt, and the success which had attended its introduction into the spinning and weaving industries gave encouragement to the paper manufacturer. Yet the entire change from a system of manufacture almost mediæval in its rudeness has been comprised within the lifetime of many persons now living.

Hand-made paper continues to be used for special purposes; but, for serviceability, machine-made paper is undoubtedly to be preferred. In Great Britain and on the continent of Europe hand-made writing paper is more largely used than in the United States, and its present employment is owing in Europe largely to the conservatism of the people; but affectation probably exerts as much influence as any other cause at home and abroad in maintaining the fashion of employing hand-made writing paper.

There is now so little hand-made paper produced in the United States that a chapter devoted to the details of its manufacture is really of no practical value; but in order that this volume may not seem incomplete a synopsis of the process of manufacturing paper by hand will be given.

The preparation of the pulp for paper to be made either by hand or by machinery is identical, but as the paper produced by the former method is usually of an expensive character, consequently only the finest qualities of rags are used. As the preparation of the pulp will be described later, we shall now only take it after it has issued from the beating engine, been stained and run into large chests from which the paper-maker's vat is supplied. The sheet of paper is moulded in the following manner: the vatman takes the mould, which consists of a framework of fine wire cloth having a "deckle" or movable rectangular frame of wood to keep the pulp from running off, and dipping the mould vertically into the pulp and bringing it up horizontally takes up a sufficient quantity to fill the deckle. The vatman then runs the pulp evenly over the mould from the front side to the back, the superfluous stuff being dropped into the vat, and then gives the mould the "shake," which motion being imparted so as to be felt in the length and across the mould causes the fibres of the stuff to intertwine and the water to pass through the openings in the wire-cloth, the sheet of paper being formed from the pulp which remains. The operation of moulding the paper requires great nicety, both in determining the thickness of the sheet and in imparting to it an uniform body throughout.

The stuff in the vat is kept at the proper temperature by a copper or other contrivance placed within the vat, steam heat being communicated through a suitable pipe, the agitation of the stuff being accomplished by means of machinery also placed within the vat.

After the sheet of paper is formed the vatman brings it to the stay, and removing the deckle applies it to a second mould and proceeds as before. An assistant, called a "coucher," takes the first mould and places it on an inclined elbow in order to cause more water to drain out of the sheet; the coucher, having by his side a heap of porous pieces of flannel called "felts," next turns the mould over on one of them, leaving the sheet of paper on the felt, and then placing another felt on the damp sheet of paper, he is in readiness to turn over the sheet from the second mould.

In this manner the two workmen proceed, the vatman moulding a sheet of paper and the coucher placing it upon a felt, until the pile, which is called a " post," contains six or eight quires.

The post is then carried to the press, where it is subjected to a powerful pressure, which causes a large quantity of water to be removed from the paper, leaving the sheets sufficiently dry to be handled by the "layer," who lays one

sheet upon another leaving out the felts, and after parting
sheet from sheet subjects the heap to a moderate pressure.

Fig. 1 shows the process of forming paper by hand, the
vatman being in the act of giving the mould the " shake,"

Fig. 1.

and the coucher being represented turning a sheet of paper
from the mould upon a felt; the " post," which is partly
built up, being so arranged as to be drawn to the press upon
a gangway of rollers.

The sheets after being pressed, as has been described, are
next parted, and then hung in the drying loft, where the
paper remains until dry, being placed in spurs five or six
sheets thick upon rope made from cow hair.

The paper is next sized by passing the spurs through a
trough containing a strong solution of gelatine, and then in

7

order to free the paper from an excess of size it is placed upon an endless felt and carried to one end of the long size trough and passed between press rollers. In order to prevent the sheets from sticking together they are separated from each other and carried to the drying loft, and finally dried at a temperature of about 75° to 80° F.

The paper is then examined, the damaged sheets being thrown out or the knots removed, and is next glazed by passing the sheets between plates, the paper being finally sorted and finished in much the same manner as machine-made paper, but with additional pains. After being neatly put up into quires, half reams, and reams, the hand-made paper is ready for market.

The water-mark observable on almost all hand-made papers when held against the light is produced by wires representing the letters or design of the water-mark raised above the other portion of the mould, thus making the paper thinner in that part covered by the marking wires by indelibly stamping the device or devices in the substance of the sheet of paper during its formation from the pulp.

Adams, Mass., is probably the only place in the United States where paper is now made by hand; the quantity produced from each vat is from 190 to 200 pounds per day. But in Great Britain, where there exists a larger demand for this class of paper, the total production averages about 60 tons per week.

CHAPTER V.

DISINFECTING RAGS—PURCHASING RAGS.

WHEN cholera, or other infectious or contagious diseases, exist in foreign countries, or in portions of the United States, the health officers in charge of the various quarantines in this country require that rags from countries and districts in which such diseases are prevalent shall be thoroughly disinfected before they are allowed to pass their stations. Rags shipped to Hull, Liverpool, London, Italian, or other ports, and re-shipped from such ports to the United States, are usually subjected to the same rule as if shipped direct from the ports of the country in which such diseases prevail.

It is usually requisite that the disinfection shall be made at the storehouses in the port of shipment by boiling the rags several hours under a proper degree of pressure, or in a tightly closed vessel, or disinfected with sulphurous acid, which is evolved by burning at least two pounds of roll sulphur to every ten cubic feet of room space; the apartment being kept closed for several hours after the rags are thus treated. Disinfection by boiling the rags is usually considered to be the best method.

In the case of rags imported from India, Egypt, Spain, and other foreign countries where cholera is liable to become epidemic, it is especially desirable that some efficient, rapid, and thorough process of disinfecting should be devised. In order to meet the quarantine requirements, it must be thorough and certain in its action, and in order that the lives of the workmen and of others in the vicinity may not be endangered by the liberating of active disease-germs, or exposure of decaying and deleterious matters, and that the delay, trouble, and expense of unbaling and rebaling may be avoided, it must be capable of use upon the rags while in the bale, and of doing its work rapidly when so used.

The object of Messrs. Parker and Blackman's invention, shown in Figs. 2, 3, and 4, is to provide such a process for

Fig. 2.

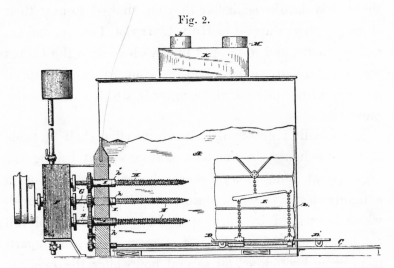

disinfecting rags and other fibrous materials while in the bale.

Figure 2 shows a view of one form of apparatus adapted for use in carrying out their process. Fig. 3 is a longitudinal sectional view, showing a bale of fibrous material as being acted upon; and Fig. 4 is an end view of the bale, showing the relative positions of the injecting-tubes.

Fig. 3.

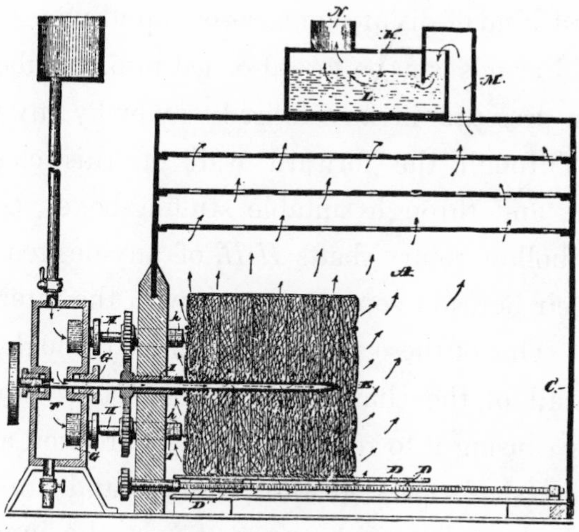

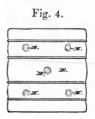

A designates a chamber provided at one end with an opening, *B*, to be closed by a door, *C*, hinged at its lower edge, and adapted, when swung outward and downward, as shown in Fig. 2, to form a continuation of the floor of the interior of the chamber. A car, *D*, supported upon suitable rollers or casters, *D'* *D'*, is adapted to be run forward and back over the floor of the chamber, and when

Fig. 4.

the door is swung down, as described above, out upon said door as a continuation of the floor. This car is shown provided with means for clamping and holding firmly a bale, *E*, of the material to be treated, and is caused to travel back and forth within the chamber and out through the door, when the latter is opened.

Just beyond the closed inner end of the chamber A is another chamber or receptacle, F, into which is to be forced any desired kind of disinfectant gas or liquid under pressure. This liquid or gas can be forced or fed into chamber F, and put under pressure in any desired way or by any preferred means. Through the forward wall of this chamber or receptacle, and through suitable stuffing-boxes, G, therein, pass the hollow rotary shafts $H H$, of any desired number, having their bores in communication with the interior of the chamber. One of these shafts is continued through the other or rear wall of the chamber, and connected with suitable means for causing it to revolve in either direction as desired. The hollow shafts pass forward through and are journaled in suitable long journal-bearings, $I I$, in the inner wall of the chamber A. The portions of the shafts within the chamber A are closed and pointed at their forward ends, screw-threaded throughout their lengths, and provided with series of small openings communicating with their central bores. Each shaft is formed with a collar, h, bearing against the inner or forward end of the shaft-bearing to prevent any backward movement of the shaft. Just outside of the wall of chamber A the shafts are provided with intermeshing gear-wheels, so that when one of them is rotated, as indicated above, the others will be revolved an equal number of times. The threads of the screws are so constructed that as the main driven screw is turned so as to screw it into anything brought against its point the other screws will by their gears be turned to screw them in also.

The mechanism for feeding forward the carriage with the bale fastened thereon is so connected with the gears on the shafts that the bale is fed against the end of the revolving screws and then continuously forward or inward as they are screwed into it. After the perforated screws have been driven fully into the bale, the door having been previously shut to close the chamber *A* tightly, disinfectant gas or liquid—preferably gas—is forced into chamber *F*, and from there out through the series of screws projecting into the bale. The gas then passes out through the openings in the screws, and is forced to pass in every direction through the mass of the bale, so as to come into intimate contact with every portion thereof, and effectually destroy every germ among the fibres of the material.

As most fibrous materials, and especially rags, are baled so as to be in layers, the inventors so place the bale to be treated on the carriage that the perforated screws shall penetrate it in directions at right angles to the layers, as shown best in Fig. 3. By so doing it is insured that the gas or liquid used, issuing through the holes in the screws, shall pass in all directions throughout the bale, so as to come in contact with and act upon every portion of the material in every layer. When the material is in layers, if the screws were inserted into the bale in directions parallel with the layers, the gas or liquid would be apt to pass out between such layers as the screws projected between, thus leaving much of the material not properly acted upon. When, however, the bale is not stratified in its formation, the screws can be forced into it in any direction desired.

Upon the top of the chamber A is a receptacle or tank, K, containing disinfecting liquid, L. At one end of this tank a pipe, M, leads upward from the interior of chamber A to a point within the tank above the level of the liquid, and is then bent over and carried down into the liquid. A discharge opening or pipe, N, is provided on top of this tank at or near its end. With this construction all the air and gas passing up out of the chamber A, as gas or liquid is forced into and through the bale, must pass through the disinfectant liquid in the tank K, so that any disease-germs contained or floating in it will be, it is claimed, effectually rendered harmless. When a sufficient amount of disinfectant has been forced into and through the bale, the disinfectant is turned off and cold dry air is forced into the chamber or receptacle F, and from thence out through the perforated screws or nozzles and the bale.

Any desired means can be used for compressing and cooling the air and forcing it into the chamber. This air, passing through the bale, cools and dries it, and then, as it issues from the bale and fills the disinfecting chamber, drives the foul air and gases from such chamber up and out through the disinfecting liquid in tank K. Only a very short time, it is claimed, is required to thus cool and dry the bale and drive out all the foul air from chamber A, so that it will be perfectly safe to open and enter the latter. The screws are caused to rotate, so as to be unscrewed or withdrawn from the bale, the carriage-moving mechanism being at the same time actuated so as to move the carriage backward and outward, so that the bale can be removed therefrom.

This process is claimed to be adapted not only to destroy all disease-germs, but also to destroy all foul and injurious gases and odors in the bale or arising therefrom. The decay of any vegetable or animal matter mingled with the material of the bale, it is claimed, will be arrested, and such deleterious matter rendered inert and incapable of injury to the health of those handling the bale or the material thereof after the bale has been opened.

Instead of using perforated hollow screws, as described and shown, the inventors contemplate using hollow perforated spindles to be thrust into the bale, either with or without rotary motion. Any suitable form of disinfectant can be used by this process, as, for instance, sulphurous acid in the gas or in solution, hot air, superheated steam, carbolic acid, any of the well-known solutions or vapors containing chlorine or sulphur, or, if desired, a solution containing a very small portion of corrosive sublimate. When the disinfectant is in the form of a solution, it can be used as a liquid or mixed with air in the form of spray or vapor, which can be forced through the perforated nozzles into the bale.

In lieu of injecting the disinfectant into the bale and causing it to permeate outward, the disinfectant can be caused to pass inward into and be drawn off from within the mass of material. In carrying out this modified process the disinfectant is fed in any desirable way into the disinfecting chamber around the bale, and the perforated tubular nozzles or screws having been previously inserted into the bale in the same manner as shown and described hereinbefore, a vacuum is produced by any desirable means in the chamber, with the interior of which the outer ends of the bores of the nozzle communicate, as described. The disinfectant within the disinfecting chamber surrounding the bale will then be drawn inward through the bale to the perforated nozzles and out through them into the vacuum chamber. From thence the gases and vapors are drawn off by any suitable means and passed through a disinfecting tank or apparatus of any well-known construction. The disinfectant, after a sufficient quantity of it has been drawn through the bale, is shut off and cool dry air admitted to the disinfecting chamber instead, and also drawn in through the bale and out through the nozzles and a disinfecting tank or apparatus.

Instead of causing the disinfectant, and subsequently the cool air, to pass inward through the bale and thence out through the nozzles by suction, the disinfecting chamber can be made strong enough to stand considerable pressure, and the disinfectant and afterward the air can be forced under pressure into it and around the bale. By such pressure the disinfectant or air can be caused to penetrate the bale from all sides inward to the perforated nozzles within it and then to pass outward through them. Either of these processes may be used, if desired.

Purchasing Rags.

The purchasing of rags for a paper-mill requires great experience, as there are so many tricks and frauds practised in making the bales, that even the most lynx-eyed and experienced buyers often find numerous weight-giving substances in the interior of the bales after they are opened. There is, of course, no way to discover these frauds until the rags are about to be used, and the only safeguard against such dishonesty is the exercise of caution in regard to the persons from whom the rags are purchased, and should the sellers refuse to make reasonable allowances for such fraudulent overweighting of the bales, good grounds would then exist for seeking to find more honorable dealers.

Such frauds will exist just so long as paper-makers submit to them without vigorous protest—but no longer.

The color and strength of the materials determine the value of the rags; city rags being easily distinguished from country rags by their respective colors and textures, city rags being fine and white, and country rags coarse and dark.

If there be ground for reasonable suspicion that the rags have been unduly weighted with water it is advisable to weigh and then dry a lot of them either by spreading them in the rays of the sun or in a heated room and afterwards reweighing them. The quantity of moisture contained in the rags is indicated by the difference between their wet and dry weights. The natural humidity of rags varies from 5 to 7 per cent., being greater in coarse rags than in fine ones, and no greater allowance than this should be made. Linen rags not uncommonly contain jute and cotton ; the jute being very undesirable as it injures the color of the paper. The presence of jute in linen may be ascertained by washing and treating with dilute chlorine, when the jute will become of a reddish color and the linen white. Cotton in linen is quickly destroyed by treating the rags with concentrated sulphuric acid, which does not injure the linen fibres but leaves them white and opaque. The gummy matter is removed by washing, and the sulphuric acid is neutralized by the addition of a small quantity of caustic potash.

By drying the rags and treating as above, and then re-drying after separating the jute or destroying the cotton the per cent. of admixtures can be readily determined.

CHAPTER VI.

SORTING RAGS — SORTING WASTE PAPER — SORTING OR " DRY PICKING" ESPARTO — MACHINE FOR FACILITATING THE SORTING OF PAPER STOCK.

SOME manufacturers after opening the bales pass the rags through a machine for the purpose of removing the dust, sand, and other adhering matters. This treatment of the rags makes them much less objectionable to the sorters and cutters whose eyesight and health are better preserved thereby.

The next operation through which the rags, waste papers, rope, and like materials used for paper stock pass is that of sorting according to fibre and color. Raw, coarse materials used for paper stock are more wasteful in treatment than fine city rags, and the same manipulations necessary to reduce the former to useful stock would prove destructive to the latter. The sorting of the raw material is, consequently, an important branch of the paper trade. In Europe a more minute classification is adopted than in the United States; but on the Continent the assortment and classification of rags are much more simple than in either Great Britain or America. This department of the work should be conducted on simple and easily-remembered principles in order to facilitate the labor and save time.

It is not possible to describe any system of assorting and classifying rags that would be generally acceptable, as every country, and almost every mill, follows a different one in order to conform to particular circumstances. The following distinctions are, however, commonly made :—

According to Fibre :—	*According to Color :—*
Linen.	White, first, second, third.
Hemp.	
Cotton.	Gray.
Manilla.	Blue.
Half wool, or woollens.	Red.
	Black, containing all dark colors.

Bagging, canvas, ropes, threads, twine, etc., are also separately classified.

After the rags are assorted, whether uncut or cut, a foreman usually inspects them after they are spread out on a large table in order to control the work of the women employed in this department.

Sometimes the labor of sorting the rags can be materially facilitated by the use of mechanical contrivances. The invention shown in Figs. 5, 6, and 7 is intended to facilitate the sorting of paper stock.

SORTING WASTE PAPER.

When waste paper has been removed from the bale it should at once be passed through a devil and duster.

Some mills use a railroad duster connected by an apron with an open cylinder duster.

After the material has been opened up and the adhering impurities removed it is ready for the sorting-rooms, which are of the same general appearance as those in which rags are sorted ; but the tables, however, are without knives, as waste papers do not require to be cut.

The sorters are required to sort out everything which has not once been white pulp, and remove all book covers, book-marks, toothpicks, matches, cigar stumps, scraps of leather, bits of wood, and other foreign substances which gravitate so naturally into the waste-paper basket and thence into the chiffonnier's bundle.

The dealers are not always to be relied upon to do the sorting properly, and should pieces of yellow straw or dark wrapping paper be reduced to pulp along with the white paper they will reappear in the finished product in the form of yellow, gray, or colored spots.

Papers made from some kinds of wood, esparto, etc., turn to a yellow-brown color during the bleaching, and as they cannot be made into first quality paper it is necessary to sort them out, and if the eye and touch are not to be relied upon such papers should be dipped into a bucket containing a strong solution of soda. If the soda does not change the color of the paper it is thrown with the No. 1 stock ; but if it turns yellow or brown under the test it is sorted out for inferior stock.

No uniform classification in the grading of waste paper is

in use in the sorting-room; the grading depending upon the custom of individual mills.

Means are shown in Figs. 68, 70, 71, and 72 whereby "imperfections" in large quantities can be assorted and delivered to the duster with the uniformity required to fully and properly supply it.

Sorting or "Dry Picking" Esparto.

The first process to which esparto is subjected after being delivered at the mill is that of "dry picking," which operation is usually performed by girls who work at separate tables placed in a long row. A coarse iron-wire screen forms a portion of the top of each table, and on this gauze each girl spreads small bunches of esparto and picks out such imperfections as pieces of weed, root-ends, etc., while the smaller and heavier impurities, such as sand, etc., fall through the openings of the wire screen into a receptacle placed under the table. "Dry picking" is a term used in contradistinction to "wet picking" which is a subsequent process employed after the boiling. In some of the mills in Great Britain a machine is used for facilitating the labor of dry picking. The esparto is first passed through a fan duster and is then carried forward on an endless belt, the roots, etc., being removed by girls stationed on each side of the machine. The root-ends, etc., are very hard to boil and bleach, and, in addition to injuring the color of the bulk of the fibre, they are liable to make their appearance in the finished paper in the form of dark-colored spots, technically known as "sheave." After the grass has been properly sorted it is carried to the boiler-house.

MACHINE FOR FACILITATING THE SORTING OF PAPER STOCK.

Messrs. Robert O. and Walter Moorhouse, of Philadelphia, are the inventors of the machine shown in Figs. 5, 6, and 7, for facilitating the sorting of materials used for paper stock, which has for its object the turning over and loosening of the same and passing them steadily upon a screen before the eyes of the operatives, whose attention being relieved of the labor of spreading, opening, and removing the paper-making materials, are enabled to more quickly and thoroughly sort them.

This invention consists in an endless travelling screen of wire-cloth supported upon and moved by suitable rollers turned by power, upon which the stock is placed near one end, and a series of rotating vanes, also turned by power, which, by sweeping over the top of the screen, rub the stock upon it and open it, thus spreading it in view of the operatives and letting much of the dirt and grit fall through the screen. The operatives, being stationed along the sides of the screen, between the several vanes, remove objectionable objects that do not pass through the screen, and the desirable stock, being carried to the end, is discharged by having several operatives each taking out a particular kind or color of material, and they are thus very expeditiously sorted.

Fig. 5 is a top view of the machine, Fig. 6 a front view, and Fig. 7 an end view.

8

Fig. 5.

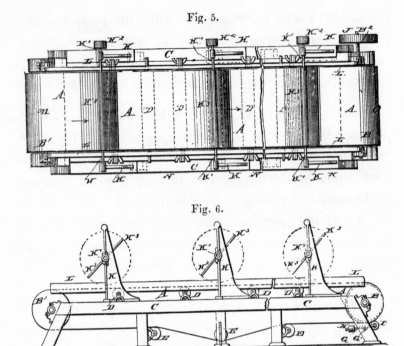

Fig. 6.

Fig. 7.

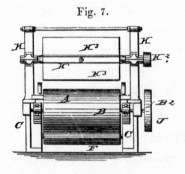

A is an endless apron of wire-cloth, supported by rollers, *B* and *B'*, at the ends of the frame *C*, and smaller intermediate rollers, *D D*, upon its upper portion, and rollers, *E E*, on the lower portion; a roller, *F*, supported and turning in bearings formed in the ends of the levers *G*, pivoted at *G'* to the frame *C*, by which the apron *A* is pressed toward the roller *B* by the action of weights, *H*, on the opposite end of the levers *G*, so that by applying power by a band to the pulley J on the shaft B^2 of the roller B the adhesion or friction of the apron A upon the roller B will be sufficient to cause the apron to move in the direction of the arrows in Fig. 5, without subjecting the entire length of the apron A to such severe strain as would be requisite were it strained by tightening the roller B'.

$K K$ are standards or frames secured to the frame C at intervals, and support shafts, K', turning in bearings in the standards K by means of pulleys, K^2, and having vanes, K^3, preferably made of sheet metal, and provided with edges of leather, rubber cloth, or similar yielding material, which are of such dimensions as to sweep the apron A.   Strips or rims, $L L$, are placed at each side of the apron A, which serve to prevent the stock spreading laterally beyond the reach of the vanes K^3 and falling off the apron A.

The operation of the machine is as follows: Power is applied to the pulleys B^2 and K^2; the materials are placed on the apron A at the place marked M, and passing, by the motion of the apron A, under the vanes K^2, are rubbed over the apron, and any grit and sand detached from them falls through the apron, and the stock, by the rubbing, becoming opened and spread out before operators stationed at $N N N$ between the reels K^3, are readily sorted by the operators removing such as is not desired to pass through the machine, and the remaining acceptable materials pass off at the end of the apron A, at the point O, ready for use as paper stock.

In the illustrations only three reels or sets of revolving vanes are shown; but in practice a larger number are used, and a greater number of operatives than three are employed, the machine being much longer than shown, as is implied by the break in Fig. 6, the increased length of machine involving a mere duplication of the parts shown.

CHAPTER VII.

CUTTING RAGS BY HAND—CUTTING RAGS BY MACHINERY—LIST
OF PATENTS FOR RAG CUTTERS AND DUSTERS—CUTTING WOOD
FOR CHEMICAL FIBRE—TREATING WOOD BEFORE GRINDING—
VOELTER'S MACHINE FOR CUTTING OR GRINDING WOOD—LIST
OF PATENTS FOR WOOD GRINDERS—CORN-HUSK CUTTER.

CUTTING RAGS BY HAND.

WHEN the stock is cut by hand the operation is usually
performed by drawing the rags against the sharp edge of a
scythe-like knife measuring about 14 inches in height above
the table. The contrivance shown in Figs. 8, 9, and 10 is
the invention of Mr. Edgar D. Aldrich, of Pittsfield, Mass.,
and consists of a device for securely holding the section of
scythe-blade used to cut rags in paper-mills, and in such a
way that it can readily be detached to be ground or reversed,
and can be tightened in place easily and quickly, and by
mechanism arranged to be entirely out of the way of the
operator, and that cannot catch in the rags.

Fig. 8 is a side elevation of the holder; Fig. 9 is a rear
elevation, and Fig. 10 is a section.

The holder proper, B, is of cast metal, in one piece, and
has a flat base to bear against the side of the table, to which
it is attached by means of bolts passing through holes left in

the flanges b b, as seen in Fig. 8. The sides of the holder, which may be made as light as is consistent with strength, support the top C and bottom D, which afford the two

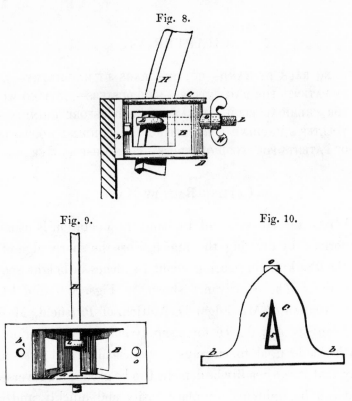

Fig. 8.

Fig. 9.   Fig. 10.

bearings, d g, to the scythe-blade H. These openings, d g, in the top C and bottom D form a triangular shape to conform to a cross-section of the blade H, and are relatively arranged to give the desired inclination to the blade when in place ; and the blade H, when in bearings in the shell, is as firm as though seated in a solid block of the same thickness as the holder B. The scythe-blade H, when in place within the holder, is grasped by the bolt hook L, the screw-shank of which passes through the holder to the outside through the nut O, and the thumb-nut W enables the scythe-blade H to be tightened or released at will. As the greatest pressure is exerted upon the blade to remove it from its top bearing the hook L is arranged, as shown in Fig. 8, to grasp the scythe at a point nearer the bearing d.

In the holder in common use, in which the blade is held in a staple by wedges, and has the staple drawn against the table by a bolt passing through the latter to the front, the nut necessary to secure the end of the bolt interferes with the operator, and the wrench to loosen it has to be detached when not in use, and fragments of wooden wedges are liable to become mixed with the rags, and the wedges require frequent adjustment; but in the above-described device all inconveniences are claimed to be done away with, and a broad flat surface flush with the surface of the work-table forms a base to the cutting edge of the blade.

In Fig. 11 is shown the manner in which the operation of cutting the rags by hand is conducted. The wire-cloth covering the top of the table is quite coarse, containing only about nine meshes to the square inch, thus allowing any dirt and fine particles from the rags to fall into a receptacle under the work-table. A casual visitor to the assorting and cutting room of a paper-mill would be impressed by the *sang-froid* exhibited by the women and girls who stand behind the scythes busily shredding handfuls of rags by drawing them down the keen edge of the blades, and the rapidity with which the handfuls of shreds are thrown into the compartments of the cutting table according to their fineness. Should a slip occur a terrible gash to the arm or hand of the operator is the result and possibly a finger is cut

Fig. 11.

off. The air of these cutting rooms is usually heavy with dust, and in order to protect the hair each operator has her head swathed in a handkerchief. In mills where fine papers are made it is necessary that every seam and hem and patch shall be ripped up in order that the dirt underneath may soak out ; and every button, hook and eye, and string must be cut off, the pins and needles picked out and any piece found to be badly stained or containing India rubber thrown out.

In preparing rags for shipment some attention is paid to their condition and classification by foreign dealers, and in the case of best cotton and linen rags they are usually perfectly clean when they arrive at the mill, and do not require the " dusting" after shredding which is given to poorer stock.

When the rags are cut by machinery it is, of course, not possible to give them the minute attention which they receive in the hand method of cutting, nor can the objectionable matters mentioned be so readily separated from them.

CUTTING RAGS BY MACHINERY.

Rags for paper stock have heretofore been "stripped"—that is cut or torn into strips—by hand. This method of stripping is both slow and expensive. Cross-cutting, which is the cutting of the strips into small pieces, has been accomplished both by hand and by machinery, the cross-cutting for the finer grades of paper being done by hand and for the coarser grades by machinery. The machinery used, however, has not been adapted to cut the materials with a sufficient degree of regularity, and after being cut the stock has not been reduced to the size which is most desirable in the various manipulations to which it is afterwards subjected, and the present construction of these cross-cutting machines is such that, however carefully they may be operated, a product of uniform size cannot be obtained. Attempts have been made to strip and cross-cut without an intermediate handling by connecting two machines. With this arrangement, however, it has been necessary to reduce the rags first by hand to a size adapted for the machine, and no positive means of conveying the strips from one cutter to the other has been provided ; but the material has been deposited upon an apron, and conveyed by it to a chute through which they pass, and are deposited upon a second apron, which conveys them to the second cutter. The manner in which the rags are thus presented to the second cutter depends upon chance, and they are as liable to be cut in one direction as another. This defect is found in almost every machine—*i. e.*, it has been necessary to first prepare the rags by hand cutting or stripping for the machine, and the feed has not been positive and has not been within the control of the operator, either as to the manner of presenting the material to the cutting device, or as to regulating the size to which the material is cut.

Taylor's Machine.

Mr. C. F. Taylor, of Springfield, Mass., has invented a process of and apparatus for stripping rags by machinery, and claims to obtain thereby a product of uniform or approximately uniform size. It is also claimed for the apparatus that it will both strip and cross-cut the stock at the same operation, and at the same time remove foreign matter from the rags.

Fig. 12 is a side view of Taylor's machine with the pulleys and gears removed, disclosing the portions of the machine which operate upon the rags. Fig. 13 is a sectional view of the same. Fig. 14 is a plan or top view. Figs. 15, 16, and 17 are detail views of the rotary cutters. Fig. 18 is a sectional view of the sieve. Fig. 19 is a top view of the sectional beater-bars. Fig. 20 is a side view of the feeding device, with parts in section, and Figs. 21, 22, 23, 24, and 25 are detail views of the parts of the feeding device.

i i represent pressure-rolls, which are also adapted to act as feed-rolls.

a represents a revolving blade, adapted to cut the rags in strips. *b b* represent rotary cutters or shears in gangs, adapted to make a shearing cut. *c* represents a beater

Fig. 12.

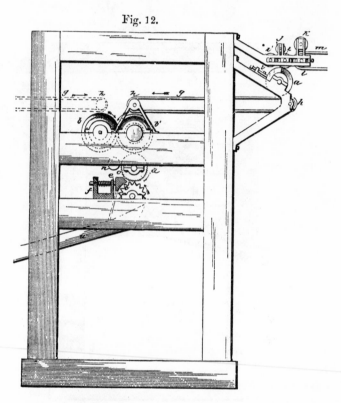

adapted to beat or pound the material against the beater-bar *e* as it is fed to it by the feed-roll *d*. The operation of the machine is as follows:—

The rags being fed upon the apron *m* are carried in the direction indicated by the arrow, and are fed to the pressure-rolls i, which are adapted to crush any hard substance. These rolls also act as feed-rolls, holding and feeding the material to the cutter a, which, revolving as indicated, carries the material against the fixed knife t, thus separating the rags into narrow strips. This operation is termed

Fig. 13.

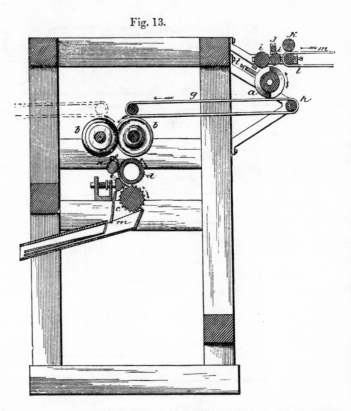

" stripping." It will readily be seen that by changing the position of the knives—*i. e.*, placing the cutters b ahead of the cutters a—the stripping will be accomplished by these cutters and the cross-cutting by the cutter a. After being cut in strips the rags fall to the apron f, which, moving in

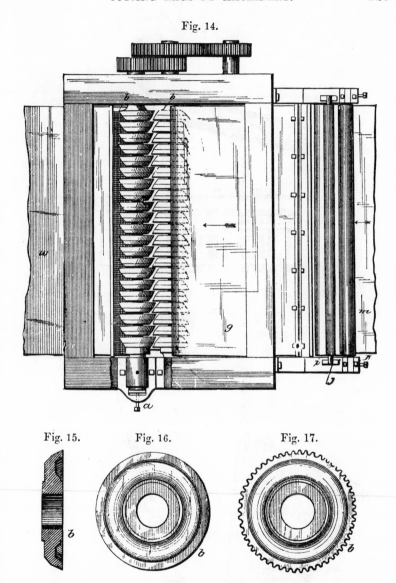

Fig. 14.

Fig. 15.   Fig. 16.   Fig. 17.

the direction indicated, carries the rags to the knives *b*, where they are cut in a direction across the cut of the first knife. It will be observed that the first knife cuts the material in a

Fig. 18.

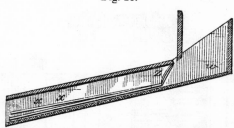

Fig. 19.

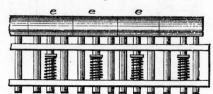

Fig. 20.

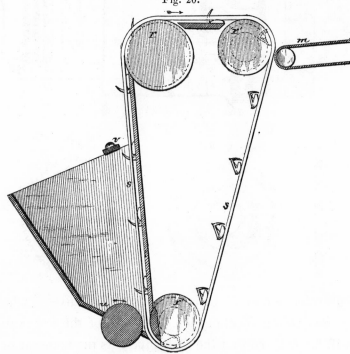

direction parallel with the axis of the revolving knife, and that the strips, falling as cut, are deposited lying in the same

direction upon the apron *f*, and are thus by a positive feed carried and delivered to the rotary cutters *b*, where the strips are separated into short pieces.

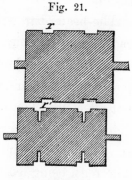

Fig. 21.

The size to which the material is cut by the first knife may be varied by varying the feed, or by varying the rapidity of the revolution of the cutter *a*, and the size to which the

Fig. 22.

material is cut by the cutters *b* is varied by varying the distance of separation of the blades. The size to which the stock is reduced may therefore be easily controlled by the operative.

To do away with the objection which might exist of the knives *b* wearing unevenly, the inventor revolves one set

Fig. 25.   Fig. 24.   Fig. 23.

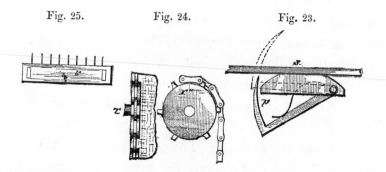

a trifle faster than the other, thus distributing the wear over the whole surface of the cutting-faces.

To assist in the feed of the rotary cutters, the edge is corrugated, as shown in Fig. 17, of either one or both sets of cutters.

The beater-bar is made in sections, as shown in Fig. 19, each section, e, being held in position by a spring. Thus, if a thick piece or bunch of cloth pass through and force a portion of the bar from the beater, the other portions are not affected. The sieve has a vibratory motion, and is provided with flails, x, which lie loosely upon the bottom of the sieve, being held in place by cords, z, or other like means. The rags, being fed in the sieve at its upper end, pass down the incline between the wire bottom and the flails. The rapid motion of the sieve causes the flails to rise and fall, thus striking the material and pounding out the dust and foreign matter in its passage through the sieve.

The feeding device illustrated in Fig. 20 consists of an endless belt provided with hooks or pins which project from the surface, and when the point of delivery is reached, retreat below the surface, thus completely freeing the rag. The rags are deposited in the hopper, and as the pins pass upward through them they catch the cloth and draw it in the direction of the moving belt. This construction—i. e., a feed taking from the bottom—is of material advantage in many respects. Clogging is avoided, as the pin being loaded at the bottom and thus covered, passes through the mass without any addition. The tendency of the mass is to roll from the apron at the top and toward the apron at the bottom. The pressure of the rags in the hopper tends to hold the rag which is being drawn from the mass, thus materially aiding in opening and straightening out the knots and bunches in which the rags are frequently found. A roll, u, located at the bottom of the hopper, revolves slowly toward the belt, thus keeping the throat filled.

The pin is constructed as shown in Fig. 23, it being pivoted to a frame, which frame is secured to the belt. The position occupied by the pin when out is shown in·dotted lines, and when withdrawn in full lines. The rolls r are grooved to permit the pins and boxes to pass, as shown in Figs. 21 and 22. The rolls r' permit the pin to pass, while the roll r is grooved only sufficiently to allow the box to pass, the pin being held projecting. The projecting hooks catch the rags on their passage through the hopper and retain their hold until reaching the roll r' at the top of the frame, where the pin is permitted to retreat, and the material being released, is deposited upon a belt or feed-rolls, as may be desired. The belt may be strained on the rolls, but it is preferable to secure the links (shown in Fig. 25) to the edges of the belt, and provide spur-wheels at each end of one or more of the rolls, which spurs will engage with the links and move the belt.

Heretofore the blades or cutters for cutting rags have either been made wholly of steel or were provided with steel facing or cutting-edges. The objection to this construction is that the steel is carried away in fibres or threads. Mr. Taylor substitutes for steel a cutter made wholly of charcoal-iron chilled. This gives a cutter of sufficient hardness, and the wear or loss is in the form of a very fine powder, leaving a sharp cutting-edge. To increase the capacity of the

9

machine a set of cutters is arranged at the opposite side of the frame, similar to the first set, and feed to the second cutters on an apron, as indicated in dotted lines.

The feed-rolls may be either smooth or roughened. It is preferable, however, to use rolls having roughened or corrugated surfaces, and provide means to force them together for the purpose of crushing and breaking such foreign matter as may be loosened in this manner.

In this machine the inventor uses two sets of cutters of different construction, for the reason that the machinery that would otherwise be required to turn the strips and present them in proper manner to the cross-cutters is avoided. A machine is thus constructed having a direct and positive feed.

Baumann's Machine.

Most rag-cutting machines in use in the mills of the United States are constructed on the shearing principle, the bed-knife presenting a cutting edge to the revolving knife. There is another form of rag-cutting machine which works with an oscillatory reciprocating movement. This is undoubtedly the best principle for a rag-cutter, as the rags are less torn than by the other methods, and machines of this character can be used for all the different grades of stock from the finest rags to the coarsest rope.

The trade objection to cutters of this character is that they cannot be made to perform as much work in a given time as those cutters having revolving knives. Rag-cutters having revolving knives cut with less intermission than those having an oscillatory reciprocating motion, as there is

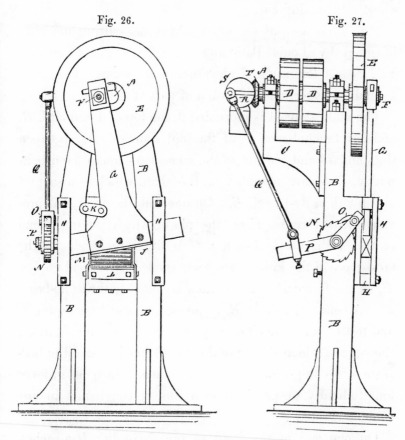

Fig. 26.

Fig. 27.

considerable time lost in the upward movement of the cutting knife in the latter class of machines.

Baumann's invention, shown in Figs. 26, 27, and 28, consists in a rag-cutting machine having a reciprocating knife which is connected by a pivoted link

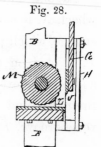

Fig. 28.

with the machine-frame, whereby an oscillatory reciprocating movement is given to the knife, which thus makes
a vertical shearing cut.

Fig. 26 is a front elevation of the rag-cutting machine
invented by Louis Baumann, of Offenburgh, Germany.
Fig. 27 is a longitudinal elevation of the same. Fig. 28 is
a detail cross-sectional elevation of part of the same.

The main shaft A is journaled in the top of a standard, B,
and in a bracket-arm, C, of the said standard, and between
the standard and the end of the arm a loose and a fixed belt-
pulley, D D', are mounted on the shaft, which is also provided with a fly-wheel, E. On one end, the shaft A is provided with a crank, F, on the pin of which the upper end
of the T-shaped knife-beam G is fitted, the ends of the
cross-piece of the knife-beam being guided by vertical guides,
H, on the frame B. The knife J is bolted to the knife-beam
G. A connecting-link, K, is pivoted to the knife-beam G
and to the frame B. The rags rest on a bed-plate, L, along
the edge of which the knife J cuts. Directly above the bed-
plate a feeding-roller, M, is journaled, which is grooved
longitudinally, or provided with ribs, or roughened in any
suitable manner.

On one end of the shaft of the feed-roller M a rachet-
wheel, N, is mounted, with which a pawl, O, is adapted to
engage, which pawl is pivoted in one end of a rocking lever,
P, pivoted on the shaft of the feed-roller. To the opposite
end of the lever P a connecting-rod, Q, is held adjustably,
the upper end of which rod is held adjustably in a transverse groove, R, in the flat surface of a bevel cog-wheel, S,

engaging with a bevelled cog-wheel, *T*, on that end of the shaft *A* opposite the one on which the crank *F* is fixed.

The operation is as follows: The crank *F* gives the knife-beam a reciprocating movement and the link *K* gives it an oscillating movement, so that the knife-beam will have an oscillatory reciprocating movement, and will make a vertical cut which at the same time is a shearing cut whereby the rags will be cut very finely and will not be torn, and the raising of dust will be avoided. The feed-roller *M* is revolved slightly after each cut, and feeds the rags the required distance to the cutter. The feeding device can easily be adjusted to feed a greater or lesser length of the rags after each cut.

Coburn's Machine for Cutting Rags, etc., or Materials containing Metallic and other Substances.

The object of the machine shown in Figs. 29, 30, 31, 32, 33, 34, and 35, the invention of Messrs. Lemuel and Jehiel E. Coburn, of Worcester, Mass., is to provide means for cutting or severing old corsets, paper-stock, rags, etc., or materials having whalebones, steels, buttons, eyelets, or other hard substances, incorporated in their structure.

Fig. 29 is a plan view of the cutting machine. Fig. 30 is a vertical section of the same at line *x x*. Fig. 31 is a rear end view of the machine. Fig. 32 is a longitudinal sectional view, showing the arrangement of the cutters upon their rotating shafts. Fig. 33 is a side view of one of the cutter disks on a larger scale. Fig. 34 is an edge view of the same, and Fig. 35 is a side view of one of the clearer-bars.

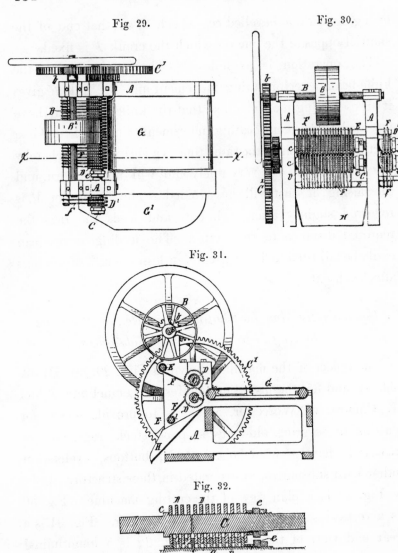

Fig 29.

Fig. 30.

Fig. 31.

Fig. 32.

A denotes the frame, of proper form and material, and provided with suitable bearings for supporting the operative parts.

B indicates the driving-shaft, mounted on the upper part of the frame, and provided with pulleys, B^2, for the driving-belt.

Fig. 33. Fig. 34.   Fig. 35.

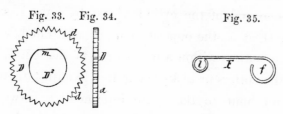

C C indicate the cutter-shaft, on which are mounted the counters or shearing-disks D, of which the working-cylinders are composed. Said shafts are journaled in bearings on the frame A, one of them being provided with a gear, C', that meshes with a pinion, b, on shaft B for operating the cylinders when shaft B is set in motion. Shafts C and C may be geared together, or one of said shafts may be left free to be revolved by the friction or interaction of the cutters D of the opposite cylinder. The cutter-disks D are made in the form shown in Figs. 33 and 34, with a central eye or opening, D^2, to fit over the shaft C, and with a series of depressions, teeth, or serrations, d, about the periphery. Said disks may be made from plate steel, or may be punched from sheet iron, and case-hardened after the teeth d have been found. The disks are preferably about one-fourth of an inch in thickness for ordinary work, although the inventors do not confine themselves to any particular dimensions, as the size of the disks may be varied as required. The cutters or disks D are slipped on to the shafts C C, alternating with each other on the respective shafts, as indicated, the cutters on the upper shaft being separated at their edges by the cutters on the lower shaft, and *vice versa*. Said cutters are retained laterally by a collar or shoulder, *c*, fixed upon the shaft at one end of the cylinder, and by nuts, *e*, screwed on to the shaft at the opposite end, as illustrated, so as to confine the cutters within a given limit of the length of the shaft *C*, the cutters or disks being free to adjust themselves within said limit to the space intermediate between the cutters of the other cylinder. Thus the edge of the cutters of the upper cylinder interact with and serve to keep separate the cutters of the lower cylinder, and *vice versa*, while all of the notched edges or peripheral angles of the several disks *D* on one cylinder or shaft *C* shear past or against the adjacent edges or angles of the disks on the opposite shaft or cylinder when the mechanism is rotated, thus effecting a series of stripping cuts corresponding in width to the respective thicknesses of the disks or cutters. The eye D^2 of the cutters *D* and the shafts *C* are made of corresponding irregular shape, or with a flattened side, as at *m*, or provided with some equivalent means for retaining them in position, and preventing any independent rotation between the shaft and cutter-disks. By turning the nuts *e* upon the shafts *C*, the shearing-edges of the cutters can be set together with greater or less force, the cutters being laterally free among themselves. Adjustment at all the cutting angles is effected simultaneously by the adjustment of nuts *e*.

F indicates clearer bars or fingers, which are arranged between the respective cutters *D* in series corresponding therewith. These clearer-bars are arranged to nearly fill the width of the spaces between the disks D. They extend from the shaft C to a position beyond the peripheral line or edges of the disks, their outer ends being retained stationary by suitable supports, E, so that any of the severed material that becomes wedged in or caught between the parallel sides of the adjacent disks will be forced outward from the spaces as the disks revolve past the clearer-bars. Thus the severed material is freed and discharged from among the cutters or prevented from winding around the cylinders. The clearers F are made as shown in Fig. 35 (preferably of round wire, although flat plates may be employed if desired), with a loop, f, at one end to fit over the shaft C, and the loop l at the opposite end, through which is passed a support-bar, E, that extends across the frame A, and by means of which the outer ends of the series of clearers are retained in a uniform line, and the respective clearers prevented from turning out of place by the strain of the work and the revolving of the cutters.

G indicates the endless travelling apron for feeding the material to the cutters. The apron may be mounted on guiding-rolls, and be operated in any suitable manner; or, if preferred, a stationary table may be used instead of the travelling apron. A stationary table is shown at G', in front of the stripping-cutters at D'.

H indicates a chute for collecting and directing the cut material as it falls from the cutter-cylinders. The machines may be made with long cutter-cylinders, as shown between the frames $A\ A$, and with bearings at each end; or they may be made with short projecting cutter-cylinders as at D', or with both the long and short cylinders, as shown in the illustrations. Also, any desired number of disks or cutters, D, may be used to compose the cylinders, and said cutters may be formed of any thickness required. Cutters of different thicknesses can be run together when desired, and one series of cutters can be readily exchanged for another of different thickness by simply raising the shafts C from the bearings, removing the nuts e, and sliding off those which are on the shaft and then sliding on the other set of cutters. When it is desired to cut the material into strips or pieces of considerable width, tubular cylindrical blanks may be arranged between two thin serrated disks, D, to form a cutter of the desired thickness, this being equivalent to making a single cutter of equal thickness to the blank and cutters as combined.

In the machine here shown the invention is embodied in a practical form for cutting up old corsets for the purpose of separating the steels, eyelets, and whalebones from the cloth fibre, the side cylinders at D' being adapted for severing the busk-steels and eyelet-strips, and the broad cylinders for cutting the body portions transversely across the whalebones.

In the operation of the machine the material to be cut is fed between the cylinders from the apron or table, and the revolving cutters sever it into uniform strips corresponding in width to the width of the cutting-disks. The teeth, d, of the disks are readily forced through the material, and the depressions between the teeth receive the hard substances, thus preventing the material from escaping or sliding in front of the contact edges of the cutters, so that the machine operates with comparative ease and effect, while the cutters, by reason of their form and arrangement it is claimed do not require to be sharpened, even though cutting hard substances, and, if broken or injured, can be readily renewed.

The cutters of case-hardened wrought-iron can be cheaply made, and are very durable, and the machine, it is claimed, can be maintained in working condition at slight expense. It is also claimed to be rapid and efficient in its operation, and is adapted to severe usage without liability of derangement.

Machine for Separating Metallic Substances from Paper-Stock, etc.

In connection with the machine of the Messrs. Colburn last described, we will here mention the invention of Mr. Charles F. Taylor, of Springfield, Mass., which has for its object the separation of small metallic substances from rags and paper-stock.

Heretofore objectionable matter in rags has been largely removed by hand, and in pulp by screens, perforated bottoms, etc. These, however, do not remove the small particles of iron and other like matter, which is very objectionable.

In reducing rags to the desired degree of fineness and purity for the bleach-boiler, no means has been devised to separate small metallic substances from the rags. A large percentage of the objectionable matter which it has heretofore been very difficult to remove is magnetic, or, in other words, is of such nature as will be attracted by a magnet. It is consequently proposed in the present invention to utilize this attractive property for the purpose of separating such substances from the rags.

It will readily be seen that many mechanical contrivances may be devised with which the attractive force of magnets may be utilized for this purpose, and that there are various stages in the process of paper-manufacture where the magnetic force may be applied to accomplish the desired result, and that both permanent and electro-magnets may be used.

In treating rags containing a large proportion of metallic substance such as are designed to be cut by the Coburn rag-cutting machine special applications of the magnetic force would have to be made; but the best result is ordinarily attained by the application of the magnetic force in two stages, the first after the rags have been reduced to the size to which they are usually cut before being placed in the bleach-boiler, and the second after the material has been reduced to pulp.

Probably the best method of applying and utilizing this force in the first instance is to provide two sets of revolving magnets, a, as shown in Figs. 36 and 37, and to feed the cut rags between these magnet-rolls, they being so adjusted that all the material which passes is caused to come in contact with or in close proximity to the magnets, when, if there be any particles in the stock, either separated from or attached to the rags of the nature above described, they will be attached to the magnets, and be carried by the revolving of the magnets away from the flow of clear material, and may be wiped from the magnets with any convenient appliance. A fixed or stationary magnet may, however, be used at this stage, and a good result attained.

Figs. 36 and 37 are side views of an arrangement for the application of the magnets for use in cut rags, and Figs. 38 and 39 are views illustrating the application to pulp.

Fig. 36 illustrates a device consisting of a hopper, *b*, having a feed-wheel, *c*, arranged to feed the stock to the magnets *a*. The feed-wheel *c* is provided with pins which catch and feed the rags through the chute.

Fig. 37 illustrates a means of feeding the rags to the magnets upon an endless apron.

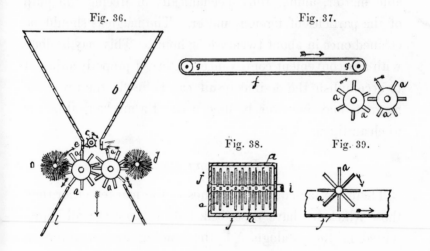

Fig. 36. Fig. 37. Fig. 38. Fig. 39.

Figs. 38 and 39 illustrate the method of application in separating the objectionable matter which may have passed the first appliance from the stock after it has assumed the form of pulp. It is preferable to apply the magnets to the pulp while it is passing through the sand catcher. The flow is here slow and shallow, and by immersing the magnets at this point all or a large proportion of the magnetic matter is caught. There is in much pulp a scaly matter, which apparently comes from the iron portion of the machinery with which the pulp comes in contact. The pulp is of such consistency that these small particles are held, and do not fall to the bottom; neither can they be caught in screens. The best method of application is probably to attach a series of magnets to an arbor, and revolve them slowly in the sand-box in a direction against the flow of the pulp. The magnets, being set near together, will thus aid in separating the globules of pulp which gather and hold the objectionable matter, aiding thus mechanically in freeing the pulp of the particles of foreign matter. The magnets should be cleaned once in about twenty-four hours. This may be done with any convenient contrivance. A brush properly adjusted to accomplish the desired result can be used; the revolving magnets may, however, be used without a mechanical means to clean them.

Other Rag-cutting Machines.

There are rag-cutting machines in the market other than those which we have described, but, as they are all represented in the catalogues of the leading manufacturers of paper-making machinery, we shall not enumerate and describe them in this volume owing to lack of space. The reader is, therefore, referred to the firms of Messrs. Cyrus Currier & Sons, Newark, N. J.; Holyoke Machine Co., Holyoke, Mass.; the South Boston Iron Works, Boston, Mass.; Messrs. La Tourrette & Co., Middleton, O.; Messrs.

Stiles & Co., Riegelsville, Warren Co., N. J.; The Black & Clauson Co., Hamilton, O., etc., from whom all desired information can be obtained.

Sizes of the Cut Rags.

In order to avoid the great waste which results from the unravelling of the rags when cut on the bias, it is desirable to make the cut in the direction of either the warp or woof. The dimensions of the cut rags vary from about two to five inches square; coarse and tough varieties being cut smaller than those which are soft or well worn.

List of Patents for Rag-cutters and Dusters, issued by the Government of the United States of America, from 1790[1] to 1885 inclusive.

No.	Date.	Inventor.
	Jan. 13, 1829.	W. Debit.
	July 27, 1831.	G. Carriel.
93	Nov. 28, 1836.	E. Burt and G. Carriel.
615	Feb. 22, 1838.	R. Carter.
920	Sept. 14, 1838.	E. Burt.
927	Sept. 19, 1838.	H. Clark and W. Albertson.
1,782	Sept. 10, 1840.	E. Smith.
11,882	Oct. 31, 1854.	A. S. Woodward and B. F. Bartlett.
23,643	April 12, 1859.	W. C. Geer.
27,167	Feb. 14, 1860.	J. Storm.
31,154	Jan. 22, 1861.	R. Daniels.
38,735	June 2, 1863.	J. Faw.
73,695	Jan. 28, 1868.	J. Collins, Jr., and N. R. Nickson.
74,506	Feb. 18, 1868.	J. Collins, Jr.
75,341	March 10, 1868.	A. Allen.
80,531	Aug. 4, 1868.	A. T. Bennett and W. O. Anderson.
85,512	Jan. 5, 1869.	} A. F. Crosby.
85,513	Jan. 5, 1869.	
98,692	Jan. 11, 1870.	J. W. Barbour.
100,718	March 15, 1870.	L. Brainard.

[1] See page 50.

No.	Date.	Inventor.
102,854	May 10, 1870.	W. E. Newton.
133,787	Dec. 10, 1872.	M. Marshall.
145,475	Dec. 16, 1873.	E. D. Aldrich.
214,185	April 8, 1879.	G. W. Patten and J. H. Knowles.
214,462	April 15, 1879.	J. T. Slack.
217,100	July 1, 1879.	T. W. Harding.
234,640	Nov. 16, 1880.	W. A. Wright.
268,075	Nov. 28, 1882.	T. W. Harding.
272,856	Feb. 27, 1883.	L. and J. C. Coburn.
280,076	June 26, 1883.	F. L. Palmer.
286,373	Oct. 9, 1883.	L. Baumann.
286,503	Oct. 9, 1883.	
287,482	Oct. 30, 1883.	} C. F. Taylor.
292,873	Feb. 5, 1884.	
298,108	May 6, 1884.	R. O. and W. Moorhouse.
299,366	May 27, 1884.	T. Ferry.
311,186 } 311,187	Jan. 27, 1885.	J. B. Hart and E. H. Walker.

STRAW CUTTERS.

The cutters used for cutting straw into the required short lengths are similar to ordinary rag cutters; the straw being spread on a table and fed to fluted feed-rolls, which push it forward over a steel bed-knife where it is cut by revolving knives.

It is usually desirable to pass the straw through a cleaner; but if this is not done it should fall from the cutting-box on a slanting rack of wire through the openings of which the chaff and grain will pass.

The straw after being cut is next thoroughly wetted and afterward thrown into a bin, where it sweats and soaks and gradually grows more pliable while waiting to be thrown into the boiler.

Cutting Wood for Chemical Fibre.

The wood is generally received at the mill as cord wood; it is freed from bark and the worst of the knots are cut out. It is cut into chips by a machine consisting of a heavy cast-iron disk, measuring about seven feet in diameter, keyed on a shaft at one end of which is a driving pulley, and all mounted in a suitable frame. Three knives are fastened to the face of the disk, in a slightly inclined position, so that the edges of the knives project about three-quarters of an inch from the face. The wood is fed to the knives through a cast-iron chute arranged in a slanting position, facing the disk in a line with the knives. The wood on coming in contact with the knives is cut into chips measuring from one-half to three-quarters of inch in length. From the cutting machine the chips fall into a pit from which they are conveyed to the digester by means of an ordinary bucket elevator or other suitable plan.

If the knives are kept true and sharp the wood will generally feed regularly and smoothly to the cutter, thus avoiding the necessity of employing either physical or mechanical force for holding the wood to the disk.

Treating Wood Before Grinding.

Various methods of treating wood previous to submitting it to the action of the grinders have been proposed and used.

By one process the pieces of wood after being cut into suitable lengths for grinding are treated by first steaming them, then removing the acids generated in the steaming operation, next treating the steamed wood with alkali, and, finally, grinding or reducing the pieces to pulp. Steaming has been resorted to for the purpose of removing the bark from wooden blocks preparatory to grinding the solid parts, and wood has also been treated with water sprinkled on it from above, and steam simultaneously applied from beneath it, in order to soften and cleanse it preparatory to grinding.

But the process which we shall now describe, which is that of Mr. George F. Cushman, of Barnet, Vermont, is intended to facilitate the disintegration of the fibres when submitted to the action of the revolving stones by a preliminary cooking of the block of wood in a bath of boiling hot water with lime, soda-ash, or equivalent chemical agent in solution, to soften the block, toughen the fibres, and lessen their lateral adhesion. By this process the block is reduced to pulp with much less power than is required to grind a block not so treated, and the pulp produced is claimed to be softer, stronger, and more desirable, since the fibres are not broken up or comminuted, but are more nearly in their natural condition, with their lateral beards or filaments preserved, so that when reunited in the paper sheet special toughness and tenacity are attained.

In carrying out this method, immerse the solid wooden blocks in a strong solution of lime, soda-ash, chloride of lime, or equivalent chemical agent, kept boiling hot by the introduction of steam or otherwise, and adapted to soften the blocks in readiness for grinding, and retain the blocks under treatment from ten to twenty-four hours, or until the liquid has had time to penetrate all parts of the block, and the lateral adhesion of the fibres is so weakened that they will readily separate by the attrition of the grinding-stone without being broken short or reduced to a mere powder; and as the chemical action is most rapid in the direction of the length of the fibres, it is desirable to cut the block much shorter than is usual, or to form transverse saw-scarfs at intervals between its ends, in order that the solution may readily penetrate from each end to the centre, so as to loosen and toughen the fibres throughout the block. The pressure of steam above the liquid in the tank tends to force the solution into all the pores of the immersed blocks. Then remove the blocks from the tank and subject them to the action of the grinders in the usual way, keeping a constant stream of water upon the stone; and the disintegration will be found to be effected with great rapidity, owing to the preliminary treatment received by the blocks, and also that no washing is required beyond what results from wetting down the stone. The pulp produced is claimed to be of superior quality, and as the blocks have absorbed only so much of the chemicals as is beneficial to the fibre, it is in condition for the successive steps in the production of various grades of paper of special strength, and for numerous other purposes in the arts. If preferred, however, this fibre may be mixed with hard stock made of other material, such mixture producing paper or board of exceptional toughness.

*Voelter's Machine for Cutting or Grinding Wood, and
Reducing it to Pulp.*

The art of reducing wood to pulp by subjecting the same
to the action of a revolving stone is not a new one, machinery
for grinding wood while a current of water was applied to
the stone having been patented in France by Christian
Voelter as early as 1847 (see 10th volume, 2d series, Brevets
d'Invention), and in England by A. A. Brooman, of London,
in 1853. (See Repertory of Patented Inventions for May,
1854, page 410.)

A large number of inventions for cutting or grinding wood
into pulp have been patented in the United States and in
Europe; but the enormous development of the paper-making
industry and the cheapening of paper in America during the
last fifteen years are largely due to the general introduction
of the machine for disintegrating blocks of wood and assorting the fibres so obtained into classes according to their
different degrees of fineness, invented by Mr. Henry Voelter,
of Heidenheim, Würtenburg, Germany, and for which
invention he received letters patent on August 10, 1858,
from the United States.

In all the processes known or used prior to Voelter's
invention the wood had been acted upon by the stone in one
of two ways, viz., either by causing the surface of the stone
to act upon the ends of the fibres, the surface of the stone
moving substantially in a plane perpendicular to the fibres
of the wood ; or, secondly, by acting upon the fibres in such
a direction that they were severed diagonally, the surface of
the stone moving diagonally across the fibres. The first

plan, in fact, made powder of the wood. The pulp had no practical length, and on trial proved worthless, or nearly so. The second plan was carried out by the use of a stone revolving like an ordinary grindstone, the wood being applied upon the cylindrical surface thereof, with the fibres perpendicular, or nearly so, to planes passing through the axis of the stone and the point or locality where the grinding was performed ; and this plan also failed, because the fibres were cut off in lines diagonal to their own length, and were consequently too short to make good pulp. There were other difficulties attending the process not necessary here to mention. Such was the state of the art prior to Voelter's invention; and his improvement in the art consists in grinding or rather tearing out the fibres from the bundle of fibres which makes up a piece of wood by acting upon them by a grinding-surface which moves substantially across the fibres and in the same plane with them. In carrying out his improvement upon the art Voelter splits a log of wood and applies the flat side upon the stone, and then so revolves the stone as to cause points upon its surface to pass the fibres in lines perpendicular, or nearly so, to the length of the fibre. By this mode of procedure it is possible to obtain a sufficiently long fibre and save much power. Voelter's improvement in the art consists, further, in regrinding the fibres by causing them, after being separated from the block, to pass under other blocks of wood, which are being reduced to pulp upon the same stone. The fibres torn out at the first operation are thus rolled over and crushed again and separated into smaller fibre.

Voelter's improvements in the machinery are in an arrangement of pockets with reference to the grinding-surface, so as

to hold the blocks of wood in such position that their fibres
may be separated from the blocks in the manner described.
and whereby fibres may be reground; and in a contrivance
for feeding up the blocks by a positive feed instead of by
force derived from weights or springs, as formerly practised;
and a contrivance for causing the feed to cease automatically.

Fig. 40.

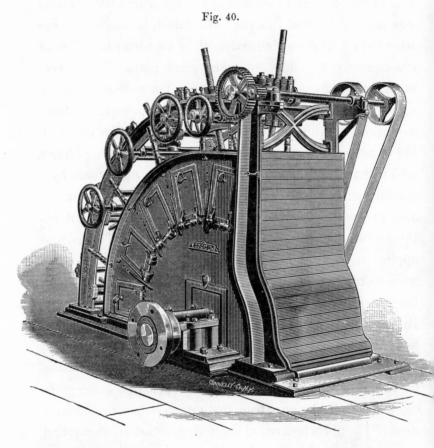

On May 22, 1866, Mr. Voelter was granted another patent
for improvement in his machine for reducing wood to paper-

pulp, which patent was reissued April 23, 1872. For the dates of the various reissues and extension of the Voelter patent see the list of patents which follow the close of the present section.

Figs. 40 and 41 show front and rear perspective views of

Fig. 41.

the improved Voelter wood-pulp machine, which is now built with either five or seven pockets.

Figs. 42 to 48 show the machine patented by Mr. Voelter

on May 22, 1866. Figs. 42 and 43 are sectional elevations of parts of Voelter's apparatus for reducing wood fibres to

Fig. 42.

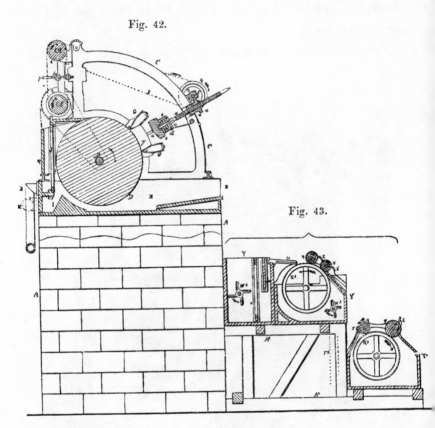

Fig. 43.

paper-pulp; Figs. 44 and 45, plan views of Figs. 42 and 43; Fig. 46, a detached view of part of the apparatus ; Fig. 47 a sectional elevation of another portion of the apparatus, and Fig. 48 a plan view of Fig. 47.

On a suitable foundation, *A*, Fig. 42, rests an oblong box, *B*, and to opposite sides of the latter are secured quadrant-shaped frames, *C C'*, in which turn the shafts *D*, *E*, *E'*, and *F*.

To the shaft D is secured a grindstone, D', and to each of the shafts E E' is secured a conical pulley, 2, a belt, 4, pass-

Fig. 44.

Fig. 45.

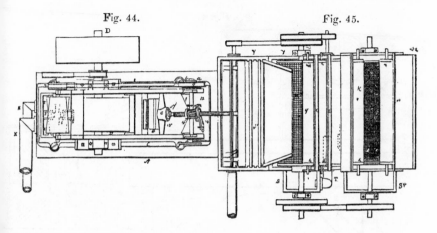

ing round both pulleys and through the forked ends of a guide, 5, which is adjustable laterally on a screw-shaft, 6.

Fig. 46.

On the shaft E is a pulley, 7, and on the shaft F turns a pulley, 9, a belt, 8, passing round both pulleys, and to the shaft F, adjacent to the pulley 9, is secured a ratchet-wheel, a, to the teeth of which is adapted the end of a spring-pawl, b, attached to the pulley 9.

Through an opening in a cross-piece, 10, extending between the side frames, passes a screw rod, 11, the rod also passing through a worm-wheel, 12, which bears against the cross-piece, and is operated by a worm, 13, on the shaft F.

To the upper side of the worm-wheel 12 are hung two jaws, which bear against opposite sides of the rod 11, and

Fig. 47.

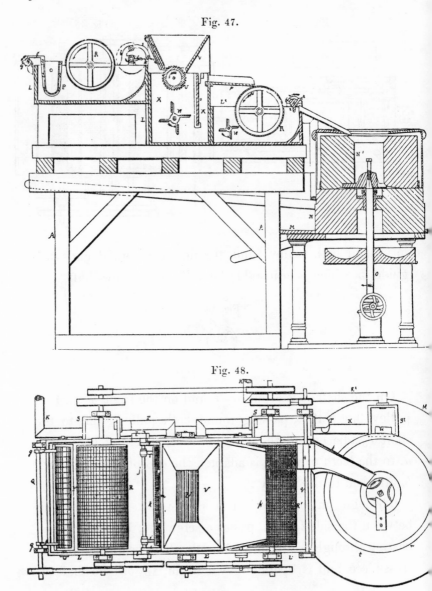

Fig. 48.

have threads cut in their edges, the side threads being adapted to the tread on the rod.

To a cross-head, 15, which slides on guides, 16, attached to the side frames, is secured a box, c, containing a rubber spring or cushion, d, and against the latter bears a disk, 17, on the rod 11, which projects through the cushion and through the bottom of the box, a nut on the lower end of the rod preventing the withdrawal of the latter.

To the lower side of the cross-head 15 is secured a wooden block, e, the face of which, near the lower edge, is cut away as shown in Fig. 42, for a purpose described hereafter.

To the side frames are secured two hollow adjustable cross-pieces or boxes, G G', each of which communicates with a water-reservoir, and in the lower edge of each box, which is nearly in contact with the face of the stone D', is a narrow slit or opening, x. The adjacent sides of the boxes G G', near their lower edges, are parallel, and are such a distance apart as to permit the ready introduction between them of the block e.

A rake, H, extends from the bottom of the box B to one side of the grindstone D', and at the bottom of the box, below the opposite side of the stone, is a projection, i, of the form shown in Fig. 42.

On a shelf or partition, 18, at the end of the box B, rests a sieve, J, so fine that fibres which can pass it do not need regrinding, the upper portion of which is inclosed by a casing, 19, secured to the side frames and to the edge of the box. Two pipes, K and K, communicate with this end of the box B, the former above and the latter below the partition 18.

On a frame-work, A', Fig. 47, rest three tanks, L, L, and L^2, and in the tank L is hung a basket, P, of wire-gauze or other suitable material, in such a manner that it receives a shaking motion by means of arms, $f f$, which project from the said basket, bearing on ratchet-wheels, g, secured to a shaft, Q, turning in brackets attached to the tank.

In the tank L revolves a cylinder, R, of wire-gauze, which communicates at one end with a reservoir, S, Fig. 48, at the side of the tank, a pipe, T, leading from the reservoir and communicating with the tank L', near the bottom of the latter.

To arms, $i i$, secured to a revolving shaft, j, turning in bearings attached to the tank L, is secured a comb, k, and to arms $i i$, hung to brackets $l l$, is secured a plate, m, for a purpose described hereafter.

In the upper portion of the tank L' turns a shaft on which is secured a fluted or serrated roller, U, and above the latter is a hopper, V, in guides, on one of the inclined sides of which slides a plate, n, the lower edge of the latter being parallel to the face of the roller. The tank L' is divided by a vertical partition, o, which extends nearly to the bottom into two unequal-sized chambers, $x x'$, and in the lower portion of the former turns a paddle-wheel, W. From the upper edge of the tank L' extends an inclined plate or chute, p, and below the latter, in the tank L^2, rotates a hollow cylinder, R', of wire-gauze, which communicates through an opening in one end with a reservoir, $S_{,}$, Fig. 48, at the side of the tank.

In the lower portion of the tank L^2, below the chute p, revolves a paddle-wheel, W, and to a shaft, q, which turns in suitable bearings secured to the tank, is attached a smaller paddle-wheel, r, the upper end of an inclined shute, s, being secured to the edge of the tank adjacent to the paddle-wheel r.

On a platform, M, supported by pillars rests the lower stone, N, of a pair of millstones, the upper millstone, N', being hung to and rotating with a vertical shaft, O, in the ordinary manner, and into the usual central opening in this stone projects the lower end of the chute s. The upper stone, N', is surrounded by a casing, t, an opening at one side of which communicates with a box or reservoir, S^2, secured to the platform M.

On the framework A^2, Fig. 43, rest the tanks Y, Y, and Y^2, and in the former is a vertical sieve, v, and a partition, w, the latter extending across the upper portion only of the tank. On one side of the sieve v revolves a paddle-wheel, W^2, and from the opposite side of the tank a chute, 31, projects over a cylinder, R^2, of wire-gauze, which revolves in the tank Y'.

The cylinder R^2 communicates, through an opening in one end, with a reservoir, S^3, Fig. 45, a pipe, T^3, communicating with the latter and with the tank Y^3, in which turns a cylinder, R^3, which communicates with a reservoir, S^4, and against both this cylinder and the cylinder R^2 bear rollers y y, on the ends of which are bands h, of leather or other suitable material, a stationary plate, z, being secured at the side of each roller. In the tank Y^2 turns a paddle-wheel, W^3

The pipe K, Figs. 42 and 48, communicates with a pipe, T', leading from the reservoir S', and also with the reservoir S^2, and from the latter extends a pipe, K^2, which communicates with the tank Y.

The material flowing through the pipe K' is discharged into the basket P, a pump or other suitable apparatus being used to elevate the material when the tank L is above the box B.

Operation.—The sections z of wood to be disintegrated are placed between the boxes G G' and against the grindstone D'. Water is admitted to each of the boxes and into the tank B, and a rotary motion in the direction of its arrow is imparted to each of the shafts D, E, E', and F. A rotary motion in the direction of its arrow is also imparted to each of the shafts Q, j, q, and O, to the paddle-wheels W, W', W^2, and W^3, to the cylinders U, R, R', R^2, and R^3, and to the rollers y y. As the worm-wheel 12 is turned the jaws 14 14, acting as a revolving nut, will cause the rod 11 to be moved forward, the block e being brought against the sections z and feeding the latter slowly toward the grindstone by which they are disintegrated, the fibrous particles thus detached being carried into the box B. The undue pressure of the wood against the stone is prevented by the elastic cushion d, which also yields slightly to permit the wood to accommodate itself to inequalities in the stone, while the wedging of the blocks between the boxes G G', which occurs when the boxes approach each other toward the bottom, is prevented by making the adjacent sides of the boxes parallel. The speed of the forward movement of the rod 11 in proportion to that of the stone is regulated by adjusting the belt 4 on the pulleys 2, the spring-pawl b, through the medium of which motion is conveyed from the pulley 9 to the shaft F, being sufficiently rigid to retain its hold on the ratchet-wheel l so long as no unusual resistance is offered to the forward movement of the rod 11 and the cross-head. When, however, the blocks of wood are not disintegrated with sufficient rapidity, or the forward movement of the rod 11 is otherwise interrupted or retarded, the pawl b will yield and slip over the teeth of the ratchet-wheel, the rattling noise thus produced informing the attendant of the necessity of readjusting the belt 4 to diminish the speed of the shaft F. As the block e is brought near the face of the stone, that portion of the wood beneath the inclined face of the block will be cut to a wedge-shape, the thick edge being toward the box G'. By this means small particles of wood are prevented from being wedged into the narrow space between the box G' and the stone, to the retardation of the revolution of the latter. The finer fibres of the wood are carried by the revolution of the stone between the teeth of the rake H, and are thrown against the sieve J, while such coarser particles as would injure the sieve are arrested by the rake. The finest filaments pass through the sieve J with the water thrown up by the stone, and are conducted through the pipe K to the reservoir S^2, Fig. 48, while larger particles fall in front of the projection I and pass with the water which flows through the pipe K' into the basket P, Fig. 47. The finer fibres pass through the meshes of the basket P, while the coarser fibres are retained and removed from time to time, such a vibrating motion being imparted to the basket by the action of the ratchet-wheels g as will prevent the meshes from becoming obstructed. The finest fibres pass with the water into the gauze cylinder R, and out of the latter into the reservoir S, and through the pipe T to the tank L', the coarser fibres being carried by the action of the cylinder R within range of the rotating comb k, by which they are caught and carried upward until the comb strikes the plate m. As the comb continues to revolve the plate m slides forward and scrapes off the adhering fibres, which fall into any suitable receptacle, the tank L being thus cleared of the useless fibres which would obstruct the action of the cylinder. After the contents of the reservoir S are introduced into the tank L', they are thoroughly agitated and mixed by the action of the paddle-wheel W, a mash being thus produced, which is directed upward through the chamber X', and on to the chute p, from which it falls on to the cylinder R'. The finer filaments, which pass through the cylinder R' are conveyed into the reservoir S', and through the pipe T' into the reservoir S^2, while the mash which remains in the tank is mashed and agitated by the paddle-wheel W', and is directed by the paddle-wheel r into the chute s, down which it flows into the opening in the upper millstone N'. As the fibres pass between the mill-stones they are split and broken into fine filaments, the stones being so prepared that the fibres may be cut rather than worn. The fibres, after being reduced to a pulpy mass, pass from the stones into the casing h and then into the reservoir S^2. The pulp flows from the reservoir S^2, through the pipe K^2, into the tank Y, where it is directed by the paddle-wheel W^2 against the sieve v, the finest fibres passing through the latter and upward to the chute 31, from which they fall on to the cylinder R^2, the fibres which pass into this cylinder being conducted to the reservoir S^3 and through the pipe T^3 to the tank Y^2. The pulp in the tank Y' is agitated by the paddle-wheel W^3, so that every portion may be brought into contact with the cylinder. The gauze on the cylinder R^3 is too fine to permit any of the fibres to pass through it. The superfluous water, however, flows into the cylinder and into the reservoir S^4, from which it is removed by a siphon or other suitable apparatus. As the cylinders $R^2\,R^3$ revolve the fibres on the surfaces of the same are transferred to the rollers $y\ y$, and after being scraped from the latter by the plates z, fall into any suitable receptacle, the leather bands $h\ h$, at the ends of the rollers, maintaining the surfaces of the same from contact with those of the cylinders, which are thus preserved from abrasion. The coarse fibres, detached by the plate z, as well as those remaining in the tanks $Y\ Y'$, are placed in the hopper V, from which they are fed into the tank L' by the fluted roller U, the sliding plate n being adjusted to regulate the passage of the fibres in such quantities as may be desired. These fibres are discharged from the tank L' into the tank L^2, and after passing between the millstones are sorted in the tanks Y, Y', and Y^2, as before. If fibres are required which are not so finely divided as those which pass into the tank Y^2, they may be removed at any stage of the process, and it will be apparent that any desired number of tanks and cylinders

11

may be employed in order to obtain a greater assortment of fibres.

Instead of arranging the cylinders as described they may be placed with their shafts inclined, and the material may be introduced into the interiors of the cylinders, the finer particles passing through the latter into the tanks, while the coarser fibres are rolled toward the lower end and discharge into any suitable receptacle.

A perforated pipe communicating with a water-reservoir may be arranged adjacent to each of the cylinders and sieves so as to throw a constant stream of water on to them, and thus maintain the meshes unobstructed.

In the process of reducing wood to fibre by a grinding operation, it always happens that slivers, chips, or small pieces of wood too large either to be used as pulp or to be reground (because they would choke the stones or lift the upper one when stones arranged as shown in the drawing are employed), are detached accidentally from the wooden blocks. These useless pieces of wood are separated from the useful fibres, first, at the rake; second, at the shaking-basket; and, third, at the first cylinder R; and it is better thus to get rid of them at three operations than to remove them all at once ; and this separation of the useless from the useful products of the first grinding, by means of sieves—for the rake, the basket, and the first cylinder are all in fact sieves —and a current of water, bearing both fibres and chips, is the separating process. The fibres which result from the action of the first stone upon the blocks are assorted at the sieve J at the cylinder R^1, in the preferred form of apparatus, and none of the fibres which pass these sieves are reground. This separation of coarser from finer fibres by sieves and a current of water is the assorting stage of the process. All the fibres might pass directly from the first to the second stone without being assorted, but in that case the finer fibres would probably be made too fine, and rendered useless; and, at any rate, the regrinding stones would be uselessly loaded with matter which did not need to be acted upon by them.

When a sieve of any kind is employed to assort fibre, the greater part of the water passes through the sieve, and the fibre which does not pass the sieve is left in a pasty state. This is the preferable state for regrinding, and the inventor therefore uses a paddle-wheel to keep the mass in motion to prevent its settling, and another wheel to produce a current to carry the mass to the chute.

The regrinding or reducing process is that which is effected by the action of the stones upon the mass of fibre introduced between them. The assorting after the regrinding is caused by the action of a fibre-bearing current and the sieves V and R^2. The fibres which do not pass V may be ground a third time. So also may those which do not pass R^2, and the pulp is deprived of the greater portion of its water, so as to fit it for transportation, by means of the cylinders $R^2 R^3$ and the rollers which gather the fibres from their surface. The cylinder R^2 is, therefore, always a part of the assorting apparatus used after regrinding, and when the fibres gathered from its surface are not ground the third time it is also, in connection with the rollers, a contrivance for partially drying the pulp.

In place of the paddle-wheel r, cords may be wound spirally around the cylinder R^1, so that as the latter revolves the material is caused to flow toward one side of the tank, and, consequently, into the chute.

Nature, etc., of the Pulp produced by Voelter's Method.

Voelter's method does not produce a real pulp, but rather a semi-flour of wood, it adds nothing to the strength of the paper, and is an injury in the sizing.

It is said with reason that the Voelter process produces little with a large amount of power; it requires, in fact, a considerable fall of water to yield 55 or 60 horse-power net to manufacture 1200 pounds of pulp in 24 hours. Furthermore, the Voelter machines can be used with profit only in proximity to large supplies of wood. Another unfavorable condition is the necessity of working the wood while green ; the sap which remains in the pulp causes it to easily ferment, to heat while piled up, and to take a reddish color.

Fig. 49 shows the arrangement of a plant for producing pulp by the Voelter process. The following letters refer to the various mechanical contrivances employed. T, elevator for hoisting the wood to the floor of the mill on which the Voelter machine is located. R, circular saw to cut the wood into blocks. B, Voelter's machine composed of a millstone mounted upon an horizontal shaft and against which the blocks are pressed by mechanical pushers, causing them to advance constantly and regularly as the blocks are ground off.

Fig. 49.

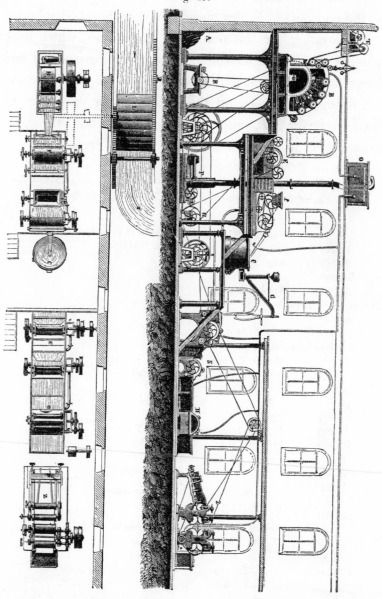

A continuous stream of water falls upon the circumference of the millstone. *E*, first sieve removing the wood splinters and separating the fine pulp from the coarser, which must pass to the refining-machine. *C*, refiner, composed of two horizontal millstones like those of a grain-mill. *G*, crane for lifting and displacing the millstones for dressing. *O*, water-reservoir. *P*, pump. *S*, sorter dividing the pulp according to its grade of fineness. *Z*, pulp press.

List of Patents for Wood Grinders, issued by the Government of the United States of America from 1790 to 1885 inclusive.

No.	Date.	Inventor.
5,251	Aug. 21, 1847.	Roberts and Hambly.
12,978	May 29, 1855	M. D. Whipple.
21,161	Aug. 10, 1858.[1]	
Reissue		
3,361	April 6, 1869.	
Extended for 7 yrs.	Aug. 29, 1870.	H. Voelter.
Reissue		
4,418	June 6, 1871.	
Extended for 7 yrs.	Aug. 29, 1877.	
37,951	March 24, 1863.	P. A. Chadburne.
40,217	Oct 6, 1863.	G. E. Sellers.
55,031	May 22, 1866.	
Reissue		H. Voelter.
4,881	April 23, 1872.	
59,042	Oct. 23, 1866.	H. and F. Marks.
77,829	May 12, 1868.	W. Miller.
84,640	Dec. 1, 1868.	H. Marks.
87,139	Feb. 23, 1869.	F. Burghardt.
89,220 89,221	April 20, 1869.	J. H. Hawes.
89,255	April 20, 1869.	J. Stutt.
97,041	Nov. 23, 1869.	F. Burghardt.
98,210	Dec. 21, 1869.	G. Vining.
99,071	Jan. 25, 1870.	H. Dodge.
101,785	April 12, 1870.	S. C. Taft.

[1] Antedated to Aug 29, 1856, so as to correspond with the date of the earliest foreign patent.

No.	Date.	Inventor.
102,239	April 26, 1870.	A. Fickett.
103,968	June 7, 1870.	Bliss and Rees.
105,622	July 26, 1870.	G. Ames.
106,710	Aug. 23, 1870.	
Reissues		
8,256		H. B. Meech.
8,257	May 28, 1878.	
8,258		
111,415	Jan. 31, 1871.	C. and C. Wolff, Jr.
111,419	Jan. 31, 1871.	Waissing and Specker.
112,733	March 14, 1871.	S. A. Perkins.
112,734		
113,297	April 4, 1871.	W. M. Howland.
113,488	April 11, 1871.	J. Bridge.
115,274	May 30, 1871.	
117,122	July 18, 1871.	
Reissue		J. Taylor.
8,845	Aug. 12, 1879.	
117,683	Aug. 1, 1871.	W. Riddell.
119,107	Sept. 19, 1871.	B. F. Barker.
119,601	Oct. 3, 1871.	J. K. Griffin.
122,353	Jan. 2, 1872.	J. Bridge.
122,581	June 9, 1872.	H. Dodge.
126,041	April 23, 1872.	J. S. Elliott and J. F. Wood.
127,337	May 28, 1872.	A. K. Gilmore.
128,788	July 9, 1872.	Burghardt and Burghardt.
130,803	Aug. 27, 1872.	H. W. Higley.
130,944	Oct. 8, 1872.	C. De Negri.
133,243	Nov. 19, 1872.	
Reissue		J. G. Moore.
5,936	June 30, 1874.	
141,206	July 29, 1873.	J. F. Daniels.
141,976	Aug. 19, 1873.	S. B. Zimmer.
144,313	Nov. 4, 1873.	J. Bridge.
144,354	Nov. 4, 1873.	M. S. and M. E. Otis.
148,452	March 10, 1874.	
Reissue		A. Harmes and A. Wagenfuer.
5,936	June 30, 1874.	
150,209	April 28, 1874.	C. W. Weld.
150,932	May 19, 1874.	B. F. Barker.
153,190	July 21, 1874.	
Reissue		F. A. Cushman.
8,198	April 23, 1878.	
155,074	Sept. 15, 1874.	L. M. Egery.

No.	Date.	Inventor.
156,355	Oct. 27, 1874.	
Reissue		} F. A. Cushman.
8,197	April 23, 1878.	
160,996	March 23, 1875.	B. F. Barker.
163,926	June 1, 1875.	J. O. Gregg.
163,958	June 1, 1875.	A. M. Zimmer.
165,706	June 20, 1875.	J. M. Burghardt.
166,835	Aug. 17, 1875.	O. Abell.
182,891	Oct. 3, 1876.	J. Chase.
183,155	Oct. 10, 1876.	J. O. Gregg.
187,292	Feb. 13, 1877.	G. H. Mallory.
191,899	June 12, 1877.	
Reissue		} J. Taylor and J. T. Outterson.
8,877	Sept. 2, 1879.	
194,591	Aug. 28, 1877.	A. Fickett.
195,478	Sept. 25, 1877.	J. W. Bowers and D. A. Curtis.
196,515	Aug. 23, 1877.	M. R. Fletcher.
196,944	Nov. 6, 1877.	E. N. Speer.
198,236	Dec. 18, 1877.	J. H. Burghardt.
198,845	Jan. 1, 1878.	W. H. Haskins.
200,540	Feb. 19, 1878.	W. W. D. Jeffers.
201,083	March 12, 1878.	S. M. Allen.
201,152	March 12, 1878.	N. Bly.
201,486	March 19, 1878.	W. J. Baxendale and D. Barry.
201,501	March 19, 1878.	F. A. Cushman.
201,550	March 19, 1878.	J. G. Moore.
202,097	April 9, 1878.	W. A. Doane.
202,185	April 9, 1878.	
Reissue		} R. D. Mossman.
8,698	May 6, 1879.	
202,698	April 23, 1878.	J. W. Brightman.
203,437	May 7, 1878.	A. H. Fisher.
203,928	May 21, 1878.	J. C. McIntyre.
204,077	May 21, 1878.	W. R. Patrick.
205,347	June 25, 1878.	B. F. Brown.
206,971	Aug. 13, 1878.	
207,553	Aug. 27, 1878.	
Reissue		} P. and G. C. Rose.
9,110	March 9, 1880.	
207,568	Aug. 27, 1878.	J. Taylor.
208,890	Oct. 15, 1878.	W. N. Cornell and C. Tollner.
209,197	Oct. 22, 1878.	W. D. and N. H. Shaw.
211,138	Jan. 7, 1879.	W. N. Cornell.
212,232	Feb. 11, 1879.	E. Johnson.

No.	Date.	Inventor.
212,782	March 14, 1879.	S. M. Allen.
217,509	July 15, 1879.	N. H. Burnhans.
218,912	Aug. 26, 1879.	S. M. Allen.
218,958	Aug. 26, 1879.	J. C. Forbes.
219,034	Aug. 26, 1879.	A. L. Sturdevant.
219,170	Sept. 2, 1879.	J. R. Moffitt.
220,808	Oct. 21, 1879.	W. N. Cornell.
220,970	Oct. 28, 1879.	} H. A. Frambach.
221,404	Nov. 11, 1879.	
221,992	Nov. 25, 1879.	
221,993	Nov. 25, 1879.	} S. M. Allen.
223,304	Jan. 6, 1880.	
223,670	Jan. 20, 1880.	W. E. Farrell.
224,002	Feb. 3, 1880.	W. A. Doane.
224,623	Feb. 17, 1880.	S. M. Allen.
225,292	March 9, 1880.	J. W. Martin.
225,988	March 30, 1880.	G. D. King.
226,013	March 30, 1880.	S. M. Allen.
228,041	May 25, 1880.	N. Cowan.
228,477	June 8, 1880.	A. W. Priest.
228,899	June 15, 1880.	P. Holmes.
229,073	June 22, 1880.	} S. M. Allen.
229,513	July 6, 1880.	
229,588	July 6, 1880.	C. W. Clark.
229,879	July 13, 1880.	H. A. Frambach.
230,471	July 27, 1880.	G. P. Enos.
231,720	May 27, 1880.	T. F. Hoxie.
231,761	Aug. 31, 1880.	C. W. Clark.
232,431	Sept. 21, 1880.	S. M. Allen.
232,480	Sept. 21, 1880.	A. Fickett.
233,014	Oct. 5, 1880.	J. C. Potter.
233,070	Oct. 12, 1880.	} J. Chase.
233,071	Oct. 12, 1880.	
233,105	Oct. 12, 1880.	R. B. Lane.
233,611	Oct. 26, 1880.	H. A. Frambach.
234,893	Nov. 30, 1880.	S. H. Scott, and Pontee and Wyman.
235,721	Nov. 26, 1880.	S. M. Allen.
236,794	Jan. 18, 1881.	G. F. Evans.
236,856	Jan. 18, 1881.	J. M. Stewart.
237,839	Feb. 15, 1881.	M. V. Eichelberger.
239,040 } 239,041	March 22, 1881.	H. A. Frambach.
239,807	April 5, 1881.	R. B. Lane.
240,027	April 19, 1881.	A. Kreider.
241,277	May 10, 1881.	E. M. Ball.

No.	Date.	Inventor.
241,311	May 10, 1881.	A. Dean.
242,138	May 31, 1881.	G. D. King.
242,308	May 31, 1881.	T. Hanvey.
243,616	June 28, 1881.	G. H. Pond.
243,965	July 5, 1881.	B. F. Perkins.
244,416	July 19, 1881.	S. M. Allen.
246,516	Aug. 30, 1881.	N. Kaiser.
247,072	Sept. 13, 1881.	R. B. Lane.
252,983	Jan. 31, 1882.	G. Werner.
253,654 } 253,655 }	Feb. 14, 1882.	S. M. Allen.
253,814	Feb. 14, 1882.	D. R. Burns.
254,327	Feb. 28, 1882.	G. L. Jaeger.
257,436	May 2, 1882.	R. Cartmell.
259,974	June 20, 1882.	D. R. Burns.
259,992	June 27, 1882.	S. M. Allen.
261,536	July 25, 1882.	A. Crosby.
263,119	Aug. 22, 1882.	W. N. Cornell.
263,250	Aug. 22, 1882.	H. P. Litus.
264,167	Sept. 12, 1882.	W. Jones.
267,715	Nov. 21, 1882.	G. H. Pond.
269,291	Dec. 19, 1882.	G. L. Huxtable.
271,409	Jan. 30, 1883.	} H. N. Brokaw.
Reissue		
10,429	Dec. 26, 1883.	
277,060	May 8, 1883.	J. Prickett.
284,433	Sept. 4, 1883.	W. Jones.
286,902	Oct. 16, 1883.	Cartmell and Ball.
287,980	Nov. 6, 1883.	E. Thompson.
289,187	Nov. 27, 1883.	F. Voith.
291,777	Jan. 8, 1884.	F. G. Ritchie.
291,848	Jan. 8, 1884.	P. H. Holmes.
293,235	Feb. 12, 1884.	G. F. Evans.
296,780	April 15, 1884.	G. H. Pond.
298,851	May 20, 1884.	Hayden and Sleeper.
298,875	May 20, 1884.	E. F. Millard.
304,182	Aug. 26, 1884.	F. A. Cushman.
305,062 } 305,063 }	Sept. 16, 1884.	E. P. Ely.
306,979	Oct. 21, 1884.	S. S. Webber.
309,532	Dec. 23, 1884.	} E. P. Ely.
310,659	Jan. 13, 1885.	
311,212	Jan. 27, 1885.	A. B. Tower.
320,574	June 23, 1885.	E. F. Millard.

CORN-HUSK CUTTER.

The machine for cutting or slicing corn-husks shown in Figs. 50, 51, and 52 is the invention of Mr. Wm. A. Wright, of Centreton, N. J.

Fig. 50.

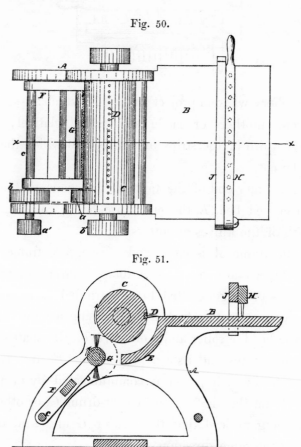

Fig. 51.

Fig. 50 is a top or plan view of the apparatus. Fig. 51 is a central vertical section thereof in line $x\,x$, Fig. 50. Fig. 52 is a front view of a portion thereof.

A represents a frame, which is provided with an apron or table, *B*, and on which is mounted a drum or roller, *C*, from

Fig. 52.

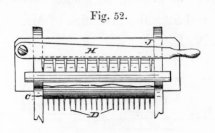

the periphery whereof project cutters, *D*, which are separated one from another, or made adjustable relatively to the required width of the strips or shreds into which the husks are to be cut.

On the inner end of the table *B*, and beneath the drum *C*, is a curved bed, *E*, the curvature being coincident with the path of the knives or cutters *D*.

To the frame *A* is connected a swinging frame, *F*, on whose upper end is mounted a rotary clearer, *G*, consisting of bristles or fingers fitted to a journaled head or roller, which receives motion from a belt passing around a pulley, *a*, on the head or roller and a pulley, *b*, on the shaft *c*, which latter constitutes the axis of the frame *F*, and carries a pulley, *a'*, to which power is communicated by means of a pulley, *b'*, on the shaft of the cutter-drum *C*, or other gearing, it being noticed that the clearer *G* and said drum *C* rotate in the same direction.

H represents a lever with downwardly-projecting teeth, pivoted to a slide *J*, whose sides are grooved or formed with guides to fit the sides of the apron or table B, so that while the slide is permitted to be moved to and from the cutters D it is prevented from vertical disengagement.

The operation is as follows: The lever H is raised and a husk, with the stalk end toward the operator, placed on the table B and held by the lever, is pushed to the cutters D by advancing the slide-clamp $H J$, power having been properly applied to the drum or roller C and the rotary clearer G. The cutters slice the husk into shreds or strips the length of the husk, the shreds or strips passing between the drum C and bed E as the husk is advanced. The clearer G forces the shreds or strips down from the cutters D, and also prevents them from winding on the drum. It will be noticed that the outer ends of the cutters describe a greater circle than the drum. Consequently when the cutters reach the clearer the latter is forced away by the former and its frame F caused to swing on the axial shaft c, so as to permit the cutters to pass the clearer, without, however, avoiding the stripping action thereof. As soon as the cutters clear the brushes or fingers of the clearer the frame G returns to its normal position and causes the brushes or fingers to sweep the circumference of the drum C unoccupied by the cutters D. When the length of the husk, excepting the stalk or stub, is entirely cut the slide J is drawn back, the lever H raised, and the husk removed, after which a fresh husk is applied and the operations of slicing, etc., are repeated.

The bed E sustains the husk during the cutting thereof, and its lower end being open permits the escape of loose cutting, dirt, etc., the rotary clearer G also serving as a fan to assist said escape.   The stalk or stub of the husk is severed preferably after the cutting operations.

Although only one row of cutters is shown, the number of them may be increased as desired.

CHAPTER VIII.

DUSTING RAGS—WET DUSTING—WASTE PAPER DUSTER AND WASHER.

Dusting Rags.

The next operation to which the rags or other paper stock are subjected is that of "dusting," and numerous mechanical contrivances for removing the dust, sand, and other foreign substances mechanically mixed with rags and paper stock have been invented in the United States and are technically known to the trade as "pin dusters," "railroad dusters," "fan dusters," etc.

The operation of dusting is commonly performed by passing the cut rags, etc., through a cylinder and sometimes through a conical drum to which a rotary movement is imparted by suitable gearing. The periphery of these dusters is formed of coarse wire-cloth having about nine meshes to the square inch. These drums are sometimes placed on an incline and at other times on a level, their disposition in this respect depending upon the arrangement of the mechanism for agitating and carrying the rags through the drum.

The rags enter at one end of the drum and issue from the other; the double motion of rotation and transmission causes the adhering impurities to separate from the rags, etc., and sift through the wire covering of the drum.

When the paper stock is coarse and very dirty and contains much straw or hemp, such as pack cloths, old ropes, etc., a " devil" is used for dusting the material. The action of the "devil" is much more severe than that of the ordinary duster, and by it the fibres are more thoroughly loosened up and put in condition for the boiling.

Some dusters which we have seen consist of a revolving drum with loose or swinging arms against which the rags are fed, and the force with which the rags are struck by these swinging arms whips them around under the drum upon a coarse wire apron through which the dust escapes. The dusted rags issue from the opposite side of the cylinder from that into which they were fed, and if not sufficiently dusted they are caused to pass through a hollow revolving cylinder, also covered with coarse wire cloth, and placed either on a line with the thresher or turned at any desired angle from it. A wooden or metal partition is placed directly against the threshing-drum, against which and the sides of the cylinder the rags are thrown with great force which assists in beating the dust out of them.

At the upper portion of the machine at a convenient point an opening is left for the purpose of ventilation and carrying off the light dust, a strong current of air being created by the rapid revolutions of the drum.

In England, Belgium, and other parts of the Continent as well as in the United States, the writer has seen rags dusted by a combination of the devil and duster; the devil being composed of two conical cylinders, the interiors of which contain projecting steel spikes. In the interior of these cylinders an iron drum, having its periphery studded with steel spikes, is made to revolve at the rate of about 300 revolutions per minute. "The rags are fed into the first cylinder by a travelling belt, and dashed through from the one to the other by the action of the revolving drum, and from the second cylinder thrown forward on the duster. This consists of a large rectangular wooden case, in the interior of which an iron cage, covered with coarse wire-cloth, revolves slowly at right angles to the devil. This cage is set at a slight incline, so that the rags which are thrown into it by the willow at one end slowly pass to the other, while the dust, etc., which have been disengaged by the action of the willow, falls through the wire-cloth, and the dusted rags pass out at the other end, now ready for the boiler."

Before describing other forms of rag duster we would call the reader's attention to the different firms mentioned on page 142, from whose catalogues all information regarding the workings of the various dusters in common use can be obtained. It would not be possible in a volume of the size of the present one to fully describe all the machines made and used in paper-making, and we shall consequently confine our attention principally to those possessing features of novelty, and the complete list of patents which this volume contains of all inventions relating to paper-making will enable the theorist or practical man to readily inform himself as to the exact state of the art up to the close of the year 1885.

12

The Waste from Dusting, etc.

The amount of loss resulting from passing the rags through the duster cannot be accurately estimated, as it varies with their nature and condition, the number and construction of the machines through which they are passed, the manner in which the rags have been cut, etc. When the cutting knives are dull a much greater loss results from ravelling than is the case when the knives are kept properly sharpened.

Prouteaux, in his *Guide Pratique de la Fabrication du Papier et du Carton*,[1] on p. 25, repeats the statement made by him nearly a quarter of a century since regarding the mean waste in the dusting, viz :—

> 1.5 per cent. for clean white rags ;
> 2.5 to 3.5 "   " hems and seams ;
> 4 to 5   "   " rags containing straw.

Owing to the greater care now given to the condition and classification of rags by foreign shippers than was formerly the case, and the improvements which have been made in the dusters now employed in the mills of the United States, it is probable that the above figures are 25 per cent. too high.

The waste of rags from the moisture which they contain, overhauling, cutting, and dusting, may be stated to be as follows :—

> 6 to 8 per cent. for fine and half-fine whites,
> 8 to 12   "   " coarse whites,
> 6 to 9   "   " white cottons,
> 8 to 12   "   " colored cottons,
> 12 to 18 "   " pack-cloths and coarse threads containing straw,
> 15 to 18 "   " ropes not of hemp,
> 18 to 25 "   " hempen ropes containing much straw.

---

[1] Paris, 1884.

The above figures are only approximate however, and are not, of course, intended to apply to every case, the only intention being to give some conception of the proportion of loss which the same materials suffer under the same treatment.

The loss is less in new and unbleached rags than in old rags.

In some paper manufactories on the Continent they have substituted for the dry dusting a washing process which gives less loss of fine thready matters.

In this connection we shall later (page 185) speak of a combined washing or cleansing and boiling process.

A Machine for Reducing the Loss in Cleaning Cut Rags.

The contrivance shown in Figs. 53 to 59 is the invention of Messrs. John B. Hart and Emory H. Walker, of Holyoke, Mass.; the object being to provide a treatment for rags and similar materials used in paper-manufacture, by which the rags, after being cut, will be separated in a degree from foreign substances without so great a waste of small rags as is common to processes now in use.

Fig. 53 is a perspective view of the invention. Fig. 54 is a diagram showing in side elevation and section the leading details of the device; and Figs. 55, 56, 57, 58, and 59 are enlarged views of certain parts in detail.

In Fig. 53, *A* is a frame or guide containing an endless belt, *B*, passing over the roll *C*, and over suitable rolls (not shown) on other portions of the frame. The roll *C* is attached to a shaft, *D*, having at one end the tight and loose pulleys *E*, for the purpose of transmitting motion to

Fig. 53.

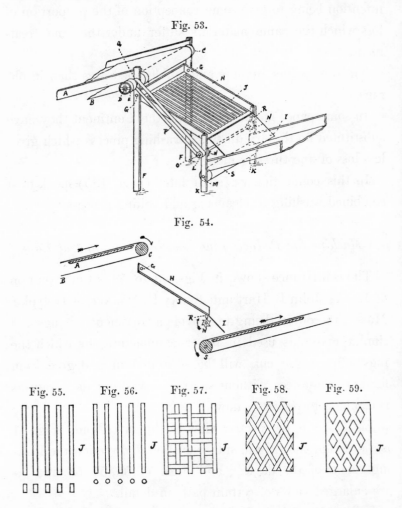

Fig. 54.

Fig. 55. Fig. 56. Fig. 57. Fig. 58. Fig. 59.

the roll *C*. The shaft *D* passes through suitable bearings in the posts *F F*. Below the roll *C*, and attached to the posts *F F* by pins *G G*, or in any other proper manner, is the tray *H*, having at its opposite end an apron, *I*, and having its bottom formed of open work, *J* (more fully described hereafter). At the end of the tray *H*, which bears the apron *I*, are two posts, *K K*, supporting the cam-shaft *L*, and a second roll-shaft, *M*. The cam-shaft *L* contains two cams, *N N*, and a pulley, *O*, as shown. This pulley *O* is connected by the belt *P* with a similar pulley, *Q*, on the first roll-shaft *D*, and by the connection described the revolution of the first roll causes revolution of the cams.

On the tray *H* are the projections *R*, resting on the cams *N*, whereby the revolution of the cams produces a vibratory motion of the tray *H*. The second roll, *S*, carries an endless belt similar to that on roll *C*, previously described, and receives motion in a suitable manner.

The arrangement of the various parts described and the direction of their motion are shown in Fig. 54.

The action of the mechanism described is as follows: The cut rags to be operated upon are delivered on to the belt or carrier *B*, moving in the direction shown, and are carried forward over the roll *C* and dropped into the tray *H*. This delivery to the tray is gradual and dependent upon the velocity of the roll. The rags, after striking the upper end of the tray, gradually slide toward its lower extremity— impelled by gravity and the vibratory motion of the tray. The bottom of the tray is formed of open-work made of bars, as shown in Fig. 55, or wires or rods, as shown in Fig. 56, arranged parallel to each other, or laid in different directions, crossing or interlacing as indicated in Figs. 57 and 58, or the openings may be perforations in a continuous sheet, as shown at Fig. 59. As the rags pass along the bottom of the tray the dust and other impurities work to the bottom and fall through. Small rags also pass through and are deposited below, where they may be collected and used, whereas, without this device, they would remain mixed with the larger pieces until blown out by a "duster," when, by their small size, they would pass through the meshes of the screen and be lost.

The object of the apron I is to carry the rags fairly on to the second belt-carrier, which moves them to the desired point of delivery.

The same gentlemen have also made another improvement in appliances for the treatment of cut rags used in the manufacture of paper; and the objects of their improvements are to secure a thorough mixing, stirring, and shaking of the rags for the purpose of separating foreign substances from the rags and disengaging small particles of cloth from the larger ones, and they claim to attain these objects by the mechanism shown in Figs. 60, 61, and 62.

Fig. 60 is a vertical section showing this second invention and parts connected therewith. Fig. 61 is a perspective view of a portion of the invention in detail, and Fig. 62 is also a perspective view of a modified form of parts represented in Fig. 61.

In Fig. 60 A is a roll or drum, which is driven by suitable mechanism, and by its revolution imparts motion to the endless belt or carrier B. The rags to be acted upon are placed upon the top side of the carrier B, and, travelling in the direction indicated by the arrow, pass over the roll A and fall therefrom.

C is a screen, made of lattice-work, interlaced wire, or other suitable material, properly arranged to form a perforated receptacle, and located, as shown, to receive the rags delivered by the carrier B.

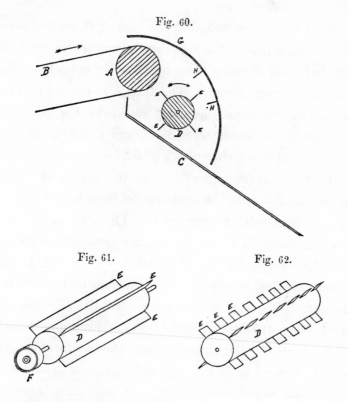

Fig. 60.

Fig. 61.   Fig. 62.

D is a shaft or cylinder having projecting arms, $E\ E$, etc. The shaft D is supported by suitable bearings, and has a pulley, F, through which rotary motion can be transmitted to the shaft or cylinder D. The cylinder D is placed beyond the carrier-roll *A*, and above and nearly in contact with the screen *c*, for the purpose of stirring and shaking the rags passing through the apparatus.

G is a bonnet or shield, which covers and closes in the space about the cylinder *D*, for the purpose of preventing the throwing off of the rags by the action of the cylinder *D* and arms *E E* of same.

H H are wings attached to the bonnet *G*, for the purpose of catching a part of the rags thrown upward by the revolving shaft, and allowing them to fall again upon the cylinder to be further acted upon. The arms *E E*, etc., may be made in single continuous pieces, as shown in Fig. 61, or they may be made up of sections of any desirable form, one arrangement being shown in Fig. 62.

In operation the rags fall upon the carrier *B*, pass over the roll *A*, fall upon the upper part of the screen *C* and upon the rapidly-revolving cylinder *D*. The arms *E E*, etc., toss the rags about in the confined space between the bonnet *G* and the screen *C*, and thoroughly stir and shake them, so that the dust is separated from the rags and caused to fall through the screen, while, in due time, the rags descend to the lower part of the inclosed space. From the lower part of the bonnet *G* the rags slide down the incline of the screen and are carried away by suitable devices.

The screen and carriers form, in combination, a device for the cleaning of rags and the saving of small rags, and are fully shown in Figs. 53 to 59.

Wet Dusting.

Combined Washing or Cleansing and Boiling Process.

The apparatus shown in Figs. 63 and 64 is the invention of Mr. W. E. Newton, of London, England, and the method employed by him is a combined washing or cleansing and boiling process, which is effected in a device in which it is claimed that the operations can be conducted with greater facility than has previously been the case.

The rags generally have adhering to them a good deal of dirt, which can be easily removed, and this is done by washing them in plain hot water in the apparatus shown in Figs. 63 and 64, before submitting them to the action of any chemical solutions, for the purpose of more thoroughly cleansing and bleaching them. A large proportion of the dirt and impurities may thus be removed in a short time, and these particles, being heavy, sink into the lower part of the apparatus, and may be drawn off in the form of mud without wasting the chemical solutions used in the subsequent processes.

Fig. 63 is a sectional plan view, and Fig. 64 a vertical section of Newton's apparatus, which consists of a cylindrical framework or cage, *a a*, covered externally with perforated metal or wire-gauze, and provided internally with any convenient number of vertical ribs, *a'*, which project radially from the circumference toward the the centre. This perforated metal cage *a* rests upon cross-bars, *b b*, or upon a ledge fixed inside the external casing *c c*, leaving an annular space between the external casing and the outside of the cage *a*. Inside this cage *a* is mounted a frame, which is fixed on the vertical spindle *e*, and consists of a series of beaters or stirrers, *d' d'*, of peculiar shape. The lower part

Fig. 63.

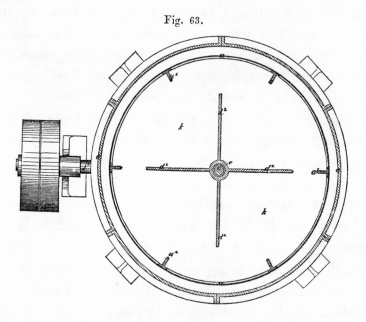

of the spindle *e* is made square at *e'*, and fits into a square socket in the end of the short shaft *f*, which passes through a stuffing-box, *g*, at the bottom of the casing *c*, and is supported in a step, *h*, below. On this shaft *f* is keyed a bevel pinion, *i*, which gears into and is driven by a similar wheel, *j*, which is actuated by a band from any prime mover to the pulleys on the end of the shaft *j'*. It will be seen that the beaters *d' d'* extend from the upper part of the spindle *e* down to the bottom of the cage *a a*, or nearly so, and that the beaters are much wider at the bottom than at the top. The effect of this peculiar shape of the beaters is that when they are in operation greater motion is given to the lower part of the water than to the upper portion, thus creating a kind of whirlpool action of the water, and a vertical or tumbling circu-

Fig. 64.

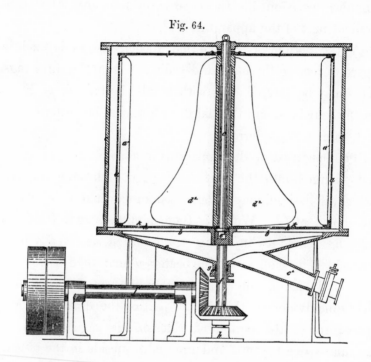

lation of the water and rags in contradistinction to a horizontal circular motion, thereby causing the dirt and impurities to be more readily removed from the rags.

After passing through the perforated sides of the cage *a* into the annular space between it and the external casing, the dirt and impurities are deposited in the form of mud at the bottom of the vessel *c*, and may be drawn off through the pipe c' before the chemical solutions are added for the subsequent processes.

If desired, the water in the apparatus may be heated by jets of steam, which may be injected from nozzles into the annular space outside the cage a, or into any other convenient part of the apparatus.

At the bottom of the external wooden vessel c is made a space to receive the mud and sediment from the dirty rags. This may be covered over with a false bottom, k, as shown in the drawings, and this false bottom may be perforated or not, as may be preferred.

The operation of the apparatus is as follows: The rags are charged into the perforated cage a through the top, which, if desired, may be inclosed or covered in by doors, as shown at $l\ l$. Water is then admitted, and, if cold, it may be heated by injecting steam into it, as already mentioned. The shaft e with the beaters d' may then be rotated, and, by knocking the rags about in the water, it is claimed will quickly detach the greater part of the dirt, which will pass through the perforated sides of the cage a into the annular space beyond, and gradually subside in the outside water, and descend into the mud space below the false bottom k, from whence it may be drawn off from time to time through the pipe c'. A solution of soda or other equivalent chemical substance may then be run into the apparatus, and the whole of the liquid with the rags in it may be boiled by means of the steam for any desired time, or until the rags are quite clean. They may then be bleached and reduced into pulp, and be converted into paper in the ordinary manner.

The steam-pipes for heating the water have not been shown, as it will be found that by simply causing steam to bubble up in the water from nozzles, in the manner already explained, the water will be sufficiently heated.

It will be noticed that as all the driving-gear is placed below, the cage *a* and the parts contained therein, together with the central shaft *e*, may be lifted out of the outer vessel *c* for the convenience of emptying the rags out of the cage, which may then be replaced in its original position.

In practice the inventor states that it will be found convenient to use three of the above-described apparatuses in combination, so arranged in reference to each other that the cage *a* with its contents may be lifted out of one apparatus and placed in the next.

The first apparatus may then be used for the preliminary washing out of the loose dirt, the second for boiling the rags in any chemical solution, and the third for the boiling or bleaching operation, or for simply washing in cold water.

WASTE-PAPER DUSTER AND WASHER.

Messrs. Hiram Allen and Lyman S. Mason, of Sandy Hill, N. Y., in their process for preparing pulp from papers, use a rotary self-discharging dusting engine by which the dry " imperfections" are subjected to the desired tumbling, beating, tearing, and screening action, from which machine the waste papers are progressively introduced in loose condition into water in a circuit vat where they are subjected in heated alkaline water to the action of a current-producing paddle-wheel for the purpose of discharging the ink and separating the fibres.

The process is intended to cover all the stages of reducing the imperfections to bleached and finished pulp; but in this section we shall deal only with the dusting and washing mechanism employed.

Fig. 65 shows, in side elevation and partial vertical longitudinal section, the dusting engine employed by Messrs.

Fig. 65.

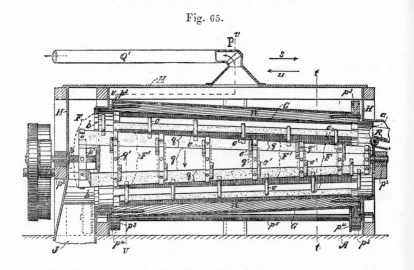

Allen and Mason, the section being at about the line *w* *w'* in Figs. 66 and 67.

Fig. 66 represents a transverse section of the same dusting-engine at about the line *v* *v* in Fig. 65, and viewed in the direction pointed by the arrow *u* in that figure, with an elevation of modified driving devices. Fig. 67 is an elevation of a section at about the line *t t* in Fig. 65, and viewed in the direction indicated by the arrow *s* in that figure. Fig. 68

Fig. 66.

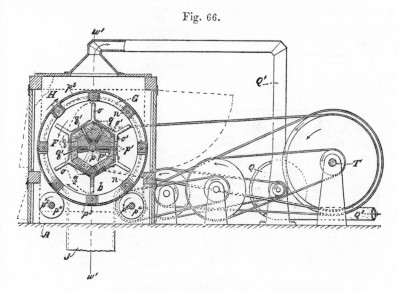

is a perspective view of the exterior of the same dusting-engine illustrated in the other figures with feeding appliances.

Fig. 67.

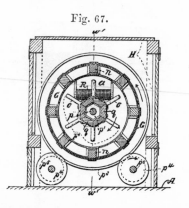

Fig. 69 is a perspective representation of a circuit-vat furnished with a paddle-wheel and a rotary washer, and arranged to receive papers in loose condition from the dusting-engine, and suitable for use in transforming the papers into washed and bleached pulp.

Fig. 68.

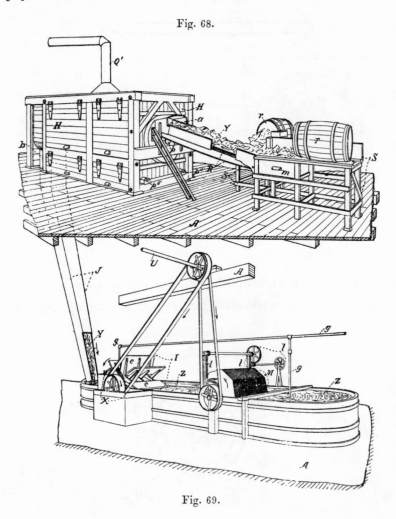

Fig. 69.

A is the stationary support for the mechanisms.

The dusting-engine has a rapidly-revolving beater, F, Figs.

65, 66, 67, within a tubular slowly-rotating screen, G, that is open at its ends and is incased closely about its ends and loosely at its sides by a cover, H, that has a feed-opening, a, in one end and a discharge-passage, b, at the other end. The general action of this dusting-engine is similar to that of some other dusting-engines heretofore used, that is, papers introduced at the feed-aperture a into the slowly-rotating screen will be therein rapidly struck, tumbled, and loosened up repeatedly by the fast-revolving beater, so as to separate from the papers dust and dirt, that may then fall down through the meshes of the screen on to the bottom of the casing as the papers are progressively beaten and tumbled along through the screen and are discharged in loose condition at b.

The circuit-vat, Fig. 69, is supplied with water, Z, with or without alkali therein, and is furnished with a paddle-wheel, I or I', constructed and arranged so that as the wheel revolves and produces a strong current in the water throughout the vat, dry papers progressively introduced in loose condition into the vat will be progressively immersed in the water by the paddles c of the wheel, and subjected to the soft disintegrating actions of the paddles in the water.

To avoid having in the paper-mill a large pile of the loose dry papers discharged from the dusting-engine, and to prevent the consequent loss of room and danger from fire, and also to avoid having persons stow away the loose papers, or feed them into the pulping-vat, as they are discharged from the dusting-engine, the inventors combine with the dusting-engine and pulping-vat an automatic transferring device, by means of which the loose papers discharged from the dusting-engine are at once progressively transferred into the pulping or circuit vat.

In carrying out this combination of the dusting-engine, circuit-vat, and device for transferring the papers from the dusting-engine into the vat, such transferring device, of course, varies in construction according to the relative positions of the circuit-vat and dusting-engine. For instance, when, as represented in Figs. 68 and 69, the dusting-engine B is nearly over the circuit-vat, the paper-transferring device J then preferably consists of a simple chute, through which the loose dusted papers will descend by their gravity from the dusting-engine into the circuit-vat.

When the circuit pulping-vat is located over, higher than, or far from the dusting-engine, the inventors combine with the dusting-engine and circuit-vat some suitable elevator or conveyer, for transferring the loose papers into the circuit-vat as fast as they are discharged from the dusting-engine. While the papers are being introduced in successive small quantities or progressively into the water in the circuit-vat by hand, or from the dusting-engine by automatic means, during one or two hours (more or less), the revolutions of the paddle-wheel I are continued, and thereafter the rotations of the paddle-wheel are continued in the mixture of papers and water for an hour or two (more or less), or until the papers become reduced to the proper pulpy condition to permit the ready washing out of the dissolved sizing and of the disintegrated ink in water, when printed papers are thus introduced and treated in alkaline water in the vat.

In order to provide means whereby the dusting-engine can be kept constantly in use without any accumulation of the dusted papers, after completing the supply of papers from the dusting-engine to the vat, and while the papers are being reduced to coarse pulp in that vat, a second vat can be suitably placed and a conduit arranged for transferring the dusted papers in loose condition from the dusting-engine into such vats at will.

While introducing the paper into the water in the pulping-vat and during the subsequent operation of reducing the papers to pulp by the rotation of the paddle-wheel in the vat, the water in the vat is kept at a temperature of from about 100° to 212° F., and preferably at about 160° to 190° F., or at whatever temperature shall secure the quick solution of the sizing of the paper and of the vehicle of the ink when printed papers are used. To quickly and cheaply accomplish the dissolution of the ink, there is added to the heated water in the vat some suitable solvent, such as caustic soda, soda-ash, or equivalent alkali. The quantity of solvent used may be sufficient to make the specific gravity of the water in the vat with the papers about one-third to two-thirds of 1° Baumé, more or less. This alkali is introduced into the water in the vat either before, during, or soon after the introduction of the printed papers. When the papers are not printed, the alkali may be omitted. The papers will be reduced to pulp in the water in the vat by the action of the paddle-wheel without having the water heated; but the moderate heating of the water hastens the pulping of the papers.

To secure the proper continued heating of the water, papers, and pulp in the vats steam is introduced by suitable means, as, for instance, by the pipes g, Fig. 69, having a stop valve or valves, and communicating at one end with a supply of steam and open at the other end or ends to the water in the vat near its bottom.

Printed papers it is claimed can be quickly and cheaply transformed into clean, refined, and bleached pulp, according to the present process, by the use of the apparatus or mechanism represented in Fig. 69, by first reducing the papers to coarse pulp and dissolving the ink and sizing therefrom by progressively introducing the papers in dry loose condition into the water in the vat, while the washer M is kept out of the water by the gearing l, and the paddle-wheel I is constantly revolved, and steam is admitted into the water for the pipes g, if desired, until sufficient papers are introduced, and thereafter continuing the rotation of the paddle-wheel, with alkali in the water, and the admission of steam, when desirable, until the papers become reduced to coarse pulp and the ink and sizing are dissolved. Then refining the coarse pulp, and removing the dissolved ink and sizing with the soiled water by revolving the paddle-wheel and the lowered washer M in the pulp and water in the vat, and adding clean water, with or without the admission of steam by the pipe g, until the reduced pulp shall be sufficiently clean for bleaching. Then further refining the washed pulp and bleaching the same in the water in the same vat by adding thereto the solution of chloride of lime or other bleaching agent, and revolving the paddle-wheel I in the pulp in water while the washer is out of action, with or without the introduction of steam by the pipe g, until the pulp shall become suitably bleached. The washer is then lowered into the water and the rotation of the washer and paddle-wheel continued with the addition of clean water, when desired, until the bleaching agent shall be sufficiently washed out and the pulp properly refined. The rotation of the paddle-wheel is then continued while subjecting the pulp in water to the agent for neutralizing the bleaching agent, and while washing out the neutralizing agent when necessary.

When practicable in carrying out this process it is preferable to have the dusting-engine above the pulping-vats, and those vats above the washing-vats, and the latter above the bleaching vats, and all connected by conduits, through which the materials are transferred by their gravity from each upper apparatus to the next one below. Whenever the vats shall not be thus arranged, a rotary fan-pump and connecting-pipes can be used, or other suitable known means employed for transferring the pulp in water from the pulping vat into the washing vat, and from the latter into the bleaching vat.

To provide excellent means whereby old folded newspapers, pamphlets, and other papers can be easily taken in hand, torn apart, assorted, freed from loose dirt, and thrown upon an endless turning feed apron, R, of the dusting-engine in the great quantities, and with the uniformity required to fully and properly supply the dusting-engine by persons seated or standing by the feeding apron, there are provided

tables *S*, Figs. 68, 70, 71, 72, each adapted to support, as
at *r*, a mass, bale, box, barrel, or package of the papers, and
each furnished at one end part with a drawer or dirt pit, *m*,
Fig. 70, preferably covered by a screen or grating, *m'*, and

Fig. 70.

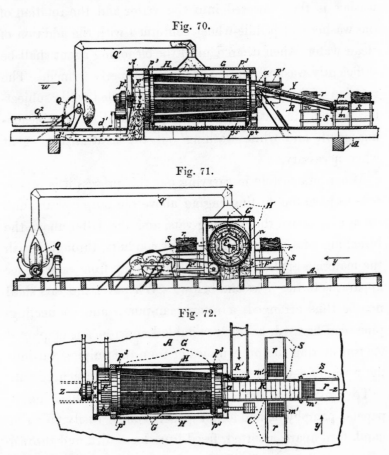

Fig. 71.

Fig. 72.

the tables are arranged with their drawer or dirt-pit ends
next to the feed-apron *R*, and at suitable distances apart to
permit a person to conveniently stand at one or either side

of each table and there handle the papers and throw them upon the feed-apron. The number of the tables, *S*, will be according to the number of persons required to supply the papers to the dusting-engine by the feed-apron. When the space in front of the dusting-engine is too limited to accommodate the required number of the tables along the main apron *R*, some or all of them can be arranged along a lateral feed-apron located as indicated at *R'*, in Figs. 70 and 72, to discharge the papers placed thereon on to the main feed-apron.

To render the dusting-engine capable of progressively and thoroughly tearing apart, beating, dusting, and freely discharging in loose condition papers as fast as they shall be properly fed into the rotary screen *G*, that screen is made with the series of lengthwise internally-projecting bars, *n*, and in a tapering form, much the largest at its discharge end, about as is illustrated by Figs. 65, 66, 67, 70, and 72, and the rotary beater *F* is also made with its body *F'* and the series of pin-teeth *o* of a similar or nearly corresponding tapering form in general outline and of suitable smaller diameter than the screen, substantially as illustrated in the drawings. The screen and beater are also furnished with driving mechanisms, whereby slow rotary motion is imparted to the hollow tapering screen and a greatly faster rotary motion is given to the tapering beater either in the same or opposite directions. By thus having the slowly-rotating hollow internally-ribbed screen *G* of a greatly tapering form, and the rapidly-revolving pin-toothed beater *F* of a suitably corresponding tapering shape, old papers fed into the small end of the screen can be thoroughly broken up, dusted, and discharged freely in loose condition at a greatly faster rate than can be done by a paper-dusting-engine of like size and movements, but having either its hollow-ribbed screen or its winged or pin-toothed beater of a general cylindrical shape or of nearly uniform diameter throughout.

The pin-teeth o of the beater can be of circular, angular, or other suitable form in cross-section, and arranged together in spiral lines, as shown, or otherwise. The body F'' of the beater may consist of a tapering log of wood mounted on a shaft, p, and have the teeth o driven or screwed into holes in the log ; or the body of the beater may be of other suitable construction, and the teeth secured thereto in any suitable manner.

In order to make the beater very durable at a cheap rate, it is composed of a rotary shaft, p, a series of cast-iron heads, p', fast on the shaft, a series of tapering staves, q, secured to the heads by bolts or screws, q', and the series of pin-teeth o, having bases, o', secured to the bars by bolts or screws, as shown by Figs. 65, 66, and 67. In the illustrations the shaft p of the beater F is shown supported by journal-bearings, p^2, Fig. 65, on the dusting-engine frame ; and the hollow screen G has circular rim bearings, p^3, that are supported by the wheels p^4, fast on the shafts p^5, that are supported in journal-bearings in the frame of the dusting-engine, so that by the rotation of one of the shafts, p^5, the screen will be revolved.

In practice the revolutions of the beater F may be about thirty to one of the screen G, or in a considerably greater or less ratio. When the screen G is about ten feet long and

about five feet in diameter at the large end, and the beater
F is of corresponding size, as illustrated in Fig. 65, a good
speed for the screen is from about eight to ten, and for the
beater from about two hundred and fifty to three hundred
revolutions in a minute; but good work may be done when
they revolve considerably faster or slower. Such different
rates of rotation are to be imparted to the screen and beater
by any suitable gearing or belting from any suitable motor
or motors. In the mechanism represented by Figs. 65, 66,
and 67 motion is imparted to the beater F, to the shaft p^5
of one of the two pairs of wheels p^4, that support the screen
G, and to the exhausting-blower Q, all from the one driving-
shaft T, by means of the pulleys, counter-shafts, and belts
clearly represented by those figures.

In carrying out this invention the paddle-wheels can have
the active faces of their paddles in planes parallel to the
axes and inclined to the radii of the wheels, as shown in
the drawings, or in planes parallel to the axes and radii; or
the paddles may be of various shapes and arrangements,
provided the wheels shall operate as already described.
The rate of rotation of the paddle-wheels can be considerably varied. A good speed is ten to twelve revolutions in a
minute, when the wheel is about seven feet in diameter and
has eight paddles, each about eight feet long and three feet
wide, arranged as in the wheel I, and the wheel is arranged
to revolve in a circuit-vat, of corresponding size, as illustrated by Fig. 69.

The paddle-wheels, rotary washers, dusting-engine and its
feed-apron, exhausting-blower, and devices for transferring

the materials may be actuated by any suitable mechanisms from any suitable motor or motors.

List of American Patents for Rag-Dusters.

For list of patents for rag-dusters issued by the government of the United States of America, see page 143.

CHAPTER IX.

BOILING RAGS—STATIONARY BOILERS—REVOLVING BOILERS—
TREATING COLORED RAGS—BOILING WASTE PAPER—BOILING
STRAW—BOILING ESPARTO—BOILING MANILLA AND JUTE—
BOILING WOOD—SODA-RECOVERY—ACID OR BISULPHITE PRO-
CESSES OF TREATING WOOD—LIST OF PATENTS FOR PREPARING
CELLULOSE FROM WOOD BY THE ACID OR BISULPHITE PRO-
CESSES—LIST OF PATENTS FOR DIGESTERS WITH LEAD LININGS
—LIST OF ALL AMERICAN PATENTS FOR DIGESTERS FOR PAPER
PULP—METHODS OTHER THAN THE MECHANICAL, SODA, AND
BISULPHITE PROCESSES FOR THE TREATMENT OF WOOD.

Boiling Rags.

AFTER being dusted, the rags are next boiled in an alkaline
lye. Prior to the substitution of the boiling process, the
rags were subjected to fermentation by piling them in stone
vats for five or six weeks, and frequently adding water,
which operation curtailed the power required for the subsequent mechanical comminution; but the process was so
wasteful that the modern method of boiling was substituted,
the invention being that of Schäuffelen, of Heilbronn.

The operation of boiling is a most important one, and
mistakes here made cannot afterwards be rectified; it is,
consequently, imperative that intelligent care should be
given to the rags during the boiling, so as not to injure their
texture or make the cost of the process too great.

The object of boiling is not only to get rid of the dirt remaining in the rags after the dusting and the removal of some of the coloring matter, but also to decompose a particular glutinous substance, which, if allowed to remain, impairs the flexibility of the fibres, and renders them too harsh and stiff to be readily fabricated into paper.

If the rags are improperly boiled, they will, in addition to consuming too much chlorine, make an inferior appearing paper.

The alkaline substances employed in boiling are fresh burned quick-lime, carbonate of soda, and caustic soda.

The quantities of chemicals used, as well as the pressure and length of time the rags remain in the boiler, vary with the character and condition of the rags, and upon the alkimetrical degree of the agent employed.

Various forms of vessels are used in which to boil the rags. Some are stationary, and others are made to revolve by means of suitable gearing; the latter class of boilers being commonly either cylindrical or spherical, cylindrical boilers being usually geared so as to revolve horizontally.

Fig. 73 shows the form of spherical rotary boiler used on the Continent, and is about the same as those used in Great Britain and in the United States. In the gearing, however, a worm is generally substituted for the small pinion shown in the illustration.

Some manufacturers of fine paper still prefer to boil the rags in stationary tubs, which are made of wood, and in such cases soda is used to saponify the fatty substances contained in the rags.

Most manufacturers use lime in the boiling process for all ordinary grades of paper; some, however, prefer soda, and others employ a mixture of lime and soda.

Fig. 73.

When rotary boilers are used, lime is probably equally as effective with uncolored rags as soda, whether used alone or mixed with lime. The quantity of lime employed varies according to its composition and the condition of the rags, and ranges from about five to fifteen pounds per hundred pounds of rags.

It is well known that the solvent properties of water far exceed those of any other known liquid. A very large proportion of all the different salts are more or less soluble in it, the solubility increasing generally as the temperature rises, so that a hot saturated solution deposits crystals on cooling. There are a few exceptions to this rule, one of the most remarkable of which is common salt, the solubility of which is nearly the same at all temperatures, the hydrate of lime (slaked lime) being more soluble in cold than in hot water, sulphate of lime being also less soluble in hot than in cold water, and insoluble at 302° F., or between 284° and 302° F. The solvent properties of water are still further increased when heated in a strong vessel under pressure; hence the greater the pressure under which the rags are boiled, the smaller the proportion of lime which it will hold in solution, and the less the quantity required to accomplish the object of the boiling.

An excess of lime will not injure the rags, but as it would be waste, it is best to employ the lime only in the necessary quantities. The pressure of the steam for any class of rags should seldom exceed thirty pounds.

Some manufacturers use a mixture of lime and soda-ash in about the following proportions:—

	Lime.	Soda.
For 100 lbs. of tarred rope or similar coarse material	15 lbs.	6 lbs.
"　" deeply dyed rags　.　.　.　.	12 "	3 "

In some cases it will be found that better results will be obtained from colored rags, such as cuttings from print cloths, by first boiling them with about 5 per cent. of lime under about 25 pounds pressure in a rotary boiler, and, after washing in the engine and put into drainers, they should be again boiled in the rotary with a solution of about 2 pounds of soda ash to each 100 pounds of rags.

The following table shows the proportions of lime and soda ash used on the Continent for boiling the various stuffs :—

	No. 1 stuff for 100 lbs. rags.	Nos. 3 and 5 stuff for 100 lbs. rags.	No. 4 stuff for 100 lbs. rags.
Lime	5.4 lbs.	8.1 lbs.	8.0
Soda-ash (48 per cent.) .	2.8 "	3.8 "	4.0

The boiling is usually continued for twelve hours under a pressure of 30 pounds of steam in a boiler revolving horizontally.

The boiling of ropes for tissue, copying, and cigarette paper is continued for twenty-four hours under a pressure of 30 pounds of steam, and for each 100 pounds of rope 16.5 pounds of lime and 11.5 pounds of soda-ash, 48 per cent., are used.

Oblong wooden boxes, measuring about fifteen by five feet, and four feet deep, are used in which to prepare the milk of lime. The boxes are divided into three compartments, and in order to retain small stones, etc., false bottoms, perforated with one-half-inch holes, are fitted to each compartment. In the third compartment a revolving drum, similar to the drum-washer of a half-stuff engine, is so arranged that the milk of lime is strained as it flows through a movable sluice in the division between the second and third compartments, and is discharged through a pipe directly into the rag-boilers. A finer wire strainer can be placed over the mouth of the pipe leading to the boilers in case it is desired to keep out finer particles of sand and grit. By means of suitable waste pipes connected with each compartment of the box all refuse can be readily carried away with the water.

The best way to introduce the soda-ash is to dissolve it, and strain through a fine wire cloth as it passes into the boiler.

Care should be exercised to use sufficient water with the rags during the boiling, as otherwise they are liable to come from the boiler with an undesirable dark appearance, which cannot be removed by either washing or bleaching.

The majority of manufacturers in the United States use lime in the boiling process, and the result is most satisfactory, but as caustic soda is often employed in Europe and sometimes in America, we will here give the quantities of caustic soda employed in boiling the different classes of rags.

$S\ P\ F\ F\ F$ are boiled with 5 per cent. of lime, and, after washing in the boiler, are afterwards boiled with 2 per cent. of soda-ash.

100 lbs.	$S\ P\ F\ F$	require	10.5	lbs. of caustic soda, 70 per cent.		
"	$S\ P\ F$	"	12.25	"	"	"
"	Fines	"	6.12	"	"	"
"	Seconds	"	5.25	"	"	"
"	$L\ F\ X$	"	17.5	"	"	"
"	$C\ L\ F\ X$	"	24.5	"	"	"
"	$C\ C\ L\ F\ X$	"	26.25	"	"	"
"	$F\ F$	"	13.0	"	"	"

STATIONARY BOILERS.

The boiling is continued for about ten or twelve hours in stationary boilers without vomit, under a steam pressure of from 20 to 25 pounds.

Fig. 74 shows a section of a form of stationary boiler much used in Great Britain. "It consists of an upright

Fig. 74.

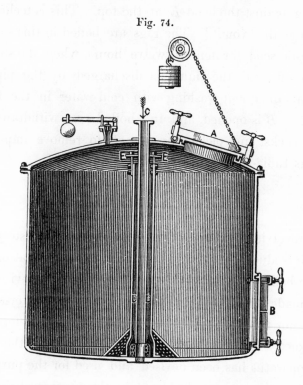

cylinder of half-inch iron, about eight feet in diameter, by six feet deep, and fitted with a perforated false bottom, on which the rags rest. The boiler is further fitted with a filling door, *A*, at the top, and an emptying door, *B*, below.

14

After being charged with rags, it is filled to about one-half its height with water; a sufficient quantity of caustic soda, varying according to the nature of the rags, is introduced; the door is then closed, and steam is then admitted by a small pipe, *C*, which is contained in, and communicates at the foot with, a larger pipe, *D*, and causes a constant circulation of hot liquid, which is dispensed all over the boiler by striking against the hood *E* at the top. This is technically known as the ' vomit.' The rags are boiled in this solution of caustic soda for ten or twelve hours, when the steam is turned off, and the liquid is discharged by the pipe *G*. After a subsequent washing with cold water in the boiler, the door *B* is opened, and the boiled rags withdrawn into small trucks, and picked by women to remove impurities, such as India-rubber, etc."

Revolving Boilers.

In revolving boilers as heretofore constructed steam has been introduced generally through the journals at one end of the boiler, sometimes at each end; but no provision has been made for determining the height of the water-level in them, nor for drawing off an excess of water, nor for knowing exactly the steam-pressure and heat inside of the boiler. An apparatus has been devised and used for the purpose of reducing the pressure of steam in the revolver below that carried in the boilers where the steam is generated for common use in paper-mills; but it is uncertain in use, and only operates when in good order by withdrawing the

steam. To accomplish these several and other desirable
results which the old form of construction failed to attain,
Mr. George F. Wilson, of East Providence, R. I., has devised
and patented the following-described mechanism and appliances.

Fig. 75.

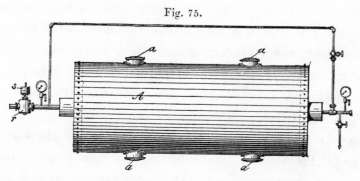

Fig. 75, letter *A*, represents a view of the revolving boiler
with its man-hole entrances used for filling the boiler, and
others at the bottom, which, by reason of the great weight
of the man-hole entrances, it has been found necessary to
use as balances for the upper ones. Fig. 76 is a vertical

Fig. 76.

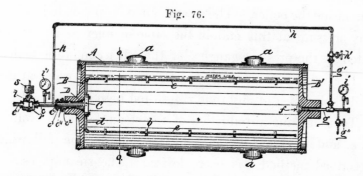

longitudinal section of the revolver, showing the proper
height of the water-level or water-line and the means which

the inventor has devised for the introduction of steam into the contents of the boiler below that water-line.

In order to determine where the water-line is inside the boiler, which, when in use, is hermetically sealed, the pipe h is inserted into the steam-pipe carrying steam into the boiler at the point c and between the boiler and the steam-gauge. The pipe h h runs around above the boiler in any convenient way for fastening it to the other or right-hand end of the boiler, through the journal of which is inserted the steam-pipe g, which has been connected with the pipe h h by a glass tube, g', Fig. 76.

At f is shown a strainer, which, in drawing off the water through the valve g'', prevents the rags from entering the pipe and stopping it up. This arrangement secures an indication not only of the pressure, and consequently the heat inside of the boiler, but enables the operator to determine the height of the water in the boiler by means of the glass tube g'. The water, being of greater specific gravity than steam, falls to the same level in the glass tube that it has inside the boiler. The steam-gauge i is put on to indicate the pressure at this end in the same manner as it is indicated by the pressure-gauge i' at the other end of the boiler.

The valve g'' is provided for drawing off the water in the boiler, should it at any time rise to too high a point.

To secure the introduction of the steam below the water-line and prevent its introduction above, the following described method has been devised: The steam-pipes e e are inserted and attached to the inner sides of the boiler by means of a hook, b, or other similar contrivance (shown in

Fig. 80), and are perforated with holes in sufficient numbers throughout the length of the pipe to secure the introduction of steam into the contents of the boiler below the water-line. The method which the inventor has devised for this purpose will be known by considering the construction of the valve *C* and the ring in which it operates, Fig. 76, and will be, perhaps, the better understood by reference to an enlarged view of the same shown in Fig. 77, in which *C* is the ring which is attached to the head of the boiler *B*, Fig. 76, which ring is fastened, as shown in Fig. 77, by the bolts *v v*.

Fig. 77.   Fig. 78.   Fig. 79.

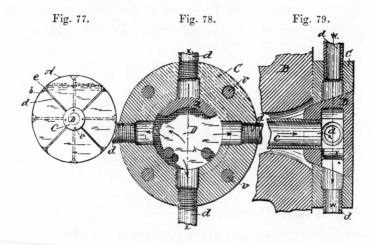

The valve *D D D* (shown in Figs. 77, 78, and 79) is stationary—that is, it does not turn with the boiler and the ring *C*, heretofore referred to—and is the means by which steam is introduced into the boiler through the pipe *e e*, Fig. 76.

In order to prevent the escape of steam into the boiler above the water-line, the valve *D* is constructed with a solid piece of metal (shown at Z, Fig. 78), which prevents the escape of steam from any of the steam-pipes while passing over this valve Z into the boiler above the water-line. The valve D is made conical in shape, and rests in a conical seat in the hollow ring C. The stuffing-box c^2 is placed within the hollow journal, and is accessible from without. By means of the screws c^3 c^4, which work against a collar, c', on pipe c and a shoulder formed by the end of the stuffing-box, the position of valve D in its seat can be regulated without stopping the rotation of the boiler. For practical operation, however, the pressure of the steam in the boiler against the head of the conical valve is sufficient to keep it tight.

Fig. 80. Fig. 81.

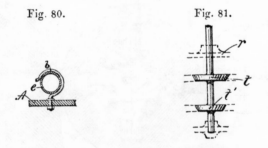

Fig. 81 shows the form of valve which it is proposed to use for reducing the pressure in the revolver below what is carried in the steam-generating boiler of the mill or works. It is a simple puppet-valve, in which the upper portion of the valve is larger than the lower portion. This valve is kept in its seat by means of weights placed on the stem outside of the valve, as shown at s s, Figs. 75 and 76. Other devices, however, for regulating the pressure of steam and water for this and other similar purposes have been made and used successfully.

Improved Strainer for the Blow-off of Paper Stock Boilers.

In Fig. 82 is shown a portion of an ordinary rotary boiler used for the treatment, under steam, of material used in the manufacture of paper.

Fig. 82.

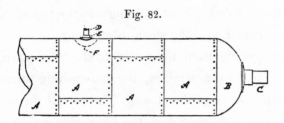

A A, etc., are sheets of iron or steel of which the cylindrical portion of the boiler is composed. *B* is a head, of which there is one for each end of the boiler. *C* is a shaft or trunnion attached to the head for the purpose of providing suitable bearings for the support and rotation of the boiler. The trunnions *C* rest in suitable journal-boxes, and gearing imparts rotary motion to the boiler. The boiler is provided with a suitable opening for the introduction and withdrawal of the stock. Steam is admitted to the boiler by a suitable pipe passing through or connecting with a passage formed in the trunnion *C*, and a pipe, *D*, attached to the boiler by the flange *E*, provides an exit for the steam. The pipes for inlet and outlet of steam are provided with suitable stop-valves.

In operation the boiler is partially filled with stock and the necessary chemicals admitted upon it, steam is turned on, and the boiler is revolved for the purpose of thoroughly stirring and mixing its contents while acted upon by the pressure and heat of the steam.   When this operation has been continued a sufficient length of time, steam is shut off and the rotary motion stopped; but before the contents of the boiler can be removed, the steam, under heavy pressure, which is in the boiler must be released.   For this purpose the blow-off pipe D is provided.   This is brought to the top of the boiler, its valve opened, and the steam blown out.

If the opening from the boiler to the blow-off pipe were not protected by a strainer, the out-rushing steam would carry with it some of the stock.   To prevent the abstraction and waste of stock described, the blow-off opening is always covered by some sort of screen or cover, as indicated by the dotted line at F, Fig. 82.

This screen as ordinarily made has the form shown in Figs. 83 and 84.   The hemispherical piece F, with a flanged edge, f, is attached to the boiler-shell A, around the blow-off opening D, by rivets, as at a, Fig. 83, or by bolts b, Fig. 84.   The spherical portion of the piece F is properly perforated, and as a separator of stock from the steam admirably serves its purpose.   In use, however, fragments of stock gradually work through the perforations, and clog the apparatus, and it must be frequently removed for repairs. To remove the strainer shown in Fig. 83, the rivets a a, etc., must be cut, which is a laborious and expensive operation. To remove the piece F, in Fig. 84, a number of bolts must be taken out, and as, under the action of heat, pressure, and chemicals, the bolts b b, etc., are likely to be corroded in place, this work will usually be found to be more expensive for removing and replacing than in the former case.

Mr. Benjamin F. Mullin, of Holyoke, Mass., has invented a strainer which he claims is readily accessible for cleaning,

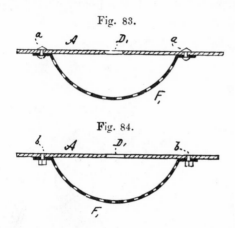

Fig. 83.

Fig. 84.

easily detachable for repairs, and firm and secure when in place. This strainer is shown in Figs. 85, 86, and 87.

The strainer proper is shown at F. It is of spherical form, of proper size, and with suitable holes, as shown, to meet the conditions respecting strength and capacity of the given service. At the line where the strainer F meets the boiler-shell A it comes to an edge (without flanging), and is given the form of the surface which it meets, for the purpose of forming a close connection. A row of half-holes is made around the edge of the strainer, as shown, for the purpose of allowing it to come to a better bearing, and also permitting particles of stock to blow back into the boiler that would otherwise collect in the corner between the strainer-edge and the boiler-shell.

The strainer F as described is attached to the boiler-shell in the following manner: To opposite sides of F are firmly united heavy ears, *G G*, as shown.   Pairs of corresponding ears, *H H*, etc., are securely riveted in the proper position on the shell *A*, as indicated, and heavy pins, *I I*, pass through

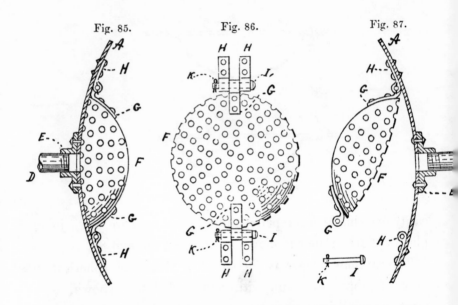

Fig. 85.   Fig. 86.   Fig. 87.

the ears, firmly holding the parts in place, accidental displacement being prevented by the split pins *K K*.

When necessary to reach the concave surface of strainer *F*, a pin, *I*, is removed, *F* being swung back on the pin as a hinge, as in Fig. 87. *F* may be entirely separated by removing both pins. The relative position of the two hinges shown gives square resistance to the action of the stock when the boiler is in motion, and prevents lateral strains.

TREATING COLORED RAGS.

The course practically pursued by paper-manufacturers in preparing their paper-stock for use is as follows:—

The stock is put into the well-known form of apparatus, and boiled for several hours in a solution of quicklime or of soda-ash, or caustic soda and quicklime, for the double purpose of removing the oils or greasy matters adherent to the stock, and for discharging the colors. It often happens that the greasy matters are rendered insoluble, or the coloring-matters are rendered more prominent, so as seriously to interfere with the action of the chloride of lime, to which the stock is subsequently subjected for bleaching purposes.

In discharging colors from misprints or calicoes, the mordants are usually removed by various acid baths adapted to the particular mordants, after which the colors can be easily removed by alkalies. This process, however, affects more or less injuriously the strength of the cloths.

Messrs. George F. Wilson and Philip O'Reilley, of Providence, R. I., have patented a process by which they claim that the mordants can be treated with chemicals in such a manner that a double decomposition in the bath will take place in contact with the colors, and the effect of this nascent action, so produced, they claim, will be to produce the oxidation or deoxidation of the mordants, and that the discharge of the coloring-matters may be brought about without injury to the fabrics or paper-stock so treated.

In Fig. 88 is shown the apparatus which the inventors

have devised to save labor and to utilize to the utmost extent all of the chemicals employed.

Fig 88.

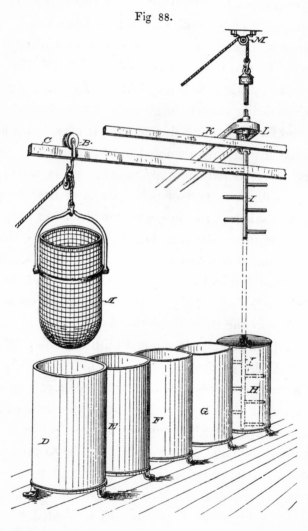

A represents an iron-wire basket with a strong iron frame, which holds the paper-stock in the several processes until it is bleached and ready to be dried. This basket is suspended by a rope passing over a pulley, so that it can be readily raised and lowered. This pulley is suspended from a carrier, *B*, which travels on a rail, *C*. Beneath this rail, and in a line with its length, is a series of keirs, *D*, *E*, *F*, *G*, *H*. The basket *A* can, by raising and lowering it and moving the carrier *B*, be readily transferred to any of the keirs.

In connection with the keir *H*, a stirring device, *I*, is employed. This is so constructed as to be raised and lowered. It is supported in a bearing attached to the bar or rail *K*, so as to be revolved by means of the belt or strap on the pulley *L*. The spindle or arbor of the stirrer passes through the hub of the pulley, and is free to move up and down therein and in its bearing; but rotary movement independent of the pulley is prevented by a spline-and-groove connection.

From the upper part of the spindle of the stirrer a rope passes over a pulley, *M*, so that the stirrer may be raised, as shown in full lines, or lowered into the keir, as shown in dotted lines, as required.

A cover is or may be employed for the keir *H*. It is made to serve as a step or bearing for the spindle of the stirrer, as well as to close the keir.

The operation of preparing the stock is as follows: The first step is to put into the wire basket a suitable quantity— say five hundred pounds — of colored rags cut into small pieces, as is the common practice, for convenience in working them. Into the keir *D* are put, for treating the weight of rags mentioned, five hundred gallons of water and fifteen pounds of caustic soda, to which is added a small quantity —say two gallons—of soft soap. The wire basket and rags are now put into the keir and boiled, preferably under atmospheric pressure, for about three hours, or sufficiently long to remove the oils and grease, which may be accomplished with frequent stirring in much less time. They may now be transferred in the basket to keir *H*, to be thoroughly washed with water. In washing the stirring apparatus is employed, it being let down into the position indicated in dotted lines in the drawing. The operation is continued until the stock is sufficiently clean, which an expert paper-maker will readily determine. The keir *E* is now supplied with two hundred and fifty gallons of water, with which have been mixed five pounds of a soluble salt of manganese, prepared for the purpose from black oxide of manganese, and muriatic and sulphuric acids, as hereafter described. The rags having been washed are now immersed in this solution, where they are to remain for about three hours, being thoroughly agitated or stirred during this whole or a portion of the time. While this process is going on put into the keir *F* two hundred gallons of water having in solution fifteen pounds of chloride of lime, where the rags are thoroughly stirred and where they remain for about two hours. If the goods have obtained or taken on a brown color they may be washed thoroughly in the keir *H*, and then removed to the keir *G*, in which have been put two hundred and fifty gallons of water and five pounds of oxalic acid. This is for the purpose of removing any traces of iron or manganese which may be left in the rags. They are allowed to remain in this solution for about three hours, when they may be again washed in keir *H*, and then thrown into a hydro-extractor of any ordinary or suitable construction, which will take out nearly all the water remaining in them; after which they may be dried and made ready for shipment. The rags coming out of this process are ready to be reduced to pulp in the ordinary engines.

Should the rags not have obtained the brown color referred to above, they may be returned to the manganese solution again, and then again to the chloride-of-lime solution, as before. The remaining part of the processes following this second treatment will be the same as before described.

Permanganate of potash could be employed for the manganese with excellent results. It would be preferable to use it but for economical reasons.

The patentees state that they have obtained nearly or quite as good results by means of the soluble salt of manganese, before referred to, which is prepared in the following way: In a suitable vessel, to one pound of black oxide of manganese add one pound of commercial muriatic acid, or other quantities in approximately the same proportion can be used. Mix them well and let the whole remain for from thirty minutes to one hour, and then add two and one-half pounds of commercial sulphuric acid; stir and heat the mass gently until the manganese is dissolved, then add about three gallons of water, stir it well, and leave the whole to settle. In charging the keir *E* only the clear liquor of the soluble salt of manganese is used.

We have indicated the proportions in the foregoing process with considerable minuteness in order that the invention can be readily carried into effect; but it is obvious that these may be varied more or less. The stirring apparatus can, if desired, be readily constructed so as to be shifted from one keir to the other by supporting the bearing for the spindle of the stirrer in a travelling bracket, and making the elevating and lowering pulley also movable, like that which supports the basket. This is not, however, considered necessary, as the stirring is principally required in the washing-keir.

Boiling Waste Paper.

Waste paper can be boiled in either stationary or rotary boilers; but in mills where its manufacture is made a specialty it is commonly boiled in stationary iron tubs.

Writing-ink can be extracted by simply boiling with water; but a solution of soda-ash is generally used for extracting printing-ink.

The tubs used for boiling waste paper vary in size in different mills, but a form of tub, very similar in design to that shown in Fig. 89, which has been found to give satisfaction is built of boiler iron and measures about eight feet in depth and eight feet in diameter at the bottom, and six inches wider at the top than at the bottom. Steam is evenly distributed to all parts of the tub through a false bottom perforated with a large number of small holes.

In order to expedite the emptying of the tubs the false bottoms have attached to them three or four iron rods to the tops of which iron chains are hooked and the false bottom and mass of boiled paper raised and deposited at any desired point by means of a steam hoisting engine or a crane.

It is preferable not to pack the waste paper while in a dry state in the tubs, as it is liable to be imperfectly boiled owing to the imperfect circulation of the liquid through dry paper.

To properly begin the boiling operation it is necessary to fill the tubs one-quarter full with a solution of soda-ash, which should then be brought to the boiling-point, when the papers should be thrown in and evenly distributed.

In order to obtain an even distribution of the boiling liquor over the surface of the tubs an iron pipe extends from the centre of the false bottom to nearly the top of the tub, and this pipe being covered with a suitable hood distributes the soda-ash solution over the whole surface of the vessel.

The iron tubs are cased with wood or covered with an asbestos coating to prevent the escape of the heat, and the top is covered with a flat iron cover in one or two pieces. The steam enters the tub at the side, near the real bottom, but under the false bottom; and the liquor is drawn off through a pipe and valve connected directly with the bottom of the tub.

In many mills the liquor is not drawn off after each boiling, the paper only being hoisted from the tub and the liquor strengthened by the addition of ten to twenty pounds of fresh soda-ash to each one hundred pounds of the paper to be next boiled. In proportion as the waste paper is more thickly covered with printing ink the more soda-ash will it require in the boiling operation. The period of boiling

15

varies from twelve to twenty-four hours, according to the nature and condition of the waste paper to be treated.

The waste steam from the engine can be profitably employed in the boiling operation, as water in becoming steam absorbs a large quantity of heat, which becomes latent, and is termed the heat of vaporization; but this heat is again given out when the steam condenses to water. The latent heat of steam by one observer is 996.4° F., and by another 998°. The latent heat of steam diminishes as the temperature of the steam rises, so that equal weights of steam thrown into cold water will have nearly the same heating power, although the temperatures may vary exceedingly. This also appears to be below the boiling-point, so that to evaporate a given quantity of water a certain amount of heat is requisite at whatever temperature the evaporation is conducted. It is for this reason that distillation *in vacuo* at a low temperature effects no saving of fuel.

The tubs which are described above will each hold about four thousand pounds of papers; heavy book papers, however, require less room than news or shavings, and the quantity which a tub will hold varies according to the class of waste papers to be treated.

Treating Waste Papers so as to make Paper entirely therefrom.

By the common process the imperfections, consisting of old letters, documents, newspapers, books, etc., are dusted, and then put into a rotary boiler, and cooked and pulped in an alkali solution. This operation produces a mass partly pulped and so conglomerated that the alkali cannot reach all of the ink. Much of the ink thus remains to form a constituent part of the pulp and paper made from it. This mass is then put into a washer, and the alkali and dissolved ink removed, after which it is beaten to the proper pulp and mixed with the other portions of the stock; usually not over fifty per cent. of imperfections being admitted into the stock for new paper, the remainder of the stock being made up of rags, etc.

The object of the process patented by Mr. J. T. Ryan, of Hamilton, Ohio, is to so treat the imperfections that a first-class clean paper may be made entirely therefrom. In executing this process first pass the imperfections through a duster, all thick old books being previously torn apart to reduce them to a few leaves. Then treat the imperfections to the action of hot alkali without pulping them. The alkali solution thus acts on the surfaces of the imperfections and dissolves off and carries away all the ink into the solution. Then drain the imperfections, which are still in sheet form, as free from the alkali solution as convenient. Then place the imperfections, still in sheet form, in the washing-engine, and wash out the alkali solution, which leaves the imperfections perfectly clean. The material is then pulped in the beating-engine, and it is claimed can be formed into first-class clean paper without the addition of any new or expensive paper-stock.

In executing this process use a common duster. Into a bucking-keir put a soda-ash solution having a density of 5° Baumé at 160° Fahrenheit. Put in the stock, and shower for eight hours at a temperature of 160° F., without pulping the imperfections; then lift, and drain and cleanse well in the washing-engine; then pulp and form into paper.

As the draining operation will always be imperfect, each charge removed will carry away some of the soda-ash solution and leave the remainder of impaired strength. After each drainage, add water to make up for loss in quantity of solution, and add enough soda-ash solution, having a density of 13° Baumé, to bring all the solution up to 5° Baumé at 160° F. In about eighteen working days the liquor will have accumulated considerable ink and other matter. Then blow off one-half of the liquor, and restore the quantity for proper working. None of the soda-ash solution is wasted, except such as fails to drain, and such as is blown out, as last mentioned.

Boiling scrap paper in alkali, then cooling it, then boiling in a new solution, then beating to pulp in alkali, then washing, etc., is an old process.

But in the present method every care is to be taken to guard against pulping before the alkali is washed out.

Other Methods of Treating Waste Papers.

Fig. 69 shows a view of a circuit-vat in which the sizing of the paper and the vehicle of the iuk can be quickly dissolved; the temperature necessary to be maintained and the quantity of alkali to be employed are given in the text describing the construction and operation of this circuit-vat.

In the section of Chapter X., devoted to " Washing Waste Paper or ' Imperfections,' " a process is described by which it is claimed that the sizing, etc., can be removed from the imperfections, and the fibres separated in the beater without breaking.

BOILING STRAW.

Mellier's Process for Treating Straw.

Mellier's process for treating straw to prepare it for use in the manufacture of paper consists in steeping the straw for a few hours in hot water after it has been cut and freed from knots and dirt.

The straw is then placed in a rotary boiler containing a weak solution of caustic alkali, and, after making it steam-tight, the pressure should be gradually raised to about 55 to 70 lbs. to the square inch, at which point it should be maintained for about three hours, the boiler in the meanwhile being made to revolve at the rate of about one or two revolutions per minute.

The solution used by Mellier is from 2 to 3 degrees Baumé, and in the proportion of about seventy gallons of such solution to each hundred weight of straw.

In order not to dilute the caustic alkaline solution by the condensation of the steam it is recommended, if the boiler is to rotate vertically on its small axis, to cover it with a jacket so that the steam may circulate from one end to the other between the two plates; but if it is to revolve horizontally, or upon its long axis, there should be fixed near each end of the boiler and inside of it a diaphragm or partition, which partition should be connected by numerous tubes arranged in a circle near the outer circumference of each partition. By this arrangement the steam introduced through the hollow axis at one end of the boiler passes through the steam-pipes, and thence into the compartment at the other end of the boiler, where it and the condensed steam are conveyed away through the other hollow axis.

By not delivering the steam directly into the boiler, in addition to not diluting the alkaline solutions, the trouble is saved of sometimes having the end of the steam-pipe in the boiler choked with straw, and there is another additional advantage of being able to quickly cool the boiler, after the pressure has been maintained for a sufficient length of time, by passing a stream of cold water through the jacket or steam-pipes.

After the apparatus and fibres under treatment have been cooled in the manner described, the manhole is opened and the materials emptied into suitable vessels, where they are washed first with hot and afterwards with cold water until the liquor runs perfectly clear.

The fibre is next steeped for about an hour in hot water acidulated with a quantity of sulphuric acid equal to about two per cent. of the fibres under treatment, and, finally, the washing is completed with cold water.

The bleaching is then done in the ordinary way, and it is claimed that it can be accomplished by the employment of a comparatively small quantity of chloride of lime.

This process was patented in America and also in Europe by Mr. M. C. Mellier, of Paris, France, the patent in the United States bearing date May 26, 1857.

Burns's Process for Treating Straw.

Heretofore in the art of making straw paper the straw has been reduced to pulp by means of beating-machines, or by means of a machine having a rotary cylinder carrying a series of cutting-knives. The objections to these machines are several, among which may be stated the fact that the fibre of the stock is sometimes completely cut to pieces, and that the bleaching material in solid particles is introduced into these machines, both of which objections, especially where straw-stock is used, tend to make the paper, straw-board, and the like extremely brittle and rotten, and consequently unsatisfactory to the trade.

To remedy these objections Mr. Daniel R. Burns, of Dayton, Ohio, has invented a process by means of which the stock is first cooked, then disintegrated by separating and tearing the fibre apart without destroying the fibre itself, thereby, it is claimed, allowing the fibre to retain all its albumen and gluten properties. Finally the stock is subjected to a bleaching process without the contact therewith of any solid-matter bleaching material. The bleaching process, as conducted by Mr. Burns, is described in Chapter XI.

Fig. 89 shows a section of the vessel used for boiling the straw; Fig. 90 shows a transverse section of the disintegrating machine, and Fig. 91 shows a face view of the stationary grinding disk.

The boiler shown in Fig. 89 consists of the vessel *A*, having a suitable cover, and in which the stock is first placed to be cooked. It is provided with an upright perforated central tube, *B*, having a deflecting-cap, *C*, at the top, and a steam-pipe entering it at the bottom. The stock is placed around this pipe *B*, and the steam from the steam-pipe, entering the pipe *B* and passing out through its perforations, thoroughly cooks the stock throughout.

Fig. 89.

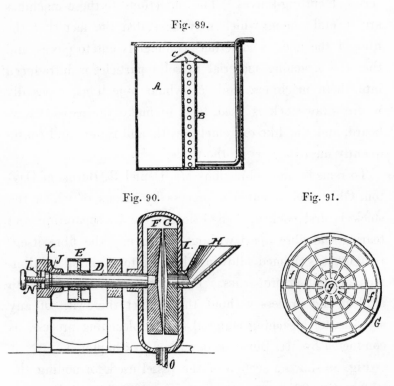

Fig. 90.   Fig. 91.

The process and manner of treating the stock when introduced into the vessel *A* is as follows: Twenty pounds of carbonate of soda are added to every one hundred gallons of water, which is not raised to a higher temperature than 212° F. at any time during the process. The straw is introduced when the water is cold, after which the heat is gradually raised to the boiling-point. In most cases it is claimed that the straw is in condition to be removed from the tub as soon as the boiling-point is reached. To ascertain if it be sufficiently cooked, a small sample is taken from the tub, and if the knots on the straw, hay, grasses, etc., can be easily crushed under the fingers by a slight pressure, it is ready to be delivered from the tub. If not, the boiling must be continued. This is intended for stock to be bleached white.

For stock which is to retain the natural color, or not beyond a buff, there are introduced into the water ten pounds of extract of hemlock or oak bark to every one hundred gallons of water, and the same amount of carbonate of soda as above (twenty pounds), and dissolved. The liquid is now the tannate of soda. The stock is then introduced as before, when the water is about 60° F. The tannic acid uniting with the gluten and albuminous properties of the stock imparts to it the properties of leather, the same as to hides or skins in tanning, which preserves it from the action of the water. The soda acts upon the fibre, softening and disintegrating it, so as to make it useful for paper-stock. This gives to the darker colored stock, paper, straw-board, etc., a greater amount of tenacity and strength than has heretofore been obtained by any other process.

After being cooked the stock is removed from the vessel in which the boiling is conducted, and is fed into the disintegrating-machine. (Shown in Fig. 90.) This machine consists of the revolving shaft *D*, driven by any suitable power by means of pulley *E*. One end of shaft *E* carries a rotary disk, *F*, of peculiar construction—that is, its grinding-

face is concaved.   The grinding-disk G is of similar construction, only it is preferably made stationary (although it may
be made to revolve), and is also provided with the central
feed-orifice g, communicating with the feed-hopper H.   The
disks F and G are inclosed by a tight casing, I, provided
with a discharge-opening, O, for the pulp after it has been
reduced.   The shaft D has a lateral adjustment for the purpose of regulating the degree to which the pulp is to be
reduced.   This adjustment is effected by means of the ad-
justing-screw N working in a nut, L.   The plate K, against
which the screw bears, is made stationary, so as not to
revolve; but has a lateral motion.

J is a loose washer free to revolve or move laterally.
This construction prevents all tendency of the adjusting-
screw to tighten or loosen due to the revolution of the shaft;
and, further, in the event of the washer J sticking to the
end of the shaft by heat, the washer would still be free to
turn against the stationary plate J.

The cooked stock is fed into the machine through hopper
H, the concave confronting faces of the grinding-disks admitting of an opening into which the stock may be easily passed.
Here, by the gradually-increased rapidity of the planes of
the disk G from the centre outward, the fibre of the stock
is completely separated and torn apart, and the whole paper
stock reduced to pulp, but it is claimed without destroying
the fibre itself.   This stock is now ready for use for the
darker-colored papers.   Where straw-stock is treated it will
be seen that all the gluten and albumen matter is retained,

which the inventor states he has found essential in order to render the paper tough and homogeneous.

Fig. 91 shows a face view of the stationary grinding-disk, which is represented with dressing. Different kinds of dressing for the disk are required, however, according to the nature of the stock to be treated.

When it is desired to have a lighter colored paper than results from the stock treated in the manner above described it is subjected to a bleaching process, which see in Chap. XI.

The inventor of the present process states that the reason it has been necessary heretofore to use caustic-alkali, acids, and the high degree of heat is that there was no machinery used that would disintegrate the stock and reduce it to pulp without being so treated, while his machine shown in Figs. 90 and 91 is so driven, that the disk has a velocity (the periphery travelling at the rate of five thousand feet per minute) which reduces the stock with great rapidity to pulp without injuring the fibre.

Boiling Coal Tar with the Alkalies employed in Treating Straw.

It has been proposed to boil coal tar with the alkalies used in the preparation of paper pulp from straw; the quantity used being about 5 pounds of coal tar to each 1000 pounds of straw or other material treated, or such quantity of coal tar as is necessary to neutralize the quantity of alkali used.

It is claimed that the paper made from straw thus treated will be much less harsh and brittle than that treated in the ordinary way.

Other Methods of Treating Straw.

For other processes of treating straw see those of Dahl, Franche, and others, described in the section of the present chapter devoted to boiling wood.

The most tender straw used in the manufacture of paper is that of oats, next that of barley, wheat, and finally rye. Maize straw is even more tender than oat straw. The time for boiling depends on the hardness of the material, as also upon the pressure under which the material is boiled and the strength of the alkaline solution, and the preliminary labor which may have been bestowed on the material.

Corn leaves and stalks are placed in a solution containing, for 100 pounds of material, 40 pounds of lime and 1 pound of potash; the straw remains under treatment about 3 hours.

Oat Straw.—For 100 pounds of straw a solution is required containing 50 pounds of lime and 2 pounds of potash. Time, 3 hours.

Barley straw is first boiled for 3 hours in water and then brought into a solution containing, for every 100 pounds of straw, 50 pounds of lime and 2 pounds of potash. It is then brought into a second solution consisting of 30 to 40 pounds of lime and 1 pound of potash. Time in each solution 3 hours.

Wheat straw is first boiled for 3 hours in water and then placed consecutively in 3 solutions, remaining in each for 3 hours. The first consists of 50 pounds of lime and 2 pounds of potash, and the last two of 30 pounds of lime and 1 pound of potash.

Rye straw, being very hard, must first be boiled in water for 3 hours, and then successively for the same time in four different alkaline solutions of the same strength as those for wheat straw.

BOILING ESPARTO.

The esparto after being sorted, as has been described on page 112, is in condition for boiling. There are various forms of boilers in use for the treatment of esparto ; but the stationary form shown in Fig. 74 is ordinarily employed in Great Britain.

The grass is filled into the boiler through the door *A* (Fig. 74), which can be firmly fastened down by the screws shown. The steam is admitted into the boiler through the pipe *C*, which extends a short distance below the perforated bottom. Surrounding the steam pipe *C* is a wider pipe, *D*, open at the top, and made slightly trumpet-shaped, also open at the bottom below the false bottom. The false bottom rests on a shoulder formed on the lower portion of the pipe *D*. The liquor passes through two or more openings in the enlarged portion of the pipe *D*, and there is thus produced a constant circulation of the hot liquor, which is dispersed all over the boiler by striking against the hood *E* at the top of the pipe *D*. This action of the hot liquor striking against the dome *E* is technically termed " vomiting." The liquor is run off from the boiler through a valve placed under the boiler, and the grass is removed through the door *B*. A safety valve is usually placed as shown in the illustration, and the chain and weights are intended to allow the door *A* to be conveniently raised and lowered. Just before starting to fill the boiler with grass the vomit is started and continued in action until the boiler is filled, as the grass is thereby softened, which allows closer packing in the boiler.

The form of boiler described cannot be used with very high pressure steam, as in such cases it requires slight modifications.

Much experience and care are necessary in boiling esparto, for if the material should be insufficiently boiled there results a loss of soda in the operation of reboiling thereby entailed, and then the result is often not satisfactory when a repetition of the boiling is made necessary.

The quality of the grass, the form of boiler, the pressure, etc., determine the quantity of caustic soda to be used. When a shipment of esparto is received at the mill, experiments should be made to find the least quantity of caustic soda necessary to properly boil that especial lot, and the quantity found necessary should be continued until the whole consignment is used.

Dunbar states : " When the necessary precautions are taken to have everything in proper order and condition, the undernoted quantities of caustic soda will generally boil the various espartos in a satisfactory manner :—

Fine Spanish esparto, boiled with 28 lbs. caustic soda (70 per cent.) per cwt.
Medium Spanish " " 24 " " " "
Fine Oran " " 30 " " " "
Medium Oran " " 28 " " " "
Fine Susa " " 28 " " " "
Tripoli " " 32 " " " "
Tripoli " " 25 " " " "

All boiled for ten hours in stationary vomiting boilers with ten pounds steam pressure, care being taken to see that the esparto is sufficiently boiled before the liquor is run off."

When the boiling of the esparto is completed, the steam is shut off, the lid lifted, and the liquor conducted to a large store-well. The esparto is then washed by running water into the boiler, fastening down the lid and turning on the steam. After a short time the steam is again shut off and the new liquor is also run into the store-well.

The door B is then opened and the esparto removed from the boiler, and if it is not to be used in fine grades of paper is carried directly to the breaking engine; but when a high degree of purity is desired the grass is first carried to the " wet-picking" department of the mill, where girls and women overhaul it and pick out all the unboiled portions.

For other methods of boiling esparto see the processes of Dahl and Franche, described in this chapter.

BOILING MANILLA AND JUTE.

Both manilla and jute may be boiled in either rotaries or open tubs, but in either case a liberal use of milk of lime will be necessary. From 15 to 30 pounds of lime per 100 pounds of material is the usual proportion; but this is largely exceeded by some manufacturers, who sometimes employ as much as 40 or 50 pounds of lime for 100 pounds of jute.

Treated by steam and water at a high temperature (248° to 266° F.) jute fibre is completely destroyed, and is then converted into soluble compounds. The same result occurs when heated with acetate of soda. In this case acetic acid distils over, thus showing that jute is decomposed, and gives rise to the formation of acids. Even a small quantity of bisulphite prevents this decomposition. This, according to Mr. Cross, is due to the known combination of this salt with the aldehydes formed by the action of oxidizing agents on jute.

Conley's Process for Boiling and Bleaching Jute.

As a preliminary step to carrying out the following process, which is the invention of Mr. Edward Conley, of Cincinnati, O., first assort, clean, and cut up the jute or jute-butts in the usual way. Then put them into a close vessel, either stationary or revolving. To every one hundred pounds of jute are added sixty gallons of caustic alkali of a strength of 7° Baumé, at a temperature of 130° F. Then boil for about ten hours under a standard pressure of eighty pounds. The boiling completed, the spent liquor is drawn off, and the stock thoroughly washed with hot water, which keeps the thick vegetable matter in solution, and carries off the black and non-fibrous substances.

Preparatory to being bleached it is washed in an ordinary rag-engine in the usual way, after which it is preferable to run it through a wet-machine, which extracts from it a large amount of water, and leaves it in a better condition to absorb the bleaching liquor.

When jute-bagging, burlaps, or gunny-bagging are used, the quantity of caustic alkali may be less by ten gallons.

The strength and quantity of caustic alkali, the pressure, and the time given above are the standards, but their equivalents may be used. For example, 7° Baumé at 130° F. has been named, but 8° Baumé at 100° F. is equivalent.

In this process there are two points of prime importance: First, boiling in alkali as strong as can be made caustic. It would be preferable to use alkali of a strength of from 9° to 10° Baumé, if it were practicable to make pure caustic at that strength. Alkali, of any greater strength than 7° or 8° Baumé, cannot be made thoroughly caustic except by evaporation. Second, a combination of a caustic liquor of a high strength and a medium pressure.

The advantages claimed to be gained by this process are as follows: First, the disintegration of the stock is effected by the single process above described. Second, this disintegration being thorough, the cellulose is left of a light brown color, and in its natural state short, with uneven ends, which is the best condition to be worked into paper. Third, another consequence of this thorough disintegration is, that all foreign matters, as silica, gluten, etc., are easily separated from the pulp, leaving pure cellulose. Fourth, owing to the absence of all foreign matters, paper made from pure cellulose dries out regularly when wet down for printing. This gives to the sheet an even surface, which enables it to pass smoothly through the press, and receive a clear and distinct impression. It is also opaque, and, when printed upon, does not permit the ink to show through from one side to the other. Moreover, pure cellulose is easily bleached, only a comparatively small amount of chemicals being required,

16

and, therefore, the paper made from it is not only of the highest grade of color, but also stronger than when the pulp is subjected to a more intense chemical action.

The cellulose obtained by this process is not chemically pure, but practically its purity is claimed to be sufficient to produce all the results claimed for clear cellulose.

It has been the practice to work jute into low grades of white paper, and in small quantities. When treated by this process, it is claimed that it can be made into the finest grades of paper, either when used alone or mixed with other stock.

Mr. Conley is not the first person to use jute in the manufacture of white paper, so called. It has been experimented upon by many persons and in many ways, such as boiling in lime, boiling in soda-ash, giving it an acid bath, etc.; but by none of such processes, however, has cellulose of the desired quality been produced.

Boiling Wood.

Chemically-Prepared Wood-Pulp.

During the past twenty years special efforts have been made with a view to dissolving the intercellular incrusting or cementing matter existing between the fibres of wood, so that the resulting cellulose might be introduced into fine papers, and latterly with considerable success.

In the earlier processes patented in England by Sinclair, in 1854, and Houghton, in 1857, wood was boiled with about twenty per cent. of real caustic soda under a pressure of from ten to fourteen atmospheres. Experience has dictated certain improvements in some of the details of these earlier methods by which so-called chemical wood-pulp is manufactured very largely on the continent of Europe from whence it is imported into England to a considerable extent; and in the United States also the preparation of cellulose from wood is receiving much attention.

It is possible to obtain a pulp of good quality suitable for some classes of paper, by boiling the chipped wood with caustic soda in the manner indicated; but when it is desired to use the pulp so prepared for papers having a perfectly white surface, it has been demonstrated in practice that the action of the caustic soda solution at the high temperature which the required pressure develops results to a certain degree in a weakening and browning of the fibres, and during the past five years much labor has been expended in the endeavor to overcome the objections named.

The outcome of these efforts has been a number of patents, having for their object to prevent oxidation and subsequent weakening of the fibres from taking place in the chemical preparation of wood-pulp. Bisulphite of lime is one of the chemical agents used to prevent oxidation and subsequent degradation of the fibres in the processes patented by Messrs. Mitscherlich and Francke, and bisulphite of magnesia is one of the agents used for the same purpose in the processes of Messrs. Ekman and Graham.

Although a common principle runs through all these methods of preparing cellulose from wood, they differ materially in detail, as to construction of the digesters employed, methods of treating the wood stock before boiling it in the sulphurous-acid solution and also as regards pressure, blowing off of the sulphurous acid gas, etc., but all these processes present a striking similarity to the method patented by Tilghmann in 1867.

The process patented by Dr. Mitscherlich, of Münden, Prussia, has been quite extensively adopted in Germany. The chemical changes which take place during the boiling process in Dr. Mitscherlich's method of preparing cellulose from wood may be explained as follows:—

The sulphurous acid is oxidized by combining with a part of the oxygen of the cellulose and of the organic substances, and formed into sulphuric acid, that under normal circumstances combines with the bases that have before been united with the sulphurous acid. When the process is not properly conducted, free acid is formed in the solution, which exerts an injurious influence upon the fibres. Besides the free acid and its combination, the incrusted substances are formed into compound combinations of tannic acid and its by-products, which are highly objectionable. For the proper carrying out of the boiling process it is an essential condition that the sulphurous-acid solution be free from polythionic salts, as by the action of such salts a brownish-black deposit is formed on the wood stock, during the boiling, so that it remains hard and causes a failure of the boiling operation. At the same time a considerable increase of temperature takes place, so that the tests taken from the boiler show an abnormally quick decrease in the proportion of sulphurous acid in the solution. These polythionic acids are generated commonly by the presence of free sulphur-fumes during the roasting process. To prevent their presence care has to be taken that sulphurous acid free from such acid and salt is obtained and used.

The great objection to the modern chemical processes of preparing pulp from wood is that, as they commonly depend on the use of bisulphite, which, being an acid salt, cannot be worked in an iron boiler, great difficulty has been encountered in practice in maintaining the lead lining of the digester in proper repair. It is probable, however, that this difficulty will be surmounted with further experience. In Dr. Mitscherlich's apparatus a thin lead lining is cemented to the inner surface of the boiler by a cement composed of common tar and pitch, and the lead lining is then faced with glazed porcelain bricks. The objection to this method is that in case of leakage or rupture in the brick facing and lead lining the pulp is liable to become injured by the tarry product. Messrs. Ritter and Kellner propose to unite the iron shell of the boiler and its lead lining by means of an interposed soft metal alloy fusible at a temperature lower than that of either. It is claimed that the shell and lining are thus securely united, while the alloy being fusible under the normal working temperature of the digester the lead lining can slide freely on the boiler shell. The objection to this method is that it is difficult to localize the creeping or sliding effect of the lead lining into adequately small portions to make the metal sufficiently durable for digesters, but this objection the same inventors claim to have overcome by a later invention. For other methods of securing the lead lining to the boiler shell, see the list of patents at the close of this section.

All the acid processes for obtaining cellulose from wood are open to the objection that the cellulose so obtained contains a considerable quantity of incrusting matter which is allowed to remain in the fibre, thus giving a harsh character to the paper manufactured solely from it, and another objection is its great transparency. In order to obtain a pure cellulose it is necessary to exhaust it in an alkaline solution subsequent to the treatment with acid.

The white woods and pines are commonly used in the manufacture of " chemical wood pulp." "The white woods, such as bass-wood, and the different varieties of poplar, are easily managed; the ' popple' or white poplar and the aspen are the least difficult to reduce, next comes the bass-wood, then the yellow poplar, but the quantity of the fibre obtained from each is inverse to this, as the yellow poplar gives the best and longest fibre, then the aspen, and lastly the ' popple.' The pines give a long fibre of considerable strength, but a smaller quantity per cord, and require severe treatment to make it white. The treatment which the different kinds and varieties of wood require is alike except in degree, and the following descriptions of boiling with soda will consequently answer for all." Marshall's boiler in which the wood chips can be digested with soda will also be described.

Boiling with Soda.

A method of cutting the wood into chips has been described on page 145. From the *Paper Trade Journal* we take the following description: The rotary boiler is generally considered preferable, and it is usually made of three-quarter inch iron, and measures about seven feet in diameter and twenty-two feet in length, with whole wrought-iron heads. Heavy cast-iron hollow trunions are riveted on each head. The digester is provided with a steam coil of $2\frac{1}{2}$ inch pipe aggregating about 700 feet in length; the steam is admitted and the water of condensation is discharged through the hollow journal at the same end. There is a short coil of $2\frac{1}{2}$ inch pipe perforated with holes $\frac{1}{16}$ of an inch in diameter, and located inside of the digester between the manholes, and connected through the shell with a blow-off valve on the outside. There is located on the opposite side, also inside of the digester, a perforated pipe which extends the whole length and is connected through the other hollow journal with a force-pump of large capacity. The chips are filled into the digester through the manholes, and after the digester is packed full, 2800 gallons of soda solution at 9° Baumé are run in; the manheads are then put in, the steam is turned on, and the digester is started. As long as a solid stream of the water of condensation flows from the discharge pipe it is allowed to run, but when the steam comes, it must be connected with a steam trap of sufficient capacity to discharge all of the water as rapidly as it is formed. The steam pressure is allowed to rise to 115 pounds by the gauge, and maintained at that point for five hours, at the end of which time the steam is shut off, and the digester stopped with the manheads at the lowest point. A blow-off pipe, which extends from under the digester to the spent liquor tanks in the evaporator room, is now connected by a flange joint to the blow-off valve, which is located between the manheads, and which connects with the perforated coil inside. The blow-off valve is opened a little at a time, to prevent the liquor from being blown out with too great violence. The valve may be opened a little more and more as the pressure decreases. At this point one of two courses may be pursued: First, all of the liquor may be blown out, thus securing the largest possible quantity of spent liquor of full strength for evaporating, and this will be found to be about two-thirds of the original quantity that was put into the digester. When this is obtained the blow-off valve is closed and the force-pump which connects with the perforated pipe inside of the digester is started and hot weak liquor is pumped in. While this is being done the blow-off pipe is disconnected, and when a sufficient quantity of weak liquor is forced into the digester, the digester is started and allowed to make a few revolutions. It is then stopped and the blow-off pipe is again attached, and the liquor which has been pumped in is blown out. Care must be taken to maintain the necessary pressure by keeping sufficient steam on the coil to accomplish this. When this is done, the force-pump is again started and a quantity of warm water is pumped in; the blow-off pipe is again disconnected, and the digester is allowed to make a few revolutions. It is then stopped with the manheads at the top, and after the manheads are taken out, the digester is rotated half way, and the contents of the digester are emptied into a tank located underneath. This tank is provided with a perforated false bottom of sheet-iron, the holes $\frac{1}{10}$ of an inch in diameter. The liquid is allowed to drain off into a cistern, and is saved to be used for the first wash. The second course that may be pursued is to blow about one-half of the liquor out of the digester, then start the force-pump in the hot weak liquor while the blowing off goes on. As soon as a sufficient quantity is obtained for evaporation the blow-off valve is closed, but the pumping is continued until there is sufficient water in the digester to give the fibre a first wash. It is then stopped, and the digester is rotated a little and emptied. Time is saved by pursuing the second course, but there is a loss of soda, and the pulp is not so thoroughly washed as in the first instance.

Dahl's Process of producing Cellulose from Wood, Straw, Esparto, or other Vegetable Matters, by boiling them under pressure in a hydrated solution containing Sulphate of Soda, Carbonate of Soda, Soda Hydrate, and Sodium Sulphide.

The process of Mr. Carl F. Dahl, of Dantzic, Germany, is applicable to the manufacture of cellulose from wood, straw, esparto, and other vegetable substances.

For the purpose of dissolving the cellular substance or fibrous mass out of the bodies incrusting them the comminuted wood, straw, esparto, and the like are boiled under pressure in wrought-iron vessels free from lead lining, containing a hydrated solution in which are contained sodium salts, partly in the form of sulphate of soda, carbonate of soda, soda hydrate, and sulphide of sodium. Two hundred and twenty pounds avoirdupois of half-dried pine wood require about fifty-seven pounds of the above-named salts in solution; straw and esparto, about twenty-two to twenty-seven pounds. Pine or fir wood requires five to ten atmospheres overpressure, the strength of the sodium solution being 6° to 14° Baumé, the time of the boiling varying from thirty to four hours. Esparto requires two to five atmospheres overpressure, the strength of the sodium solution being 5° to 8° Baumé, and duration of boiling eight to three hours. By the boiling process the incrustations combine with the sodium solutions; the cellular matter, it is claimed, remains uninjured of a loose consistency. After the boiling is completed, the brownish-black lye is blown off into iron basins for the purpose of afterward recovering the sodium salts. The remaining cell matter is washed with warm water in the boiler or other suitable receptacle, and is then manufactured into paper-pulp, in the well-known way, by means of a Hollander or pulp-engine, and bleached with a solution of chloride of lime. The color of the unbleached mass is yellowish-gray; that of the bleached mass pure white or slightly yellowish, according to the degree of bleaching.

Sulphate of soda serves for the production of the sodium solution. The sulphate dissolved in water is boiled with about twenty to thirty per cent. of burnt lime. The lye thus prepared is already serviceable for boiling; but it receives its proper composition by the addition of the salts regained from the sulphate solution after the boiling process. The lye, after being used, is forced into an evaporating-oven for the purpose of regaining the salts, is strongly calcined, and after thus being deprived of gas is drawn from the oven as a cake-like mass, washed, and the resulting solution used for the preparation of fresh lye.

For obtaining pure salts without the admixture of carbon, the thickened lye is drawn from the evaporating-oven and is fused in a calcining-oven at a dark-red heat. The fused mass, after cooling, assumes a reddish-brown color, is readily soluble in water, and has approximately the following composition: sixteen per cent. sulphate of soda, fifty per cent. carbonate of soda, twenty per cent. sodium sulphide, four per cent. diverse non-essential matters. This composition is very variable, according to the properties of the boiled matter, but without influencing the dissolving power of the solution afterward prepared therefrom. The salt which is regained should be dissolved as soon as possible, or guarded against the influence of atmospheric air. By the process of boiling and regaining, about ten to fifteen per cent. of the salts in the solution is lost. In general practice the losses are replaced in the preparation of the lye by sulphate of soda. For the solution are taken eighty-five per cent. of regained salt and fifteen per cent. of sulphate, which mixture, boiled with twenty to twenty-three per cent. of burnt lime, yields the proper lye. With ten per cent. of loss a clear watery solution is taken for the lye, in which are contained one hundred and ninety-eight pounds of regained salt. Twenty-two pounds of sulphate are added, and the solution, in which are contained two hundred and twenty pounds of salts in the already-named proportion, is boiled with forty-four pounds of burnt lime. If the loss amounts to fifteen per cent., then thirty-three pounds of sulphate and one hundred and eighty-seven pounds of regained salt are taken, the whole being boiled with fifty pounds of lime. In the case of twenty per cent. loss, one hundred and seventy-six pounds of regained salt and forty-four pounds of sulphate are taken for the solution, and are boiled together with fifty-two pounds of lime. If twenty-five per cent. of sulphate is to be added, one hundred and sixty-five pounds of regained salts, fifty-five pounds of sulphate, and sixty-one pounds of burnt lime are taken for the solution.

In regular operation the utmost limit of sulphate addition should be thirty per cent., one hundred and fifty-four pounds of regained salts, sixty-six pounds of sulphate, and seventy pounds of burnt lime. The proportion of salts, contained in the boiling solution is, on an average, thirty-seven per cent. sulphate of soda. eight per cent. carbonate of soda, twenty-four per cent. soda hydrate, twenty-eight per cent. sodium sulphide, three per cent. diverse combinations. This composition is very varying according to the qualities of the materials to be boiled.

The transfer from the soda hydrate treatment to the described treatment with sulphate is accomplished in the preparation of the lye by replacing the loss of soda hydrate by the sulphate, instead of by soda, and then, with the disappearance of the hydrate, gradually reducing the addition of lime during the boiling of the solution from forty-five per cent. to about from twenty to twenty-three per cent.

Defects of Boilers for Digesting Wood by the Soda Process.

The manufacture of wood-pulp by the chemical process has, from its invention down to the present time, been attended by very disagreeable and expensive features. It is a well-known fact that all wood-pulp digesters, of whatever description, leak more or less when a certain pressure has been attained. Generally, when the steam pressure has reached sixty pounds per square inch, the digester begins to leak. Then as the pressure is increased to one hundred or more pounds the leakage is increased. It frequently happens that as much as one-fourth or one-third of all the boiling liquors in the digester are lost in this manner, and necessitates charging the digester with an excess of alkaline liquor. The leakage is forced out through the riveted seams of the digester in the form of a fine spray and charges the surrounding atmosphere with an exceedingly offensive and suffocating vapor, which is unhealthy and injurious to the men working in the vicinity, who are obliged to inhale some portions of it. It frequently happens that the leakage is so great as to prevent the disintegration of the wood, thereby causing the total loss of the charge. Various methods of riveting digesters have been adopted. Plates have been planed to a true surface at the laps of the seams. Rivets have been even screwed in through the laps and headed cold. Digesters have been made of steel in order to have a close-grained surface at the seams. But none of these various methods have proved successful in preventing the leakage; and it is a well-known fact that wood digesters can be operated but a few weeks without recalking and replacing rivets, and the utmost care and skill of the boiler-maker have failed to produce digesters that will not leak at or through the seams. The filling and discharging of the digester two or three times daily, thus exposing it to temperatures varying from 150° to 320° F., produces expansion and contraction sufficient to cause the iron or steel plates to "creep" at the seams, and thus wear away the calking. When the pressure in the digester has reached a certain point, it opens the seams sufficiently to allow the escape of the very volatile liquor composed of the caustic alkali and other products generated in the process of the disintegration of the wood. Even in the pulping of wood plain, without chemicals, the escape of the pinic, pyroligneous, and other acids is annoying and injurious. By the invention shown in Fig. 92, Mr. George E. Marshall, of Turner's Falls, Mass., claims to have overcome all these difficulties, and he states that he has constructed and successfully operated for months wood digesters that are perfectly tight, not leaking at all, and, therefore, he claims the honor of having changed the manufacture of chemical wood pulp from the most disagreeable, offensive, and wasteful process known in the whole art of paper-making to a pleasant, safe, and economical system, always producing sure results, and worked with more ease and comfort than the ordinary process of boiling rag stock.

Marshall's Boiler for Digesting Wood by the Soda Process.

Fig. 92 represents Marshall's stationary upright wood-digester with its appurtenances.

A, the digester, is six feet in diameter, and sixteen feet long, made from half-inch iron or steel boiler-plates—all seams double riveted.

Fig. 92.

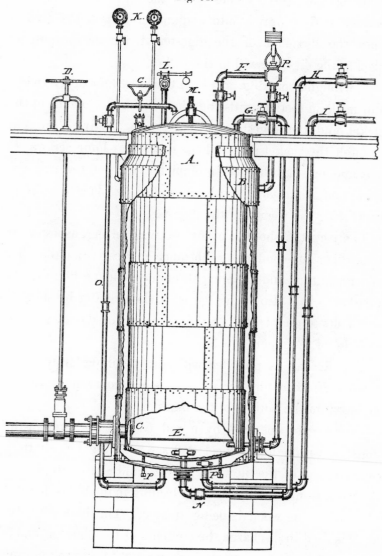

B, the outer shell or jacket, is six feet and eight inches in diameter, and connected to the digester, *A*, at a distance of from eight to twelve inches below the top. The upper end of jacket B is drawn into a diameter two inches greater than the diameter of the digester A, for the purpose of placing the digester within the jacket.

A wrought-iron ring, one inch thick by four or five wide, and made in sections, fills the space between the ends of the jacket and the digester, and two rows of rivets are put through the jacket-ring and the digester. Four screws, P, one and one-eighth inch in diameter, are put through the lower end of the jacket with a re-enforcing plate to assist in supporting the weight of the digester.

The jacket B should be covered with felting, asbestos, or some other non-conducting substance, to preserve a uniformity of temperature and to prevent undue condensation in cold weather. It is also connected with the digester by suitable stay bolts to guard against the explosion of the one or the collapse of the other.

The digester is provided with the usual perforated false bottom, E, extending across at a height from ten to twelve inches from the bottom of the digester.

C is a gate operated by a rod passing up through the digester, with a hand-wheel and screw at the top, shutting over the end of the blow-off pipe D, and is used to prevent the wood from entering the pipe leading to the blow-off valve at D while the wood is being treated. The blow-off valve at D is used to discharge the contents of the digester, and is connected to a heavy eight-inch wrought-iron pipe passing through a stuffing-box in the jacket B, and screwed into a heavy wrought-iron ring riveted inside of the digester, as shown in the drawing.

G is a two-inch iron pipe connecting the top and bottom of the digester, and is used while the steam-pressure is being raised to conduct the liquor from the bottom to the top of the digester for the purpose of removing any pulp that may have passed through the perforated false bottom, it being important to keep the space under the false bottom clear of fibre. The pressure is greater at the bottom of the digester than at the top until the full pressure required is attained. The opening of the valve in the pipe G will cause the liquor to circulate from the bottom to the top of the digester, carrying with it any fibre remaining in the liquor. This pipe may also be used to circulate the hot liquor from the bottom to the top of the digester while the pressure is being raised.

F is a two-inch pipe extending upward from near the top of the jacket, having an arm extending across and down into the top of the digester, and contains a pressure-valve, P, for the purpose of regulating the steam-pressure in the space between the jacket and the digester, and for relieving the pressure on the outside of the digester when its contents are being discharged through the blow-off valve D, insuring the digester from the danger of collapsing from outside strain by pressure in the jacket. This pressure-valve P is controlled by weights, which can be regulated according to the indications of the steam-gauges $K\ K$, one of which connects with the jacket and the other with the disgester.

The steam for treating the stock in the digester is admitted

17

through a pipe, *H*, directly to the alkaline liquor, all attempts to treat the stock by the heat of the steam within the jacket proving insufficient, it not penetrating to the centre of the digester, and leaving a core of " uncooked" wood in the middle.

I is an inch-and-a-half pipe for conveying steam into the space between the jacket and the digester, for the purpose of maintaining a pressure equal to or greater than the pressure within the digester, thus balancing the pressure inside the digester with an outside pressure, and thereby preventing leakage of liquor through the seams of the digester.

L is a safety-valve.

M is a manhole.

N is a check-valve.

O is a hot-water pipe to be turned into the manhole for washing out the digester with the hot water obtained by the condensation of the steam in the jacket, which is a great convenience over the former way of washing the digester with cold water.

SODA RECOVERY.

Fig. 93 shows a sectional elevation and plan of a Porrion oven, the various parts being indicated by the following letters: *D*, grate.   *C*, oven where the incineration takes place.   *H*, conduit which brings the hot air from the furnace.   *A*, evaporator with its paddles.   *B*, large flue.

The liquor is conveyed to the Porrion evaporating and incinerating oven, which differs from an ordinary reverberatory oven in its being provided with paddle-agitators which project the liquid, making it come down in sprays; this increases the surfaces coming in contact with the hot air and the smoke current traversing the oven. The expense of fuel is greatly reduced by this method.

Fig. 93.

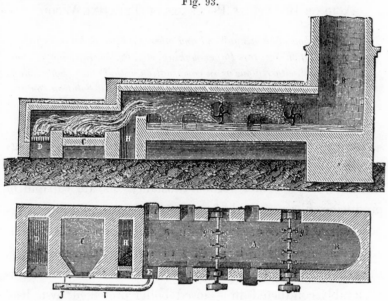

The residue is in combustion when it comes from the oven; and is disposed in a pile so that it may burn slowly. The combustion terminated, the carbonates are rendered caustic like those for the trade; they contain at an average 42 per cent. of anhydrous soda. Two-thirds of the soda used are thus recovered, the lost third being mostly retained by the lime, notwithstanding whatever care might be taken in the washing.

The economical regeneration of the carbonate of sodium by means of the evaporation and the incineration of the black liquors produced by the soda treatment of straw, wood, etc., have been a progress of the highest importance in these industries.

ACID OR BISULPHITE PROCESSES OF TREATING WOOD.

Graham's Method of treating Wood and other Fibrous Substances for the production of Fibre, for Paper-making, etc., by the Injection of Sulphurous Acid, either alone or in combination with Potash, Soda, Magnesia, Lime, or other suitable Base in the form of a Solution containing an excess of Acid, into a closed or open Vessel or Digester during the operation of Boiling.

The process of Mr. James A. Graham, of London, England, which we will now describe, relates to the treatment of wood and such other fibrous substances as are capable of producing fibres suitable for paper-making, and other purposes when boiled or steeped in a solution of sulphurous acid, or a sulphite or bisulphite of soda, potash, magnesia, lime, or other suitable base and water. The operation is preferably conducted in a closed boiler protected by a lead lining or otherwise from the action of the chemicals used, and fitted with a valve which can be opened, so as to allow the gases and volatile hydrocarbons contained naturally within and around the fibres (either in chemical combination or mechanical adhesion therewith) to escape.

We will first explain the manner of carrying out Graham's invention in a closed boiler.

In carrying out the process there is a constant loss of sulphurous acid gas going on, and consequently a continual weakening of the solution employed, in order to avoid which it is preferable to employ the monosulphite of potash, soda, magnesia, lime, or other suitable base and water. Either of these substances, or a suitable combination of them, and water are placed in the boiler with the fibrous substances to be treated, and the temperature raised to or above boiling-point. After the hydrocarbons, air, and gases natural to the fibrous substances have been driven out by the heat and allowed to escape there is pumped or injected into the vessel or boiler sulphurous acid, either in its gaseous or liquid state, or in combination with potash, soda, magnesia, lime, or other suitable base and water, or a solution of sulphurous acid. There is thus forming in the closed boiler a solution containing an excess of sulphurous acid above that required to form, in combination with the base, a monosulphite. The operation of injecting sulphurous acid or its combinations with potash, soda, magnesia, lime, or other suitable base, as above described, may be repeated from time to time during the boiling, so as fully to maintain, and, if necessary, increase, the strength and efficiency of the chemical solution employed.

According to this mode of treatment a saving of the chemical employed is claimed to be effected, as little or no sulphurous acid gas is lost during the time the gaseous hydrocarbons, air, and other gaseous or volatile matters are being driven out of the fibrous materials.

It will be readily understood that in the case where there is employed an open vessel or boiler the operation will naturally be carried on at the temperature of the boiling-point of the solution employed; but the mode of keeping such solution at a fairly uniform strength, or, if necessary, increasing its strength, will be substantially the same as that above described when using a closed vessel or boiler, in which latter case the operation may be carried on either at or above the boiling-point of the solution. When using an open boiler, it is evident that the excess of sulphurous acid supplied during the boiling will be constantly given off in a gaseous state from the surface of the liquid, and must consequently be replaced by further injections, while the acid given off can be led away and condensed, so as to enable it to be again used, if desired. In cases where the vegetable substances are boiled with water alone, or in conjunction with potash, soda, magnesia, lime, or other suitable base in the form of an oxide or an acid sulphite, the injection of sulphurous acid or its combinations with potash, soda, magnesia, lime, or other suitable base during the boiling will be equally beneficial. The inventor prefers to inject the sulphurous acid, or its combinations, as above described, into the vessel or boiler at the bottom, and to cause it to come in contact with the solution therein, before reaching the fibrous material. For this purpose there is formed a kind of chamber beneath the boiler, and separated therefrom by a perforated disk or diaphragm of lead or other suitable material capable of resisting the action of the solution, so as to allow the latter to fill the chamber. To this chamber a pipe is connected, through which the sulphurous acid, or a combination of it with a suitable base, as described, is forced or injected by any suitable apparatus.

It will, of course, be necessary to coat with lead the interior of the vessel or boiler and the parts with which the sulphurous acid or its combinations described come in contact.

Mitscherlich's Processes of Preparing Cellulose from Wood.

In the process patented September 5, 1882, by Dr. A. Mitscherlich, of Münden, Germany, cellulose is obtained as a by-product in the manufacture of tannic acid.

In carrying out this invention the wood is freed of bark and cut into pieces of convenient size for handling without removing the knots and small limbs. These pieces of wood are placed in a boiler which has an interior lining of lead, and which is provided with heating-tubes and with other accessories required for the induction and eduction of steam, etc. As soon as the boiler is charged with wood the digester is hermetically closed and the wood treated with steam and then with the aqueous solution of bisulphite of lime, according to the size of the wood, at a temperature of 226° F., for a certain length of time, preferably for somewhat more than eight hours. By the action of the steam and bisulphite of lime upon the wood all the soluble substances which surround and permeate the fibres of the wood are dissolved, while the cellulose remains as a soft mass in the liquid. The contents of the boiler are raised to boiling-heat, which is maintained for such a length of time as the vapors which are conducted off to the tower or shaft employed continue to have a strong smell of sulphurous acid. Instead of conducting the vapors into the tower or shaft, they may also be conveyed to a tank containing slaked lime (milk of lime). A concentrated solution of bisulphite of lime is obtained in the tank, which is extensively used in the trades for preventing the formation of acetic acid in solutions while in the process of fermentation. If in place of the slaked lime, carbonate of soda or other salts are placed in the tank, the different sulphites can thus be readily produced. The solution remaining in the boiler is then run off and separated from the cellulose. It contains, besides salts of lime (sulphate of lime, etc.), essentially tannic acid, also adhesive substances, acetic acid, and a small quantity of sulphurous acid, which latter is retained in the solution.

The solution can be utilized, first, as a new tanning material; secondly, for the manufacture of adhesive substances; and, thirdly, for the manufacture of vinegar.

The insoluble residue in the boiler consists of cellulose with the knotty parts of the wood, which knots are not changed by the boiling process while in the boiler, owing to their greater consistency. The fibres, knots, together with particles of bark, are finally removed from the boiler.

The white or nearly white cellulose is obtained by the above process in considerably larger quantity than was supposed to be obtainable from the wood. For instance, from air-dried spruce it is claimed that over sixty-six per cent. of dry cellulose can be obtained. This cellulose may be utilized, either directly or by bleaching with chloride of lime, in the manufacture of paper, and in case of longer fibres, even in the manufacture of textile fabrics.

Dr. Mitscherlich's invention, patented May 4, 1883, has reference to certain improvements in the apparatus for and the method of making cellulose, whereby it is claimed a perfectly

white and tough cellulose is obtained at a considerably reduced cost, to be used as a substitute for the best rags, in
the manufacture of paper.

The invention consists, first, of certain improvements in
the apparatus for boiling the wood with the sulphurous-acid
solution, and, secondly, of a method of treating the wood
stock by first steaming the stock, so as to expel the air from
the pores, then boiling it with the sulphurous-acid solution,
first at a temperature of about 226° F., which is gradually
raised to about 244° F., and finally lowered until the sulphurous acid is entirely driven off.

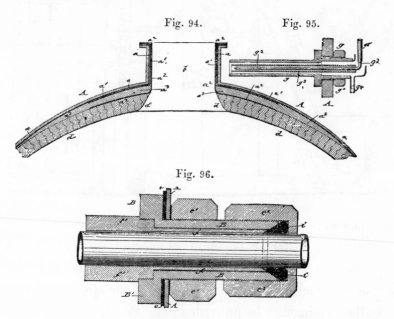

Fig. 94.   Fig. 95.

Fig. 96.

Fig. 94 represents a vertical central section of a part of a
boiler for the wood stock and the sulphurous-acid solution.
Fig. 95 is a detailed vertical transverse section, showing the

connection of the steam-heating pipes with the wall of the
boiler. Fig. 96 is a vertical longitudinal section of a device
for testing the contents and indicating the temperature; and
Figs. 97 and 98 are respectively vertical and transverse sections
of a stamp for disintegrating the boiled wood stock.

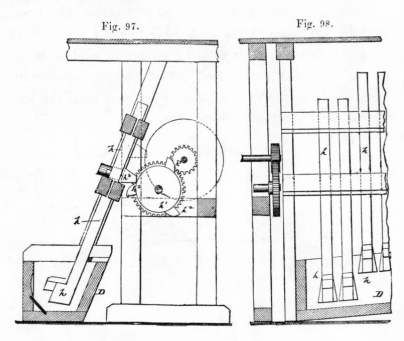

Fig. 97.   Fig. 98.

The wood stock employed for making cellulose is boiled
in an iron vessel or boiler of cylindrical shape, which is
provided with interior protective layers of special construction, by which the corrosive influence of the acid on the iron
walls is claimed to be prevented.

The boiler *A* is made of a very large size, preferably
about twelve feet in diameter and thirty-six feet in length,
so that large quantities of wood stock can be treated therein.

The interior surface of the boiler A is covered with a thin layer, a, of sheet-lead, which is applied at any ordinary temperature to the iron walls upon a layer of cement composed of common tar and pitch. This cement is heated and the lead placed thereon and rubbed down smoothly.

The lead lining should not be too thick, as in that case it cannot be made to adhere properly to the upper part of the boiler-wall, nor can it be properly worked into the different indentations of the iron. A too thick lead lining would form air-spaces between it and the boiler-wall, which would weaken the lining at that point, and would destroy large portions of the boiler-wall by the action of the acid in case of a leak. The boiler would thereby become unfit for use and some of the iron dissolved, whereby the contents would be discolored and otherwise injuriously affected.

By carefully cementing the thin lead lining to the inner surface of the boiler the expensive soldering on of the lead lining, which had heretofore to be resorted to, is done away with, and thereby it is claimed a more efficient boiler lining obtained, by which the security and durability of the boiler are claimed to be considerably increased.

The manholes b of the boiler, which serve for the introduction of the wood stock as well as for the removal of the product obtained therefrom, are closed by suitable lids and are covered, besides the interior layer a, with a second layer, a', of sheet-lead, which is cemented in the same manner over the manhole and extended to some distance from the manhole over the interior surface of the boiler. A third layer, a^2, of sheet-lead is next applied over the second layer a', and extended from the manhole and over the adjoining part of the boiler, a thick layer, a^3, of cement being interposed between it and the second layer a'. These different layers of lead are required at the manholes, as the lead lining a at that point is not protected by the glazed porcelain bricks d, which cover the interior surface of the boiler, as shown in Fig. 94. As soon as the innermost covering a^2 shows signs of wear it is renewed, and thereby the interior surface of the manhole is fully protected.

The contents of the boiler are heated by coils of lead pipes, which extend from the bottom to about half the height of the boiler. Several parallel systems of lead coils are preferably used, so that in case of leakage one or the other coil can be shut off and the boiling operation continued with the remaining coils. For the heating-pipes an alloy of lead and antimony is used, as this resists in a higher degree than pure lead the action of the mechanical and chemical agencies to which the pipes are exposed. The lead coils are connected with the steam-boiler at one end, and at the other end with condensing-chambers, through which at high pressure the water of condensation is forced out. These features are not shown in the drawings but they are mentioned, as thereby temperatures considerably above 212° F. can be readily obtained.

The lead pipes are connected to the boiler-wall by the coupling shown in detail in Fig. 95. A cast-iron sleeve, B, which is flanged at one end and threaded at its exterior surface, is inserted through an opening in the boiler-wall to the interior of the boiler, until the interior flange B' abuts against the wall of the boiler, between which and the flange a lead lining, e, is interposed.   A screw-nut, e', is screwed over the outer threaded shank of the sleeve B tightly against the boiler-wall.   A second flanged nut, e^2, is screwed over the end of the outer sleeve B, its flange pressing tightly on a lead or asbestus ring, i, and forcing it against the beveled end of the sleeve B, as shown clearly in Fig. 95.   The sleeve B is covered at its inside by a lead lining, f, which is screwed by its threaded thicker end f' to the flange B'.   By this construction a very reliable steam-tight coupling of the heating-pipe and boiler is obtained, which is capable of resisting the acid in the boiler.   The boiler is next provided with a device for testing the contents of the digester and for readily observing the temperature and pressure.   These objects are combined in one attachment (shown in Fig. 96), so that only one opening has to be made through the boiler-wall.   The device consists of a fixed tubular socket, g, which is firmly secured by a threaded portion and exterior screw-nut, g', to the boiler-wall.   A thermometer tube, g^2, passes centrally through the socket-tube g, and is protected by a metal sleeve, g^3, between which latter and the tubular socket f sufficient space is left for drawing off a small portion of the contents of the boiler through a suitable valve, g^4.   A thermometer, g^5, is applied to the upwardly-bent outer end of the pipe g^2.   Besides the thermometer g^5 and test-valve g^4, a pressure-gauge may be arranged, and also a gauge for indicating the level of the solution in the boiler.   After the wood has been treated for the proper length of time in the boiler the stock is removed and passed through a stamp-battery, which has for its object to separate the fibres of the cellulose from each other, as well as to wash out the substances incrusting them. This separating and washing apparatus is shown in Figs. 97 and 98, and consists of a number of inclined stamps, h, that are successively actuated by the cams h^2 of the revolving cylinder a', which cams engage projections, h^3, on the shanks of the stamps. The stamps h h move up and down in a trough, D, having a longitudinally-inclined bottom, so as not only to separate the fibres of the wood stock from each other by repeated blows thereon, but also to separate the softer parts of the stock from the harder parts—such as knots—which latter would otherwise be broken up by the stamps, and would deteriorate the pulp. To prevent the breaking of the knots the stamps h are not allowed to touch the bottom of the trough, and are also guided at some distance from the inclined side walls of the trough, as shown in Fig. 97. By this arrangement the stamps exert a pressing and rubbing action on the mass without breaking up the knots. The boiled wood stock is introduced in the trough at the lower point, and the water at the highest point of the trough, so that they pass in counter-directions to each other in the trough. The stamps are dropped alternately in such a manner that the mass is moved in an upward direction over the inclined body of the trough. In this manner the proper separating and washing of the fibres is obtained without injuring them, as with each blow of the stamp the water is pressed out of the fibres directly affected by the blow, which quickly again absorb the moisture, and so on throughout the stamping operation. After the stock is thus properly broken up and washed it is bleached and treated in the usual manner, so as to bring it into proper marketable form.

Operation.—To boil the wood stock the following process is employed: The wood is cleared of bark, cut into small pieces, and charged in a proper quantity into the boiler. It is then steamed in the digester, which steaming, however, has to be carefully watched, as on it depends, in a high degree, the success of the subsequent boiling. The object of the steaming is not to prepare the stock for the chemical action of the acid solution, but to drive out the atmospheric air from the pores of the wood, and give easy access of the solution into the cells of the wood. In this manner not only a more rapid chemical action upon the wood stock takes place, but also, owing to the increased absorption of the solution into the stock, a larger quantity of wood can be charged and treated in the boiler, as the space is more advantageously utilized, and thereby the output of the boiler is considerably increased. The steaming is continued for a greater or less length of time, according to the condition of the wood. If the wood is freshly cut and moist, the atmospheric air is expelled therefrom in a comparatively short time; but if it is dry and hard a longer exposure to the steam is necessary. The expulsion of the air is further· accelerated by the cold acid solution when admitted into the boiler, which causes the quick condensation of the steam and reduces thereby the pressure in the boiler. Care has to be taken that the temperature during the steaming process does not rise above $212°$ F., as practical tests have proved that at a higher temperature the steaming of the wood does not take place in so perfect a manner. This steaming of the wood stock at this stage is entirely different from steaming of the wood before it is ground up, as in the latter case a chemical change is effected and a brownish color imparted to the stock. After the wood has been properly steamed the boiler is supplied with the sulphurous-acid solution. The quantity of organic substances to be worked up has to be in proportion to the degree of concentration of the sulphurous-acid solution, which varies and is dependent upon certain conditions connected with the preparation of the solution. If the proper proportion between the organic substances and the sulphurous-acid solution is not established, and, for instance, an insufficient quantity of organic substances be present, then insoluble salts are deposited in the fibres, which can be washed out only with difficulty. For instance, when bisulphite of lime is used in the solution, sulphite of lime may be formed, which is only soluble with difficulty. If such fibres be worked into pulp, it would produce so-called "knots" in the paper. Furthermore, in bleaching such fibres with chloride of lime much larger quantities of chlorine would be required, so that the expense of the bleaching process is considerably increased. If, however, a too great quantity of organic substances is present in the acid solution, the product does not become sufficiently soft or "opened." The proper proportions have to be determined by a series of tests, which are made by drawing off from time to time small quantities of the contents through the testing-tube described on page 269.

During the boiling process it is necessary to carefully observe the temperature as well as the duration of the boiling. The first stage consists in the slow and gradual chemical action of the solution on the wood, so that that part which has been absorbed by the stock can be replaced. This takes place best at a temperature not exceeding 226° F. After this a quick reaction at a temperature which is gradually increased to about 244° F. has to be produced. Special care is necessary toward the end of the reaction, as this has to go hand in hand with the driving off of the sulphurous acid. By the driving off of the sulphurous-acid solution the chemical action is gradually retarded in the same manner as by lowering the temperature. By properly observing the different stages of the process, by taking small test quantities from time to time from the boiler and mixing them with suitable reagents, the quantity of active solution still present can be readily determined. If, for instance, bisulphite of lime is uséd, and the testing-solution is mixed with ammonia and the precipitate carefully observed, by the ammonia or similar chemical substances a part of the sulphurous acid is retained in solution, while the sulphite of lime in the acid solution is precipitated. The salts which are formed in the due course of the process are not precipitated. From the precipitate the proportion of the effective solution can be readily determined. When the precipitate is only equal to one-sixteenth of the volume of testing-solution the proper time has arrived when the driving off of the sulphurous-acid solution has to be commenced. As the solution passes off with the steam forced into the boiler, the temperature is low-

18

ered to about 223° F., whereby also a decrease of pressure is obtained. If the precipitate in the testing-tube is only one thirty-second part of the testing-solution, the process is fully completed, and the solution has to be quickly drawn off. A still smaller precipitate will prove that the process has gone too far, and that no further organic substances are present, in which case free acid, probably sulphuric acid, would be formed that would impart an injurious brownish color to the organic mass.

To carry on this process with such a certainty through the different stages and temperatures described, which are of considerable importance, a boiler of large dimensions, with the different accessories described, is of special advantage. By a higher temperature the boiling process may be accelerated; but there would also result a higher pressure, and the cellulose obtained thereby would not only be inferior in quality, but also in toughness and quantity.

The chemical change which takes place during the boiling process has already been described.

Francke's Process of Manufacturing Paper-Pulp from Wood, Esparto, Straw, etc.

The invention of Mr. David O. Francke, of Korndal Mölndal, Sweden, relates to the manufacture of wood-pulp from wood, esparto, wheat, maize, or other straw, or from other suitable vegetable fibre. For this purpose Mr. Francke prepares a solvent, which is the acid sulphite of an alkaline earth or of an alkali—that is to say, a solution of such sulphite with an excess of sulphurous acid. As the cheapest and most accessible base, the inventor prefers lime. It has long been known that a solution of sulphite of lime combined with free sulphurous acid will at a high temperature tend to dissolve the undesirable portions of vegetable structures, and leave the fibres in a fit condition for paper manufacture. Mr. Francke, after prolonged and careful experiments on a large scale, claims to have determined the conditions for effecting this with rapidity, and so as to preserve the strength of the resulting fibres and attain a practical and successful method of manufacturing paper-pulp by this means. He employs only a moderate strength of the solution with a high temperature and gentle but constant mechanical agitation. Mr. Francke has devised a method of producing the acid sulphite in large quantities at small cost, and supplying it at a temperature nearly up to that required with agitation for its most effective use. He charges a tower or column with fragments of limestone, which he keeps wetted by a shower of water, and passes through the tower sulphurous-acid fumes produced by burning sulphur or by roasting or calcining sulphides, such as pyrites. The liquid which collects at the bottom of the tower is the solvent required which should have a strength of 4° to 5° Baumé. It is not essential that the limestone should be pure, as mineral containing a proportion of magnesia, or of other alkaline earth or mineral—such as witherite—will answer well; also, minerals consisting, principally, of magnesia or of alkaline earths other than lime may be employed, their treatment being the same as for limestone. The soluble alkalies, soda and potassa, may also be used when their greater cost is not objectionable. For these alkalies the treatment has to be modified as follows: The tower or column is charged with fragments of inert porous material—such as coke or bricks or porous stone—and these are kept wetted by a shower of a solution of the caustic alkali, which solution should have a strength of 1° to 2° Baumé, while the sulphurous-acid fumes are passed through the tower. In like manner the carbonates of soda or potassa may be treated; but when they are employed the solution showered on the porous material should be stronger than that of the caustic alkali, so that it may contain approximately the same amount of actual alkali. Whatever be the alkaline base employed, the liquid collected at the bottom of the tower, having, as stated above, a strength of 4° or 5° Baumé, and being the acid sulphite of the base, or a solution of the sulphite with excess of sulphurous acid, is the solvent which Mr. Francke employs for the manufacture of pulp, as we will now describe: When wood is the material to be treated for pulping, it is freed as much as possible from resinous knots by boring or cutting them out, and is then cut, by preference, obliquely into chips or fragments, which may be from one-quarter to three-quarters of an inch thick. When esparto, straw, or analogous fibre is to be treated, it is cut or chopped into fragments. The fibrous material is charged, along with the solvent, into a strong vessel or boiler, which is heated by a steam casing or coil or by steam-tubes, the steam employed being at a pressure of four to five atmospheres, and consequently capable of raising the solution to the temperature of about 300°

Fahrenheit.   As agitation greatly promotes the pulping
action, Mr. Francke employs a vessel or boiler of cylindrical
form, which is caused to revolve while its charge is under
treatment.   Such a vessel is conveniently made with a
steam-jacket at each end, connected by longitudinal tubes,
the steam being supplied through its trunnions, the steam
entering through one end, and the water of condensation
being removed through the other.

Fig. 99.   Fig. 100.

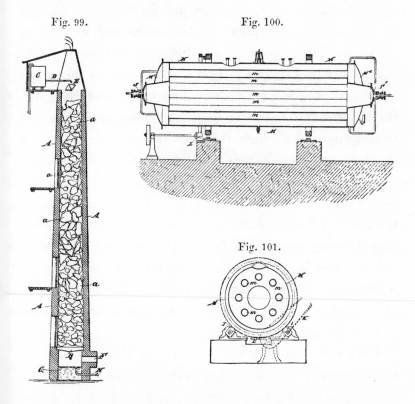

Fig. 101.

Figs. 99, 100, and 101 show one means by which Mr.
Francke, can carry out his process.

Figure 99 is a central vertical section through the tower and its contents employed for the production of the solvent. Fig. 100 is a longitudinal section through the vessel in which the pulp is treated, and Fig. 101 is a cross-section of the same.

A is a tower loosely filled with irregular lumps, a, of carbonate of lime (ordinary limestone), supported on a grate, B.

E is a tilting distributing-vessel, to which water is supplied through a pipe, D, from an elevated tank, C, kept filled by a pump or other suitable means, and allowed to trickle down over the surfaces of the limestone below. There is a furnace (not represented) in which iron or copper pyrites are roasted, the sulphurous-acid fumes therefrom being led through the flue F into the base of the tower A and allowed to move upward through the interstices between the pieces of limestone a, such portion of the fumes as are not absorbed escaping freely at the top. The sulphurous acid is absorbed by the water, which, becoming acid, attacks the carbonate of lime, setting free the carbonic acid and combining with the lime, forming sulphite of lime. The conditions are such that just a sufficient quantity of free acid will remain in the solution, which will accumulate in the tank G at the bottom, and ultimately flow out through the pipe H into any suitable retaining-reservoir. (Not shown.)

I represents anti-friction supporting-rollers mounted in fixed bearings. M is a cylindrical vessel resting thereon, and revolved slowly by a screw, L, operated by gears driven by a steam-engine or other suitable power through a belt, K. The vessel M has tubes, m, communicating between chambers,

M' M^2, in the ends thereof. Steam at a pressure of four or five atmospheres is supplied from a boiler (not represented) through a tightly-packed swivelling connection, J, at one end, and the water of condensation, with a small quantity of the steam, is allowed to escape at the other end, controlled by a suitable valve. (Not represented.) The acid sulphite thus cheaply formed is pumped or otherwise supplied into the vessel M in such quantities that, with the wood, straw, or other material also inserted, the vessel shall be about three-quarters full. Then the orifice through which it is charged and removed being tightly closed by a suitable cover and secured, so as to allow a considerable pressure within, the vessel M, with its contents, is rotated by the gearing, making, preferably, one revolution in about ten minutes. The solution is received warm, and the steam in the pipes m rapidly raises it to a temperature of about 300° F., with the corresponding pressure. The rotation of the vessel gives an efficient but moderate agitation. The proportion of solvent required varies according to the character of the material treated. Mr. Francke states that he found that from two thousand to twenty-five hundred gallons of the solvent would generally suffice for the production of one ton of wood pulp. For esparto, straw, and the like, the quantity of solvent may be somewhat less; but the best proportions are soon learned by experience. The material, having thus in presence of the solvent been subjected to heat and pressure with agitation from twelve to fifteen hours, is withdrawn from the vessel, and, being well washed with water, is in the condition of pulp which is ready for paper-making, but which, when great whiteness is required, may be bleached like ordinary pulp.

Figs. 102 to 105 show a later form of boiler invented by Mr. Francke.

Fig. 102.

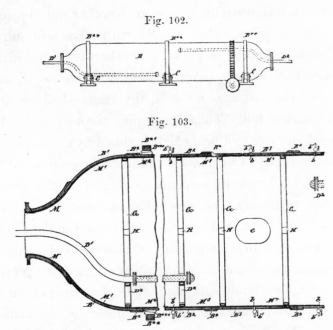

Fig. 103.

Fig. 102 is a side elevation of the boiler. Fig. 103 is a longitudinal section of portions of the boiler on a larger scale. Fig. 104 is a corresponding transverse section. Fig. 105 is a partial section and elevation of a detail. It is on a still larger scale.

The outer shell, B, of the boiler is of steel, made in several distinct lengths or sections (indicated by additional marks, as B' B^2), each united to the next by what is sometimes called a "jump-joint," that is to say, the edge of one section abuts

directly against the edge of an adjacent section, and both
sections are secured by riveting, brazing, welding, or other

Fig. 104.   Fig. 105.

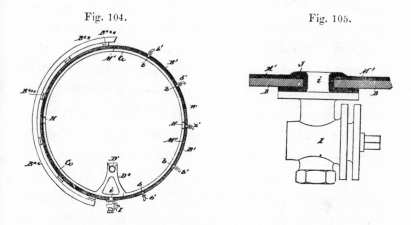

efficient means to an inclosing-ring B^*, of such width as to
lap sufficiently upon each section. The inner shell, M, is of
thick lead, applied in distinct lengths or sections (indicated
by additional marks, as M' M^2, etc.), united along their
several edges. The boiler is mainly cylindrical, but the
ends are partially spheroidal, terminating in trunnions (not
represented), which are hollow and equipped with suitable
stuffing-boxes and steam-pipe connections leading from one
or more boilers (not shown), which supply steam at a high
pressure.

D' D^2 are perforated pipes extending from the trunnions
inward to points near the centre of the boiler, where their
ends are each nearly closed by a perforated plate. These
pipes allow the steam received through the trunnions to be
introduced directly into the contents of the boiler.

Manholes, e, properly re-enforced and equipped with

strongly-secured covers, allow a man to enter the boiler, when required, to effect repairs; but Mr. Francke claims that his present invention reduces the liability of the lead to require examination, and in case of failure at any point directs the workman to the point requiring attention. The lead need not be pure. Ordinary lead of commerce will suffice. It is preferable that the lead shell have a thickness of about one-quarter of an inch. From some cause, probably the difference in expansion and contraction between the lead shell and the outer shell, the lead shell is likely to fail after a period. One steel shell will outlast many lead shells. When it is required to change the lead shell, it is cut out and removed through the manholes or trunnions without difficulty. The introduction of the new shell requires some care, but it can be effected by rolling up the new lead and introducing it through the manholes and carefully unrolling and fitting it to its place and joining the edges. Between the lead M and the steel B is a thin space. The steel shell affords the requisite strength for the entire structure, and the lead shell alone comes in contact with the contents of the boiler. As heretofore worked, the lead lining of boilers is liable to failure by cracking so minutely as to be difficult of detection, but a small quantity of fluid leaking from the interior of the boiler into the space between the head lining and the steel causes great mischief. When the inner lead boiler is made, air will remain between the lead and the steel. When the solution at a high temperature fills the lead boiler, this air expands and after one or two operations breaks the lead M. To avoid this Mr. Francke provides a great number of small orifices, *b*, through the steel plate, to connect with the space between the lead and steel boilers, which orifices let out the expanded air as soon as the heat rises inside the lead boilers, instead of breaking the lead. The orifices *b* serve a double purpose. Sooner or later the lead boiler must commence to give way by cracking. This cracking commences so minutely that it is difficult to detect, and the solution or fluid which escapes through these minute cracks would in a short time consume the steel. Steam penetrates more easily through these cracks than the solution. Consequently when a crack in the lead under the rotation is turned upward, which part of the boiler is empty, the nearest orifice gives or throws out escaping steam, and when turned down shows small drops of solution, in the latter part of the operation, mixed with the dissolved parts of the wood By close observation of the orifices, when steam from inside the lead has shown itself, there is no difficulty in fixing the whereabouts of the crack within a narrow space. If the steam is from the inside of the lead boiler, it smells of the solution. This it does not do if the escape depends on moisture or air between lead and steel.

G are internal bracing-rings, made each in two pieces, and held distended by wedges, *H*, introduced at the joints. These bracing-rings and wedges are made of a composition— as brass—that will be unaffected by the solution, and which will have a greater expansion and contraction than iron. Care is taken to envelop each bracing-ring with a thick covering of lead. These internal bracing-rings are arranged about three feet apart. They press outwardly with sufficient force to support the weight of the upper portion of the lead shell when the structure is empty and cold, and press it firmly against the interior of the iron or steel shell when it is filled and hot.

The apparatus will serve best when the lead shell is made in a continuous band extending quite around the interior of the boiler, and of the proper width to reach from one of the internal braces, G, to the next.

The wedges H, which distend the internal braces G, may be made of considerable length, and, after being driven to set the braces tightly out against the inner face of the lead, their ends should be smoothly cut off by any suitable cutting instrument. It is important to leave no recesses to form a lodgment for the pulp, as a retention and reboiling of the pulp injures its color, and on mixing it with the next batch will injure the whole.

The steam-pipes D' D^2 are supported by brackets, D^*, extending inward from the internal braces G. The latter, being made in sections, can be easily removed through the manholes and new ones introduced and brought to the proper positions when required.

Those portions of the outer steel shell B, which run upon the supporting-rollers C are thickened by the addition of re-enforcing material on the exterior. Mr. Francke states that in his experiments he has used boilers having a length of forty feet with a diameter of seven feet. Such a boiler complete and fully charged weighs about forty tons. He re-enforces successfully at the supporting-point by rings, B^{**}, of cast-iron, made each in a single piece and centered exactly on the boiler by means of wedges, B^{***}

Each of the orifices, b, is provided with a stopcock, b'. Under ordinary conditions these cocks, b', are all open and each is ready to discharge any fluid, whether hot air, steam, or solution, which may seek to issue through the orifice b, which it controls.

I is a sample-cock controlling an orifice, i, made at a convenient point in the boiler, and through which small quantities of the contents of the boiler may be drawn from time to time to examine its condition.   The surfaces of the cock I which are exposed to the solution are lead.   The orifice i is re-enforced by a bushing of lead, J, having a head on the inner face and one on the outer face.   The cock is secured by bolts tapped in holes in the steel shell B.

Operation.—As the boiler is rotated by mechanism (not shown), the contents of the boiler are gently agitated.   The proper valves (not shown) being operated, steam at a sufficient pressure is allowed to flow inward through the pipes D' D^2.   These pipes agitate the contents of the boiler by being traversed through the same as the boiler slowly revolves, and deliver steam through orifices distributed along their whole length.   Thus the heat is delivered in the form of steam, mingling directly with the contents of the boiler and imparting all its caloric thereto, rapidly raising the temperature of the boiler until it very nearly corresponds to that of the steam-pressure employed.   The transfer of heat from the steam to the contents of the boiler results in the production of considerable quantities of water due to the condensation of the steam.   This water becomes added to the contents of the boiler.   It is therefore important at the commencement that the solution be strong, though not stronger than $4\frac{1}{2}°$ to $5°$ Baumé, and not of sufficient quantity to fill the boiler. As the work proceeds, the water, added by the condensation of the steam, increases the volume and weakens the strength of the fluid contents of the boiler. At the close of the operation the boiler will be nearly full. The solution—the acid sulphite of lime—is produced, as has been explained on page 275, by causing sulphurous acid-fumes to pass up through a tower containing carbonate of lime kept wetted with water. The sulphurous fumes are absorbed by the water, making the water acidulous, which then attacks the lime, and in trickling down the tower obtains nearly its equivalent of alkali, leaving just a sufficient excess of the acid. This gives the desired acid sulphite of lime for the proper treatment of the woody matter in the boiler.

By the ordinary methods of treating wood with acid sulphite a large quantity of sulphate forms and remains attached to the fibres of the pulp, which sulphate is practically insoluble, and, adhering to the fibres, remains in the pulp. All known methods of extracting it tend to darken the pulp and make it more difficult to bleach. Mr. Francke is not confident as to the precise chemical reactions occurring, but he claims to have discovered that the absence of highly-heated surfaces reduces the evil. The formation of the sulphate and its disposition upon the fibres depends, probably, on a high temperature throughout the solution; such as is required to effect the heating by metallic surfaces. Mr. Francke's method of heating by direct steam avoids the necessity for any particles of the material under treatment being heated much above the mean temperature of the solution. The steam-pipes are liberally perforated, and allow a perfectly free discharge of the steam. The apparatus, by thus avoiding the presence of any surfaces much hotter than the solution, produces, it is claimed, a pulp having, when dried, not more than one per cent. of sulphate.

Lead is of such a nature that even with the care taken in the present instance to give it expansion and contraction it is still shorter lived than the steel. When a flaw occurs in any portion of the lead lining of sufficient magnitude to induce a visible escape of steam from the nearest stopcock, b', the attendant marks that stopcock and then closes all the stopcocks, thus preventing any serious loss of the contents. As soon thereafter as practicable, a workman enters the boiler and, knowing by the marked stopcock what part of the lead shell is defective, solders or otherwise repairs the defect, introducing a new sheet of lead if required.

We have in Figs. 100 and 101 described a boiler having a portion of each end occupied as steam space with tubes connecting such spaces, and in which steam flowed from one chamber to another. Such apparatus imparted the heat of the steam to the contents of the boiler only through the medium of the metal of the tubes and of the tube-sheets. A portion of the steam was necessarily allowed to escape at the opposite end of the boiler from that through which it was received, in order to maintain the presence of steam on all the surfaces. Mr. Francke's present invention, by introducing the steam directly, imparts the heat of the steam fully and effects the heating more rapidly and with less consumption of steam. It also economizes room by dispensing with the considerable steam-spaces at each end and by dispensing with a large proportion of the tube-spaces. It also (and to this the inventor attaches the most importance) conveys the heat to the solution directly without being transmitted through metal. It thus avoids the presentation to the pulp of any surfaces materially hotter than itself.

It will be understood that the wood or analogous material to be treated, previously made quite fine by mechanical means, is introduced into the boiler with the solution, so as to fill the boiler about two-thirds full, and then it is heated by the direct application of steam to about 300° F., or somewhat more or less, and rotated slowly from ten to fifteen hours, as described on page 279. After being discharged from the boiler, the dissolved material may be removed by washing in a common rag-engine. The pulp may then be either made into paper directly by any ordinary or suitable process, or it may be dried and stored or transported to distant points.

The paper-pulp produced in the present boiler is claimed to be not only relatively free from gypsum, but easily bleached, and even without bleaching, it is said to be of a very light color. It is claimed that it may be used without bleaching for many purposes requiring white or nearly white paper.

Some of the Defects of the Acid or Bisulphite Processes of Treating Wood.

In the acid treatment of wood for the purpose of converting the fibres into pulp for use in the manufacture of paper the general practice has been to use alkaline solutions of soda, combined in various proportions with certain acids—such, for instance, as sulphurous acid, hydrochloric acid, etc. These solutions have been heated in digesting-vessels, and the high temperature resulting from this process of heating developing a pressure of from six to seven atmospheres, the wood being disintegrated by the action of the boiling solution. The gum, resinous constituents, and other incrustating or cementing substances that bind the fibres together are decomposed, destroyed, or dissolved, while the pure cellulose, which constitutes the essential elements of the ligneous fibres, is separated therefrom. To this end high temperatures had to be employed, otherwise the disintegration was found to be only partial, the wood remaining in a condition unfit for further treatment. The high temperature not unfrequently converts a large proportion of the resinous and gummy constituents of the wood into tar and pitch—that is to say, carbonaceous bodies that penetrate into the fibre and render its bleaching difficult, laborious, and costly, while the frequent washing and lixiviation necessary to bleach such products seriously affect the strength of the fibre, its whiteness, and also materially reduce the percentage of the product in some instances as much as eighteen per cent. These difficulties and detrimental results necessarily materially

19

enhance the cost of production, while the fibre itself suffers considerably in strength from the repeated action of the chloride of lime employed in the process of bleaching.

The difficulties are due chiefly to the carbonization of certain constituent parts of the fibres under temperatures exceeding 212° F., such carbonized constituents being insoluble and incapable of being bleached, and as they permeate the fibres cannot be entirely removed.

To overcome these difficulties, the wood should be chemically treated at a temperature sufficiently low to insure that in the solution and decomposition of the cementing substances of the fibres the carbon will remain chemically combined with other elements — such as the hydrogen, oxygen, and nitrogen—in order to obtain an increased product of superior quality, and render the process more economical.

Pictet and Brélaz's Process of treating Wood for conversion into Paper-Pulp, which consists in first subjecting the same to the action of a Vacuum and to that of a sursaturated Solution of Sulphurous Acid at a temperature not exceeding 212° F.

The process invented by Mr. Raul P. Pictet, of Geneva, and Mr. George L. Brélaz, of Lausanne, Switzerland, consists, essentially, in the use of sursaturated solutions of sulphurous acid—say from $\frac{1}{5}$ to $\frac{1}{3}$ lb. avoirdupois of sulphurous acid to a quart of water—employed under a pressure of from three to six atmospheres, and at a temperature not exceeding 212° F. Under these conditions the cementing substances of the wood fibre retain their chemical character without a trace of decomposition of a nature to show carbonization, while the liquor completely permeates the wood and dissolves out all the cementing constituents that envelop the fibres.

In carrying into practice the process invented by Messrs. Pictet and Brélaz the wood is cut into small blocks, as usual, and charged into a digesting-vessel of such strength as will resist the necessary pressure, and of any desired or usual form and material—as, for instance, of iron or steel lined with lead. Water is then admitted to the vessel, and afterward the sulphurous acid from a suitable receiver, in which it is stored in a liquid form, until the proportion of acid has reached that above indicated—namely, from one hundred to one hundred and fifty quarts of acid to one thousand quarts of water. The volume of the bath will be determined by the absorbing capacity of the wood, and is preferably so regulated as not to materially exceed that capacity.

In practice it is preferable to form a partial vacuum in the digesting-vessel, whereby the pores of the wood are opened, when it will be in a condition to more readily absorb the solution, thereby accelerating the process of disintegration. When disintegration has resulted, which generally occurs in from twelve to twenty-four hours, according to the nature of the wood treated, the liquor, which is usually not quite spent in one operation, is transferred to another digester, a sufficient quantity of water and acid being added to complete the change.

In order to remove the liquor absorbed by the wood, the latter is compressed, the digester being connected with a gas-receiver, into which the free gas escapes, and in which it is collected for use again in the operation of disintegration. The bath is heated and kept at a temperature of from 177° to 194° F. by means of a coil in the digester supplied with steam from a suitable generator. The wood, after disintegration, undergoes the usual treatment for converting it into paper-pulp, which may thus be readily bleached by means of chloride of lime.

The unaltered by-products contained in the bath may be recovered and treated for use in various branches of the arts by well-known methods and means.

Marshall's Boiler for Treating Wood for Paper Pulp by the Acid or Bisulphite Processes.

The boiler shown in Figs. 106 to 108 is intended to be used in the manufacture of wood pulp according to the acid or bisulphite processes, and is the invention of Mr. James F. Marshall, of Rumford, Rhode Island.

When wood is boiled with sulphurous acid or similar agents for separating the fibres of the wood, the boilers employed require to be lined with lead in order to protect the iron shell from the action of the acid, and as usually made the sections are transverse and united by horizontal flanges, so that there are about five joints to each boiler.

The object of Marshall's invention is to reduce the jointed surfaces and consequently lessen the liability to leakage, and to that end he forms the boiler with vertical flanges, and packs the joints as hereafter described.

Fig. 106 is a side elevation, partly sectional, of a boiler constructed after Marshall's idea. Fig. 107 is a cross section of the same; and Fig. 108 is a detailed section in larger size of the flange-joint.

The two sections of the boiler A are connected together by means of their flanges, a, which extend lengthwise of the

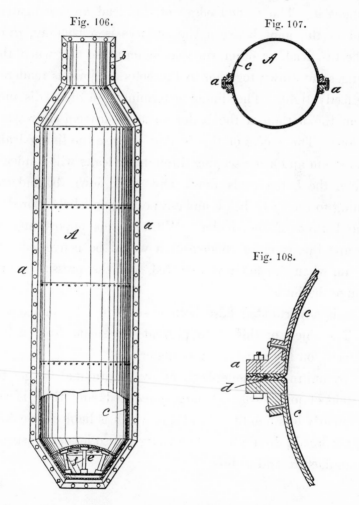

Fig. 106.

Fig. 107.

Fig. 108.

boiler. The upper part of the boiler is drawn inward to form a steam-dome, b, and the bottom is also drawn inward to give support to the lining and also to reduce the area of the false bottom, described hereafter.

c is the lead lining of the boiler, attached and held in place by its edges, that are bent out to pass between the flanges a. The turned edges of the lead are corrugated, and in the joint between the surfaces is a packing, d, of asbestos, lead, or other suitable material, so that when the flanges are drawn together by the bolts the joint is rendered perfectly tight. The lining c terminates a short distance from the bottom of the boiler, so as to leave a clear space below. The object of this is that in case the lining leaks the steam and acid escaping through the leaks will condense when the boiler cools down and work down behind the lining to the space below and escape by small holes bored in the bottom of the boiler. Without this opportunity to escape the water of condensation would be converted into steam when the boiler is reheated, and the pressure would bulge the lining.

e is a perforated false bottom supported by brackets f. The object of this is to prevent the steam from acting directly on the wood or other material.

By uniting the boiler-sections by longitudinal flanges the extent of joint surface is largely reduced, and there is consequently less liability of leakage, which is liable to weaken, if not break, the lining. The boiler is also less expensive to manufacture and to line.

List of Patents for preparing Cellulose from Wood by the Acid or Bisulphite Processes, issued by the Government of the United States of America from 1790 to 1885 inclusive.

No.	Date.	Inventor.
67,941	Aug. 20, 1867.	J. B. Biron.
70,485	Nov. 5, 1867.	} B. C. Tilghman.
92,229	July 6, 1869.	
119,224	Sept. 26, 1871.	A. K. Eaton.
253,357	Feb. 7, 1882.	
Reissue		} C. D. Ekman.
10,131	June 6, 1882.	
263,797	Sept. 5, 1882.	A. Mitscherlich.
274,250	March 20, 1883.	
Reissue		} G. Archibold.
10,328	May 22, 1883.	
280,171	June 26, 1883.	J. A. Graham.
284,319	May 4, 1883.	A. Mitscherlich.
295,865	March 25, 1884.	D. O. Franche.
296,935	April 5, 1884.	C. F. Dahl.
306,476	Oct. 14, 1884.	F Fremerey.
307,972	Nov. 11, 1884.	D. Minthorn.
310,753	Jan. 13, 1885.	G. B. Walker.
329,215 } 329,216	Oct. 27, 1885.	E. B. Ritter and C. Kellner.
331,323	Dec. 1, 1885.	R. P. Pictet and G. L. Brélaz.

List of Patents for Digesters with Lead Linings to be used in the Preparation of Cellulose, issued by the Government of the United States of America, from 1790 to 1885 inclusive.

No.	Date.	Inventor.
238,227	March 1, 1881.	H. H. Furbish.
259,206	June 6, 1882.	A. H. Pond.
265,649	Oct. 10, 1882.	G. F. Wilson.
284,319	Sept. 4, 1883.	A. Mitscherlich.
298,602	May 13, 1884.	J. S. McDougal.
300,778	June 24, 1884.	J A. Hitter.
304,092	Aug. 26, 1884.	D. O. Francke.
304,674 } 304,675	Sept. 2, 1884.	J. A. Southmayd.
305,740	Sept. 30, 1884.	E. H. Clapp.
307,587	Nov. 4, 1884.	G. R. Philippe.

No.	Date.	Inventor.
307,608 } 307,609	Nov. 4, 1884.	C. L. Wheelwright *et al.*
312,485	Feb. 17, 1885.	J. Makin.
312,875	Feb. 24, 1885.	J. F. Marshall.
314,643	March 31, 1885.	T. Alcheson.
328,812	Oct. 20, 1885.	} E. B. Ritter and C. Kellner.
329,214	Oct. 27, 1885.	

List of all Patents for Digesters for Paper Pulp, issued by the Government of the United States of America, from 1790 to 1885 inclusive.

No.	Date.	Inventor.
1,753	Sept. 2, 1840.	} G. Spafford.
Reissue		
171	June 11, 1850.	
4,093	June 25, 1845.	R. Deering, Sr.
6,980	Dec. 25, 1849.	L. W. Wright.
7,497	July 9, 1850.	H. Pohls.
9,910	Aug. 2, 1853.	} J. T. Coupier and M. A. C. Mellier.
Reissue		
1,295	March 25, 1862.	
11,981	Nov. 21, 1854.	W. Watt.
17,387	May 26, 1857.	M. A. C. Mellier.
20,294	May 18, 1858.	M. Nixon.
21,077	Aug. 3, 1858.	A. S. Lyman.
24,484	June 21, 1859.	} J. B. Palser and G. Howland.
Reissues		
996 } 997	July 3, 1860.	
1,590	Dec. 15, 1863.	
2,730 } 2,731	Aug. 15, 1867.	
24,819	July 19, 1859.	A. S. Pitkin.
25,418	Sept. 13, 1859.	M. L. Keen.
26,199	Nov. 22, 1859.	M. Nixon.
27,564	March 20, 1860.	G. Howland and J. B. Palser.
28,062	May 1, 1860.	C. S. Buchanan.
37,846	March 10, 1863.	S. M. Allen.
38,901	June 16, 1863.	M. L. Keen.
40,659	Nov. 17, 1863.	J. B. Fuller.
40,696	Nov. 24, 1863.	A. S. Lyman.
41,812	March 1, 1864.	J. B. Fuller.
42,319	April 12, 1864.	J. Stover.
43,015	Jan. 7, 1864.	J. B. Fuller and J. P. Upham.

No.	Date.	Inventor.
43,073	Jan. 7, 1864.	J. B. Fuller.
44,209	Sept. 13, 1864.	H. B. Meech.
45,791	Jan. 3, 1865.	W. Deltour.
45,849	Jan. 10, 1865.	H. B. Meech.
47,217	April 11, 1865.	T. A. Nixon.
47,539	May 2, 1865.	J. B. Fuller and J. B. Upham.
50,108	Sept. 26, 1865.	J. Evans.
50,266	Oct. 3, 1865.	T. A. Nixon.
50,835	Nov. 7, 1865.	H. B. Meech.
51,430 51,431 51,432 51,433	Dec. 12, 1865.	
51,570 51,571	Dec. 19, 1865.	
51,704 51,705 51,706	Dec. 26, 1865.	J. W. Dixon.
51,813	Jan. 2, 1866.	
52,543 52,544	Feb. 13, 1866.	
52,694	Feb. 20, 1866.	
52,941	Feb. 27, 1866.	J. Eastor, Jr., and F. Thiry.
52,994	March 6, 1866.	A. K. Haxtun.
54,308 54,309	May 1, 1866.	J. W. Dixon.
54,510	May 8, 1866.	J. W. Dixon and G. Harding.
54,932	May 22, 1866.	H. B. Meech.
55,031	May 22, 1866.	H. Voelter.
55,253	June 5, 1866.	J. W. Dixon.
55,418 Reissue 2,383	June 5, 1866. June 23, 1866.	H. L. Jones and D. S. Farquharson.
55,835	June 26, 1866.	J. W. Dixon.
56,832	July 31, 1866.	J. Tiffany.
57,947 61,848	Sept. 11, 1866. Feb. 5, 1867.	H. B. Meech.
63,044	March 19, 1867.	J. R. Haskell.
71,728	Dec. 3, 1867.	A. Fickett.
73,138	Jan. 7, 1868.	J. Tiffany.
80,737	Aug. 4, 1868.	W. Holdman.
84,850	Dec. 8, 1868.	Geo. L. Witsil.
90,566 94,228	May 25, 1869. Aug. 31, 1869.	G. E. Marshall.

No.	Date.	Inventor.
96,237	Oct. 26, 1869.	V. E. Keegan.
106,135	Aug. 9, 1870.	L. Dean.
108,241	Oct. 11, 1870.	A. H. F. Deininger.
108,487	Oct. 18, 1870.	M. L. Keen.
109,595	Nov. 29, 1870.	L. Dean.
110,873	Jan. 10, 1871.	} G. Sinclair.
Reissue		
4,771	Feb. 25, 1872.	
113,502	April 11, 1871.	J. Denis.
114,301	May 2, 1871.	M. L. Keen.
115,327	May 30, 1871.	W. F. Ladd.
116,980	July 11, 1871.	H. B. Meech.
117,427	July 25, 1871.	M. L. Keen.
117,683	Aug. 1, 1871.	W. Riddell.
119,107	Sept. 19, 1871.	B. F. Barker.
119,465	Oct. 3, 1871.	M. L. Keen.
123,757	Feb. 13, 1871.	F. W. Zanders.
124,196	March 5, 1872.	G. Demailly.
128,732	July 9, 1872.	M. L. Keen.
131,794	Oct. 1, 1872.	D. A. Fyfe.
137,484	April 1, 1873.	L. Routledge.
140,333	June 24, 1873.	W. E. Woodbridge.
141,016	July 22, 1873.	L. Routledge.
143,546	Oct. 7, 1873.	A. Ungerer.
148,125	March 3, 1874.	H. J. Lanhouse.
151,127	May 19, 1874.	J. P. Herron.
151,991	June 16, 1874.	A. S. Lyman.
155,836	Oct. 13, 1874.	W. F. Ladd.
166,117	July 27, 1875.	H. Loring.
168,382	Oct. 5, 1875.	J. W. Dixon.
196,965	Nov. 13, 1877.	H. Allen and L. S. Mason.
197,850	Dec. 4, 1877.	W. W. Harding.
206,277	July 23, 1878.	J. Thorpe.
209,179	Oct. 22, 1878.	G. Miles.
212,447	Feb. 18, 1879.	S. and J. Deacon.
234,144	Nov. 9, 1880.	W. R. Patrick.
234,431	Nov. 16, 1880.	J. Saunders.
238,227	March 1, 1881.	H. H. Furbish.
240,318	April 19, 1881.	M. L. Keen.
241,815	May 24, 1881.	H. B. Meech.
246,083	Aug. 23, 1881.	H. Coker.
258,400	May 23, 1882.	H. A. Frambach.
259,206	June 6, 1882.	G. H. Pond.

No.	Date.	Inventor.
259,658	June 20, 1882.	T. Atcheson.
269,649	Oct. 10, 1882.	G. T. Wilson.
276,163	April 24, 1883.	J. W. Dixon.
284,319	Sept. 4, 1883.	A. Mitscherlich.
286,031	Oct. 2, 1883.	G. E. Marshall.
298,602	May 13, 1884.	J. S. McDougall.
300,778	June 24, 1884.	J. A. Hitter.
304,092	Aug. 26, 1884.	D. O. Francke.
304,674 } 304,675 }	Sept. 2, 1884.	J. A. Southmayd.
305,740	Sept. 30, 1884.	E. H. Clapp.
307,587	Nov. 4, 1884.	G. R. Phillips.
307,608	Nov. 4, 1884.	C. S. Weelwright.
307,609	Nov. 4, 1884.	C. S. Weelwright and G. E. Marshall.
312,875	Feb. 24, 1885.	J. F. Marshall.
313,011	Feb. 24, 1885.	B. F. Mullin.
314,643	March 31, 1885.	T. Atcheson.
328,812	Oct. 20, 1885.	} E. B. Ritter and C. Killner.
329,214 } 329,217 }	Oct. 27, 1885.	
329,949	Nov. 10, 1885.	J. F. Quinn.
333,105	Dec. 29, 1885.	C. Bremaker and M. Zier, Sr.

METHODS OTHER THAN THE MECHANICAL, SODA, AND BISULPHITE PROCESSES FOR THE TREATMENT OF WOOD.

Aussedat's Process of Treating Wood.[1]

By this process the wood is disintegrated by means of an injection of steam. The apparatus consists of a vertical boiler, tested at six atmospheres, four and one-half feet in diameter, and about ten feet high; it is closed at its upper part by a manhole used to fill and discharge the wood; it is provided at the lower part with a perforated false bottom upon which the wood rests, so that there remains between the

---

[1] Dictionnaire de Chemie, Wurtz, tome ii. p. 749 *et seq.*

false and true bottoms of the boiler a space sufficient to contain the condensed steam, which can be discharged by means of a cock or valve, according to the requirements of the work.

Another valve is placed upon the manhole, to be used at the end of the operation to discharge the non-condensed steam. The boiler is supported upon two hollow lateral axles, resting in suitable boxes, and used for the admission of steam. The wood is piled in the boiler so as to occupy the least possible space; the filling of the boiler being completed the manhole is closed. The discharge valves being closed, the injection of steam is proceeded with. The steam must be as dry as possible, and it must be admitted gradually. This last condition is essential, if a supple and strong pulp is desired, and to prevent the wood from getting black. For this reason the steam-cock is only one-third opened, and it is then regulated so as to attain, after three or four hours, a temperature corresponding to five atmospheres, which temperature must be maintained for about one hour.

During the operation the condensation chamber must often be emptied, as the least contact of the wood with this water would blacken the wood; furthermore, the rush of steam which takes place at the moment of the opening of the discharge valve facilitates the disintegration, in removing, either by dissolution or by displacement, all the gummy and resinous matters which fill the cells of the wood. These condensed liquids are heavily colored during the first part of the operation. It is advisable to open for a few minutes the valve placed on the manhole, so as to insure a convenient distribution of the heat; the air thus being evacuated a free circulation of steam is insured.

The time required for the injection varies according to the wood to be treated; it takes three hours for white woods, poplar, aspen, birch, etc., and five hours for hard and resinous woods.

The steam valve being closed, the two valves on the boiler are opened in order to completely discharge the steam and not to allow it to settle upon the wood. Everything being cooled the discharging of the wood is proceeded with.

The wood is then of a reddish color, more or less dark, according to the nature of the wood treated and to the pressure of the steam injected. The higher the pressure used, the darker will be the wood.

A much clearer shade in the wood is obtained if a pressure of three atmospheres is not exceeded and the duration of the injection prolonged; but the disintegration of the wood is not so complete and the subsequent operations for the trituration will be rendered longer and more difficult.

When, for the injection, steam is introduced in a boiler the pressure of which is already at five atmospheres, great precautions must be taken to prevent the too prompt heating of the wood. The admission valve must be operated with intelligence. It is best to make steam in proportion to the consumption and to start the fire under the steam generator when the filling of the boiler is commenced; the greatest drawback to be feared in the management of the operation is thus avoided.

The wood is charged into the boiler for injection in any shape, with or without the bark, without it being necessary to remove the knots or the rotten parts; the bark is sufficiently softened by the injection, while the knots, etc., are removed by the condensing water. Still, it is better to remove the bark than to allow it to remain, but, then, its removal is done so much more easily after the injection than before. The wood is generally used in the shape of logs three feet, of any diameter; chips, shavings, wastes from saw-mills, packing-shops, floor-joiners, etc., can all be utilized.

The trituration succeeding the disintegration is subdivided into two parts: the crushing and the refining.

The wood, which has been injected in the shape of logs, is cross cut in sections $\frac{3}{4}$ of an inch thick by means of a circular saw. The sawdust produced by this operation is very fibrous: it may be worked into pulp, but as its refining takes a long time it is generally burned.

The production of this sawdust may be avoided by submitting the log at its end to the action of a chipper of the style used to crush resinous barks or tincture woods; but its reduction in disks gives a more uniform pulp, while the products obtained through the use of a chipper are very uneven. Furthermore, we may, with the disk, and when the disintegration has been well performed, obtain long or short fibres by varying the thickness of this disk, which is of great advantage, considering the commercial value of the product.

The disks are crushed by means of the crushing-machine

invented and patented in France by Mr. Iwan Koechlin and operated for the first time at the Isle Saint Martin.   This apparatus is essentially composed of a vertical shaft upon which a burr-pestle of a special pattern is mounted, which drags, breaks, and crushes the wood against the sides of a fixed envelope or shell, the interior of which represents, in inverse sense, the relief of the burr-pestle.   It is, in one word, a kind of coffee-mill.   One of these crushers, requiring three horse-power, will prepare about 75 lbs. of wood per hour.   According to the opinion of manufacturers who have these machines in use, they answer the purpose very successfully.   Mr. Roger, machinist at Epinal, France, built them.

The crushed wood is mixed with water in sufficient quantity, in the agitating boxes, before its passage through the mill.

The mills used and patented in France by Mr. Aussedat, are provided with conical millstones.   The opening in the lower millstone is more inclined towards the horizontal line, in order to give passage to the material.   Burr-stones give excellent results, but they should be of one piece, as the water rapidly disintegrates the plaster which is commonly used to bind the fragments together ; when whole stones cannot be had, the fragments should be united with cement. Granite-stones yield less product, and furthermore, they wear off rapidly and require frequent dressing.

These pulps are especially used for fine card-board and wall papers.   The cream tint of the prepared wood presents to the eye a very agreeable ground, much appreciated by the wall-paper manufacturers, as the colors show upon this ground in fresher and sharper lines than upon any other paper. Still, it is possible, that some economical method may be found to bleach the product of the Aussedat process. The problem does not offer as many difficulties as the bleaching of the Voelter pulp, as the larger part of the sap and the rosin have been displaced by the dissolving action of the steam.

It results from experiments made at the Onnonay laboratory, and which Mr. Bourdillat has kindly communicated, that the manipulation by steam causes a mechanical and a chemical action; it traverses the cellular tissue of the wood, dissolving and expelling the larger portion of the gummy and resinous substances which fill the cell. Furthermore, the heat disengages a certain quantity of acetic acid, the action of which adds to that of the steam, acting especially upon the incrustating matter.

The following experiments have been made to bleach the more or less deep reddish pulp:—

1. The use of hydrochloric or azotic acids in the boiling removes from the pulp a large portion of the incrustating matter: thus treated, the pulp presents a clearer tint after washing, but it is susceptible to the action of hypochlorites only when operated upon in highly concentrated baths and under the influence of heat. The hypochlorite baths used must be at least of 400 chlorometric degrees. (French.)

2. The fermentation, produced by the addition of a certain quantity of beer yeast to the pulp, previously slightly acidulated by sulphuric acid, produces a clear enough pulp, if the

fermentation lasts only a few days; the pulp will be gray if this action is unduly prolonged.   In the first form the pulp is not sensibly susceptible to the hypochlorites, while in the latter the bleaching is effected tolerably with 400° (French) baths.   The large waste resulting from this method, and the considerable time needed to complete the action of the fermentation, make this process impracticable.

The caustic alkalies have not given good results, the coloring matter becomes of a darker hue, and resists more energetically the action of the hypochlorites than when the pulp has been previously treated.

Bachet-Machard Process of Disintegrating Wood.[1]

Messrs. Iwan Koechlin & Co. have carried on the Bachet-Machard patent at the Isle Saint Martin, near Chatel (Vosges), France, and it has also been experimented with on a large scale at Bex and at Saint Tryphon, Switzerland.   At the start, the inventors had in view the saccharification of wood, the paper-pulp being intended to be only a secondary product of the manufacture of alcohol, but in practice the inverse result has been obtained; the paper-pulp becoming the principal product and alcohol the secondary one.

The wood, previously sawed in thin disks, was thrown in tubs, the filling of which was then completed with water and sulphuric acid; the latter in the proportion of one-tenth. Each tub would contain 188 cubic feet; 18 hours' boiling was needed; the disks were then washed as well as possible

---

[1] Dictionnaire de Chemie.   Wurtz, tome ii. p. 749 *et seq.*

20

in order to eliminate the acid, then passed through the crushers and the mills.   Each $31\frac{1}{3}$ cubic feet produced about 330 lbs. of dry pulp; 65 lbs. of acid and 136 lbs. of coal were used for the production of 220 lbs. of pulp.   Calculating the value of the wood at $3\frac{8}{10}$ cents per cubic foot, the cost of production of 220 lbs. of pulp would be $1.95.

With the Bachet-Machard method a brown pulp is obtained, producing a good brown folding paper costing about 90 cents per 100 lbs. dry pulp.   This brown pulp is easily transformed, by a half bleaching, into a blond pulp costing about $2 per 100 lbs., and which can be utilized, with or without mixing, for the manufacture of wrapping-paper and of all the colored papers.   Up to this time a method for economically transforming this blond pulp into white pulp has not been found.

The inventors think that the tenth of acid, which they cause to react at 212° F. upon the wood, saccharifies the ligneous, or rather the incrustating substance, without touching the cellulose fibres.   Thus the cellulose becomes easily separated into fibres by mechanical means.   It is probable that the acids modify the incrustating substance and render it friable, and that at the same time certain principles of the wood are converted into glucose.

The process is the same as with straw and esparto, when alkaline washes are used, but it requires more energetic boiling; the proportion of alkali is doubled and the boiling done at a pressure of 165 lbs.

A little more chlorine is also required for the bleaching.

The yield of esparto is . . . . 48 per cent.
    "   rye-straw . . . . 42 "
    "   wheat-straw . . 40 "
    "   oats-straw . . . . 36 "
    "   barley-straw . . . 32 "
    "   buckwheat-straw . . 26 "
    "   pine-wood (the most used) . 30 "

As per quality, the succedaneous pulps may be classified in the following order:—

1, Esparto-pulp; 2, wood-pulp; 3, rye-pulp; 4, wheat-straw-pulp; 5, oats-straw-pulp; 6, barley-straw-pulp; 7, buckwheat-straw-pulp.

Treating Wood with Aqua Regia.

A. Poncharac, near Grenoble, uses aqua regia (nitro-hydrochloric acid), cold, for disintegrating the wood; 94 parts of ordinary hydrochloric acid and 6 parts of azotic acid are employed in earthen vessels of a capacity of 175 gallons. It is allowed to soak from 6 to 12 hours. 132 pounds of aqua regia are required for 220 pounds of wood. It is a dangerous process. When it is desired to operate with hot liquids, 6 parts of hydrochloric acid, 4 parts of azotic acid, and 240 parts of water are used, in granite tubs, provided with a double bottom; it is then heated by the admission of steam during 12 hours, then washed and crushed.

Treating Wood with Ammonia, etc.

It has also been proposed to boil the wood in ammonia, in a closed vessel, heating by steam by means of a worm. 900 gallons of ordinary ammonia are used to disintegrate

3300 pounds of wood.   65 to 75 pounds of caustic soda are added, and the operation is followed by bleaching.

There has been mentioned a process patented in France by Mr. Tessié du Motay.   The wood is treated under pressure with alkaline liquids, then bleached by means of manganate of sodium.   The incrustating substances dissolved in the lye may be separated from it by a stream of carbonic acid; the alkali is thus regenerated at little cost.

It has also been proposed to replace the caustic alkalies with sulphuret of sodium, which would act like them, but which would offer the advantage of an easy regeneration of the active agent, as it would be sufficient to evaporate and calcine the lyes in order to destroy the organic matters and to recover the primitive alkaline sulphuret.

CHAPTER X.

WASHING RAGS—WASHING WASTE PAPER OR "IMPERFECTIONS"
—WASHING STRAW—WASHING WOOD PULP—WASHING AND
POACHING ESPARTO—WASH WATER—LIST OF PATENTS FOR
PULP-WASHING AND STRAINING.

WASHING RAGS.

THE next process to which the material to be reduced to
paper pulp is subjected is that of washing.

For lower grades of paper, such as wrapping, etc., the
rags (of which we shall first particularly speak) or other
materials are washed and beaten into pulp in one engine;
but for the production of "half-stuff" or bleached pulp
separate engines are commonly employed in the United
States in which to accomplish the washing and the beating.

The rag engine is commonly known as "The Hollander,"
from the fact that it was invented in Holland about the
middle of the eighteenth century. Prior to the invention
of the rag-engine, rags were reduced to pulp by stamps or
beaters acting in mortars, the contrivance being not unlike
the stamping mills used for reducing ores to powder; but
as it would require about five thousand of these stamps to
supply a modern Fourdrinier machine of average width and
speed, it will readily be seen that the enormous modern
development of the paper-making industry is largely owing

to the Dutch invention which made it possible to use the
paper machine invented by the French workman Robert,
who was employed in the paper-mill at Essone in 1798,
and which machine is now commonly known as the
"Fourdrinier."

The machines used for washing and beating are almost
similar in their construction; in the rolls of the washing
engine, however, there are usually only two bars to the
bunch, while in the rolls of the beater there are usually
three bars to the bunch.

Fig. 109.

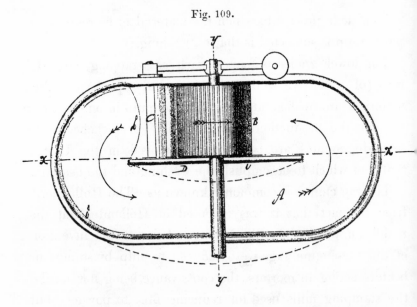

The rag engines employed in the mills of the United
States vary in size and details of construction. In Figs.
109 to 111 the principal parts of the rag-engine are shown.
Figure 109 represents a top or plan view of the engine.

Figure 110 is a vertical longitudinal section of Fig. 109, through the line *x x*. Figure 111 is a vertical cross-section of Fig. 109, through the line *y y*.

Fig 110.

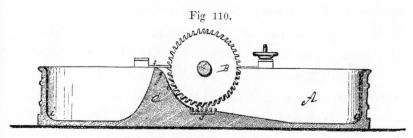

Fig. 111.

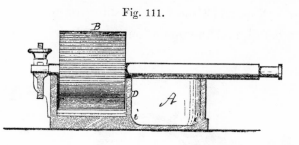

A represents the tub, trough, tank, or vat, which may measure about 12 feet long by 6 feet wide and 2 feet deep.

B is the cylinder or roll fitted with bars which revolve at a high rate of speed on the plate *J*, also furnished with bars; the term "bars" being the technical name for knives. *C* is the back-fall; and *D* the mid-fellow.

As these engines were made about fifteen years ago, the back-fall *C* was carried up to an angle at its top by continuing the curve *e*, and then dropping back directly, as at *f*. From the point *f* the back-fall sloped down to the bottom of the tub, leaving sharp angles on each side. These sharp or right angles were continued around the mid-fellow and

around the tub, allowing the fibre of the half-ground pulp
to catch and hang in the angles or corners, thereby obstruct-
ing the current, while the fibre which thus caught in the
corners would not receive its proper share of grinding, but
being mixed with the rest of the pulp, the result was pulp
of uneven fineness, which is very detrimental to good paper.

To remedy these evils, the top of the back-fall was curved
or rounded off as seen at g, and the back slope curved transversely, as seen at h, and the corner or angle of the tub and
that of the mid-fellow were rounded, as seen at $i\ i$.

Figs. 112 and 113 show the washing or breaking engine
more in detail, and the operation is as follows :—

Fig. 112.

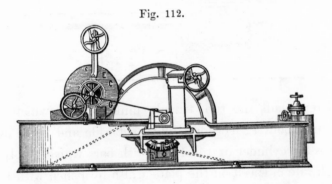

The tub of the engine should be half filled with water,
which is admitted through the valve D, after which the rags
or other materials are taken from the trucks on which they
come from the boiler and are gradually introduced into the
engine. When the proper quantity of boiled rags or other
material has been placed in the engine the operation of
washing is commenced, and the roll B is let down just
sufficiently to open up the rags and allow the dirt to escape.

The rags or other material to be washed should not be introduced into the trough of the engine in such quantities

Fig. 113.

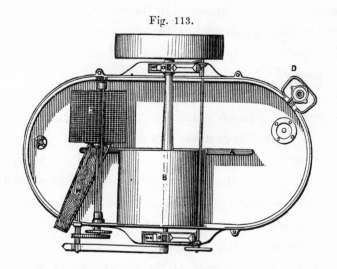

as to be so thick that difficulty will be experienced in turning it.

When the engine is started the stirring stick should be used directly above the sand-trap, and around the sides and back-fall of the engine. The object in thus using the stirring stick is to prevent "lodgers," or pieces of rag not reduced to half-stuff, from hanging or catching about, which pieces if not forced to travel with the current cause knots and gray specks in the finished paper.

The water introduced into the trough of the engine is withdrawn by the washer E, which consists of a drum about three feet in diameter and of such width as to allow a space of about two inches on each side between it and the sides of the engine. The periphery of the washer E is covered with fine wire-cloth, and in the interior of the drum there are arranged buckets indicated by the dotted lines *G*. The washer *E* is partly immersed in the water and material contained in the trough, and as the drum revolves the buckets *G* lift the water into a conical pipe and discharge it through the spout *H*.

In regard to the time which the rags or other materials are treated in the washing and breaking engine, it is of course not possible to apply any fixed rule, as the duration of the treatment varies with the capacity of the trough, the weight of the roll, the number of its revolutions per minute, the extent to which its bars and those of the bed-plate are worn, the quantity and nature of the water used in washing, the nature of the rags or other material being washed, and the skill and experience of the workman who directs the operation.

The quality of the paper depends largely upon the knowledge possessed by the workman having charge of the washing and breaking department of the mill, and an experienced man is generally known by the cleanliness of his surroundings.

When the water from the washing engine runs off clean the roll is lowered upon the bed-plate so as to disintegrate the rags or other material being operated upon, which gradually lose their compact or textile appearance, and are converted into a substance greatly resembling that of a fine, long-fibred lint saturated with water, the new substance being more or less white, according to the nature of the material from which it has been produced.

In mills where the number of washing and beating engines is limited it is often necessary to hasten the process of reducing the rags to half-stuff, but such haste must necessarily be at the expense of the quality of the product, and it is seldom that the work of reducing rags can be properly accomplished in less than two or three hours.

It is necessary to continue the reduction of the rags for a longer time when the bleaching is to be accomplished with chlorine gas than when liquid chlorine is to be employed.

The old custom of bleaching in the washing and beating engine is almost exclusively employed in the mills of the United States; but the plan is not a good one, as a larger number of engines are required, and the metallic tubs are more or less corroded by the action of the chlorine and sulphuric acid employed to neutralize the agent used to accomplish the bleaching, and this is especially true of the bars of the roll and the bed-plates which are not protected by a coating of paint as is the case with the interior of the trough.

When the stuff is in condition for emptying into the drainers or into the bleaching cisterns, the discharge valve should be carefully drawn and deposited on the floor until the engine is empty.

Before replacing the valve the workman in charge throws a few buckets of water under the cylinder in order to remove any of the half-stuff remaining adhering to the back-fall.

The valve should also be carefully washed before it is replaced, as dirt and sand are always lodged in the hole on the top of the valve, and when the valve is carelessly drawn these impurities escape with the stuff.

The sand-trap plate should next be lifted and all impurities carefully removed, after which the plate is replaced and the engine is ready for the reduction of a new lot.

If the paper is to possess the requisite strength care must be observed not to reduce the rags to half-stuff too quickly, as in such case the washing will not be properly done and the material will be weakened; but if attention and time are given for properly drawing the stuff into fibre there results no injury to its texture, and a stronger and better paper can be produced.

It is desirable to keep a register of the daily operations in the engine-room, in which the time of commencing and completing each operation may be recorded, and for night-work, when the superintendence is less regular than during the day, such a record is especially desirable.

The average waste resulting from washing, boiling, and reduction of the rags to half-stuff can be only approximated; Prouteaux gives the following figures, which are probably from 10 to 15 per cent. too high:—

Whites, fine, half-fine	7 to 10 per cent.
" coarse	9 to 13 "
Cottons, white	7 to 9 "
" colored	8 to 14 "
Thirds and pack-cloths	18 to 26 "
Ropes	20 to 25 "
" tarred and containing much straw .	20 to 35 "

The pipe which supplies the water to the washing or breaking engine varies in size from 3 to 6 inches in diameter, the diameter depending upon the pressure of the water, the size of the engine, and the capacity of the washers, and, as it is necessary to mix the water with the pulp as quickly as the action of the cylinder will allow, the water is admitted at the end where the rags ascend.

There are two systems in use in regard to the manner of admitting the water to the tub of the engine; one method being to allow the water to flow in from the top, and the other plan being to let it in at the lower part of the tub.

When the supply of water is introduced from the top the flow can always be observed, and flannel bags or other additional filtering arrangements can be more conveniently attached to the mouth of the supply pipe.

But if the water is naturally very pure or has been carefully filtered, its admission from the bottom of the tub furnishes an easy mechanical means for keeping up the stirring of the contents of the tub, removing "lodgers," and forcing the dirty water to the top, where it is removed by the washers.

Washing Waste Paper or "Imperfections."

Scrap-paper, be it newspaper, letter-paper, or book-paper, when used as a stock from which to make paper, is by the paper-maker termed "imperfections."

The boiled papers are conveyed on trucks from the open boilers, from which they are lifted on the false bottom and are supplied to the washing engine, which should be provided with blunt bars, and abundantly supplied with clear wash water.

When the tub of the engine has been properly supplied

with the desired quantity of imperfections the cylinder is raised and the washers made to revolve until the water runs off clear, when the cylinder is lowered and the papers brushed out.

When the imperfections are boiled, rags and threads always make their appearance in the stock, no matter how carefully the papers have been sorted. Sometimes papers are pasted on one or both sides of a body of cloth, which does not make its appearance until the paper has been separated from it, either by the action of the water when heated in the boiler, or by the brushing in the engine, and small threads are often overlooked on account of their fineness.

A rack, constructed similar to those used for water-wheels, and placed across the tub of the engine, between the mid-fellow and front side, where the pulp begins to ascend the cylinder, has been recommended for catching the rags and strings.

The upright teeth which form the rack are fastened in a frame of hard wood which hangs in boxes on the mid-fellow and front side of the engine. The lower part of the frame rests on the floor of the engine.

The teeth should be sufficiently strong to resist the pressure of the pulp, and are arranged about $2\frac{1}{2}$ inches apart, and should be about $\frac{3}{8}$ of an inch thick at the back end.

It is desirable to have the edges of the teeth which come in contact with the pulp either sharp or pointed, as the strings then catch on easier. Metal teeth, which become polished and allow the strings to slip off, are not as good as wooden ones.

When it is not desired to use the rack it is raised and held above the tub by placing under the rack a stick or paddle, one end of which rests on the mid-fellow and the other end on the front side of the tub.   But when the imperfections have been sufficiently reduced to pass through the rack, the stick or paddle is removed and the rack lowered into working position, and it must be frequently raised to remove the rags and strings from the teeth.

The pulp, after being properly reduced, can be bleached in the engine in the same way as rags, or by such other methods as may be desirable.

Low-priced paper is produced from one class of imperfections, and the entire operation can be finished in one engine; but if it is desired to produce a paper of the best quality from different classes of waste papers by mixing the various pulps, it is much better to empty the bleached pulp into drainers.

When it is possible to obtain fine blue letter-paper in large lots such material should not be bleached, but only washed, as its pulp furnishes a coloring material which can often be used in lieu of ultramarine.

The difficulty commonly experienced in repulping paper-stock—that is, clippings and scraps of paper—arises from the breaking or shortening of the fibres, which are thus made so short that they will not unite to form a sheet of adequate strength.   The paper having been previously hardened and toughened by the admixture with the pulp of size and other

substances, it is found that the necessary grinding and disintegrating required to pulp it break and destroy the fibre.
It has been customary, therefore, to mingle with the paper-
stock a quantity of rags, and to reduce the whole to pulp by
grinding them together. Thus the paper-stock is employed
merely as a filler, the fibre being supplied by the rags.

Mr. Charles Coon, of Saugerties, N. Y., by a process invented by him, claims to preserve the fibre of the paper-stock
in repulping it, so that the stock may be employed alone or
without the admixture of rags or other fibrous material in
the manufacture of paper. To this end it is necessary to remove the sizing which firmly binds the fibres together, after
which it is claimed they will readily separate without breaking under the subsequent operation of the beater or pulper.

When proceeding according to Mr. Coon's method, first
place the paper-stock, which usually consists of cuttings,
clippings, and waste, in an upright tank or vessel having a
perforated false bottom, and add to it a solution of pearlash
of about six degrees strength, in the proportion of about two
gallons to six hundred pounds of the dry stock. Water,
either hot or cold, may be added to the dry stock before the
steam is admitted, if desired. The vessel is then covered
and hot steam admitted under the false bottom of the vessel
for about four hours. This treatment causes the sizing and
other substances to separate from the stock and to rise to the
surface of the water of condensation which will have accumulated in the vessel. The steam is now shut off, and water
(either hot or cold) is admitted at the bottom of the vessel
until the water in the latter rises, bearing the sizing, etc.,

on its surface, and overflows the top of the vessel or passes off at a waste outlet. This the inventor terms "floating" the size. When the surface impurities are thus removed the incoming water is cut off and the water in the vessel is allowed to drain off at the bottom. The stock thus treated is termed "water-leaf," and contains little or no sizing or other substances which would cause the fibres to adhere.

In floating off the size after the steaming operation, it is preferable to admit warm or hot water at the bottom of the tank, although the hot stock may be sufficient to warm it. Care should be taken that the size be not chilled, as it will set and be difficult to remove.

The next step consists in removing the water-leaf to the beater, where it is placed in water having a temperature of from 120° to 150° F., and to which has previously been added two gallons of solution of pearlash of about 6° strength, which serves to remove or destroy all the size that may remain in the stock, and leaves the latter in condition to receive the bleach, color, etc.

The temperature of the water or solution in the beater should not exceed 150° F., as a higher temperature is liable to thicken or cook the sizing materials employed in preparing the pulp for use. For pulping soft material 120° F. is sufficient. It is preferable to employ this temperature for rag-stock fibre, while for grass or wood fibre 140° F. is preferable.

It is best to employ pearlash as the best form of potash, as caustic potash appears to weaken, rot, or burn the fibres, and they break and become too short in the operation of

21

pulping. The pearlash solution should be of sufficient strength to remove the sizing and free the fibres, and this will vary somewhat with the kind of stock employed, the softer kinds requiring a little weaker, and the harder kinds a little stronger solution than that named; but a slight experience will enable the operator to readily determine the proper strength for his purpose.

When the water-leaf is placed in the beater the pulping proceeds, and while it is in progress the chlorine is added, which bleaches the mass in about fifteen minutes. The chlorine being employed while the stock is hot, the bath in the beater being kept at from 120° to 150° F., it accomplishes its work and passes off with the steam and vapor, leaving no traces behind. Consequently it is not necessary to employ anti-chlorine to remove it, as is ordinarily done.

By the ordinary method the bleaching is commonly done while the stock is in the washer and known as " half-stuff," and from the washer it is let down into the drainers, where it is allowed to remain eight or ten days. The anti-chlorine is added when this half-stuff is removed to the beater, and as cold water is usually employed in beating or pulping, the chlorine is generally only partially removed, and the paper made from the stock is apt to turn yellow with age.

In the present process the hot pearlash solution acts in concert with the chlorine, so as to accelerate the bleaching of the stock or pulp and to dissipate the chlorine, as above stated.

After the stock has been reduced to pulp in the beater it may be sized, colored, etc., in the usual way.

Waste papers are sometimes washed in a circuit-vat furnished with a paddle-wheel and a rotary washer such as is shown in Fig. 69.

WASHING STRAW.

After the straw has been boiled it is sometimes washed by emptying it from the boiler into a vat or tub of suitable capacity, which should have a false perforated bottom. Before the straw is emptied into the vat the perforated bottom should be covered with a piece of coarse bagging or cocoa-matting, which will allow the liquid to escape. Hot water under a strong pressure should be introduced beneath the false bottom, and after the water has risen and become sufficiently mixed with the pulp the contents of the tub should be stirred with paddles and then allowed to drain ; the operation being repeated until the pulp has been sufficiently washed.

If possible, the straw pulp should be washed in the washing engine in preference to the tubs just described ; but in some mills the tubs and washing engine are both used, the pulp being pumped from the tubs to the washing engines while it is in a fluid state, or if it is allowed to drain in the tubs the pulp is conveyed on trucks to the engine.

If the straw has been properly digested the bed-plate of the washing engine should be smooth and the bars of the cylinder should be blunt, as it will not be necessary to subject the pulp to any further disintegrating action.

The first washing of the pulp in the washing engine

should be done with hot water in order to thoroughly wash out any alkali so as to avoid loss of chlorine in the bleaching. After being washed the pulp is either emptied into drainers, and after being removed treated like rag pulp, or it may be emptied into a large stuff-chest provided with a suitable agitator from whence it is run over a wet machine for the purpose of removing sand, knots, and other parts of the straw which have not been thoroughly digested.

The bleaching of straw pulp is usually accomplished in the washing engine in the same manner as rags.

Washing Wood Pulp.

Mechanically prepared wood pulp is simply added to and thoroughly incorporated with the rag pulp in the beating engine; but as it is necessary to reduce it to the finest fibres it is subjected to the action of the cylinder for about one and one-half hour.

Chemically prepared wood pulp after being emptied out of the digester into the discharge tank is allowed to drain, and the liquor is saved to be used for the first wash to which the pulp is subjected while in the digester, as has been explained on page 249. When the fibre has been allowed to properly drain in the tank the drain-cock is closed and warm water is run in until the tank is full, and after it has been allowed sufficient time to penetrate the fibres the drain valve is again opened and the water drained off, and afterwards the tank is again run full of water and then drained as before.

After being washed in the tank as has been described, the pulp is allowed to drain as dry as possible after the second washing and is then transferred, either by a centrifugal pump or by other suitable means, to the washing engine, which should have a smooth bed-plate and blunt bars on the cylinder.

If the pulp has been properly prepared but little washing will be necessary in the engine, and any chips which are in the pulp will be readily reduced to fibre by the action of the cylinder.

After the washing is completed the valve on the water-supply pipe is closed and a sufficient quantity of water is removed by the washer to make room for the bleaching liquor, which is then introduced at a temperature of about 100° F., and the engine run until the desired color is obtained, after which the pulp is run into the drainers.

If the pulp is to be used at the mill where the fibre is produced it is desirable to run it from the washing engine into a large stuff-chest, such as has been mentioned for straw, from whence it is passed over a wet machine before bleaching; but if it is intended to make the fibre into dry rolls for transportation the screening can be accomplished during the passage over the machine as it is being fabricated into rolls.

WASHING AND " POACHING" ESPARTO.

The washing of esparto is a simple operation if the treatment of the grass in the boiler has received proper attention. The tub of the washing engine is half-filled with water, after which the grass is introduced and run for about twenty or thirty minutes.

Esparto is not commonly bleached in the washing engine,

but is passed in a "poacher," which is arranged on a slightly lower level than the washer.

The poacher or the "potching engine," as it is also termed, is larger than the washer, and instead of the cylinder and bars it has a hollow drum which carries on its periphery a number of cast-iron paddles which thoroughly agitate the pulp.

In introducing the half-stuff from the washer to the poacher care should be exercised to keep the quantities as near uniform as possible, as irregular bleaching will result if the quantity of stuff filled into the poacher is changed and the bleach is not varied accordingly.

A finer wire is used on the washer of the poacher than on that of the washing engine; the washing in the poacher being continued until the esparto is thoroughly washed, after which the bleaching liquor is introduced in the desired quantity, the washer of the poacher being raised before the bleach is put in.

After remaining in the poacher usually for about two hours the pulp is emptied into stone chests or drainers, which have each a capacity for containing two engines of the bleached stuff, where it remains usually for about eight hours.

The stone chests are commonly placed in an exposed position, as it is thought that the action of light assists the bleaching.

The chests are supplied with two perforated zinc drainers, one extending up the back of the chest and connecting with the second, which is placed on the bottom of the drainer.

In some cases the washing and bleaching are done in the washing engine, in which case it is provided with two drums, the peripheries of which are covered with wires of different fineness, the coarser being used for removing the water from the washing and the finer for removing the excess of bleach, at which stage much of the pulp is so fine that it would pass through the coarse wire-cloth of the washer. When this method is employed the pulp is run directly into the beating engine without pressing.

List of American Patents for Washing Engines.

The distinction between washing engines and beating engines is so slight that it would be impracticable to separate the two classes, and both varieties are consolidated in the "List of Patents for Pulp Engines and Bed Plates" at the close of Chapter XIV.

WASH WATER.

An abundance of pure, clear water is one of the first conditions in the manufacture of fine white papers; for lower classes of papers, such as wrapping, etc., it is not a matter of such vital importance.

When we consider that each one pound of rags or other material to be converted into paper will be brought in intimate association with from 100 to 200 times their own weight of water, it becomes manifest that even the smallest proportion of certain impurities which such water may contain will result in serious injury to the pulp.

The value of the soda, bleach liquor, alum, sulphuric acid, and coloring matters neutralized by impure waters aggregate an important sum of money in a short time, and every paper manufacturer should know the exact constituents of the water which he employs.

Water is composed of the two gases, oxygen and hydrogen, in the proportion by weight of 88.9 parts of the former and 11.1 parts of the latter, or 1 volume of oxygen to 2 volumes of hydrogen in chemical combination.

The composition of water can be proved analytically as well as synthetically, a current of electricity decomposing it into its constituent gases, twice as much hydrogen as oxygen, by volume, being produced.

Water, when pure, is colorless (in small quantities) and transparent, without taste or odor, and a bad conductor of heat and electricity. It is slightly elastic; under a pressure of 30,000 pounds to the square inch 14 volumes may be condensed into 13 volumes. It is 815 times heavier than atmospheric air, an imperial gallon weighing (at 62° F. and barometric pressure at 29.92 in.) 70,000 grains, or 10 pounds avoirdupois; but being the standard to which the gravities of solids and liquids are referred, its specific weight is usually said to be 1.0.

It is proper that we should give a description of the different sources from which natural waters are obtained, and also the properties of the water in each case when they are used in paper-making. We will divide the several natural waters into rain, river, and well waters, and the principal source of these is rain, snow, or hail.

It is probable that rain as it leaves the clouds is almost pure, but in its passage through the air it absorbs certain gases, and carries with it small particles of organic matter

which are floating about in the air. The substances thus
dissolved by the rain in its passage to the earth, *i. e.*, in the
open country, are the gases, oxygen, nitrogen, and carbonic
acid, a little carbonate of ammonia, nitric acid, this latter
more especially after a thunder storm, it being formed from
ammonia and water by the passage of the electric spark
through the air. In or near large manufacturing towns
several other substances are found in rain water, such as
sulphurous acid, sulphuretted hydrogen, etc., varying with
the kind of manufacture carried on near the spot. Again,
if rain water is collected after having fallen upon the roofs
of houses it will be further contaminated by various substances with which it comes in contact. Rain water from
the absence of earthy salts is very soft, and on that account
is preferable to hard waters. Rain, after it reaches the
earth, soaks down into it, and during its passage through
the various strata dissolves certain salts, etc., the quantity
and quality of which vary with the nature of the strata with
which it comes in contact. When this takes place on high
ground the water percolates the strata, and very frequently
finds an outlet at some lower point, as a spring. One or
more of these springs is generally the source of commencement of rivers, which, as they flow on in their course, become
increased in size by the various additions of water received
from rain, drainage from the surface of the earth, etc. The
springs above mentioned generally yield hard waters, that is,
water containing earthy salts in solution, the most frequent
of which are carbonate of lime, carbonate of magnesia,
sulphates of lime and magnesia, common salt, and organic

matter. These are the substances which the rain, containing a considerable quantity of carbonic acid in solution, dissolves in its passage through the earth. Spring waters resemble well waters. The river water, receiving supplies from those other sources which do not contain earthy matters, is, of course, softer than spring water. River water usually contains from 10 to 20 or 25 grains of solid matter per imperial gallon of 70,000 grains. The quantity, however, varies with the time of the year and the dryness of the season. Carbonate of lime, carbonate of magnesia, sulphate of lime, chloride of sodium, and organic matter are the substances most generally found in river water, the quantities per gallon and the relative proportions of the constituents varying according to circumstances. The hardness of water is generally determined by a solution of soap in proof spirit, made of such a strength that every degree of hardness shall be equivalent to one grain of carbonate of lime in a gallon. This simple method is known as Dr. Clark's soap test.

In water the carbonate of lime is held in solution by the presence of free carbonic acid. When the water is boiled this carbonic acid escapes, and the carbonate of lime is deposited; and it is this deposit which forms the principal incrustation in steam boilers. The removal of this carbonate of lime, or the greater portion of it, of course renders the water softer than before boiling.

If carbonic acid gas be passed through lime-water until the precipitate first formed is dissolved, the resulting liquid is a solution of carbonate of calcium in carbonic acid water.

When the solution is boiled carbonic acid escapes, and the carbonate is again precipitated.

Such an experiment will serve to show how chalk is kept in solution in ordinary well waters, giving the property of "hardness" and the manner in which the incrustation of boilers is formed. It may here be stated that sulphate of calcium produces similar hardness, and that these, with small quantities of the sulphate and carbonate of magnesium, constitute the hardening constituents of well waters.

The waters from wells differ from each other much more than do river waters, from the fact of the waters passing through different strata in different spots, and having no direct addition of rain water.

Determination of Constituents and Hardness of Water.

We cannot enter upon a full description of the different qualitative and quantitative methods for determining the constituents of water, but will only briefly describe a few examinations of importance, and refer those of our readers who may desire more minute information, concerning methods and apparatus employed, to the treatises of Wanklyn and Frankland on Water Analysis.

The qualitative examination of water as to its admixtures of iron, lime, magnesia, alkalies, chlorine combinations, sulphuric and carbonic acids, the larger or smaller quantity of which generally determines its character, can be executed in the following manner :—

1. The presence of iron can be readily discovered by the

addition of a solution of yellow prussiate of potash to the water; the iron salts will form with it Prussian blue.

2. The chlorine combinations are shown by the formation of a white precipitate when treated with nitrate of silver in nitrate solution.[1]

3. Sulphuric acid and sulphates are recognized by the formation of a white precipitate with chloride of barium.[2]

4. Carbonic acid is present when the addition of clear lime-water gives a white precipitate.

5. The presence of silicic acid, lime, and magnesia, by evaporating to dryness, with an addition of hydrochloric acid, in a platinum dish of a capacity of about one litre. The residue is taken up with hydrochloric acid and water, the portion remaining undissolved being silicic acid. The lime can be separated as calcium oxalate from the filtrate with ammonium oxalate. After removing the calcium oxalate by filtration and evaporation of the filtrate, the magnesia is precipitated with ammonium phosphate, as ammonium magnesium phosphate.

6. Organic substances are shown by adding a few drops

---

[1] Numerous apparatus containing chemical tests for water have been contrived, and without wishing to disparage such apparatus, it is probably best to state that without a knowledge of chemistry those who use them will be worse off with than without them. For instance, nitrate of silver is usually provided to determine the presence of chloride and chlorine; but if carbonate of soda should be present in the water under examination, carbonate of silver would be formed as well as chloride. Before the nitrate of silver could be applied the water should be acidulated with nitric acid to remove the carbonates, and then the nitrate of silver would throw down the chloride.

[2] 200–300 c. c. of clear water is heated to boiling, and then heated with a slight excess of solution of chloride of barium and a few drops of hydrochloric acid, boiled and filtered. The precipitate is washed, ignited, and weighed. Good filter-paper is essential for this determination.

of potassium permanganate and some pure sulphuric acid. If organic substances are present the potassium permanganate, added drop by drop, is decolorized until all the organic substances are completely oxidized.

7. Determination of the entire residue. One litre is carefully evaporated to dryness in a platinum dish, the weight of which has been previously determined. The residue is dried at 356° F. until a decrease in weight no longer takes place. It not uncommonly happens that the solid residue is exceedingly deliquescent; in such a case it must be rapidly weighed.

8. A determination of hardness with alcoholic soap solution serves in most cases as a substitute for a quantitative analysis. We give, therefore, a short description of it. The process of determining the hardness of water by a soap solution of a determined percentage, which was introduced by Clark, is a very simple one. By an addition of soap solution to water containing too much lime or magnesia a white precipitate of lime or magnesia soap insoluble in water is formed as long as calcium or magnesium salts are present.

A distinction is made between "total hardness" and "permanent hardness." The hardness of water not boiled is termed total hardness, and the hardness produced by the earthy sulphates is termed "permanent hardness," because unaffected by ebullition; the term "temporary or changeable hardness" being also frequently used to denote the hardness produced by the earthy carbonates, because removable by ebullition.

The process of determining the total hardness is as follows: 50 c. c. of water are measured with a pipette into a bottle having a capacity of about 8 ozs., and provided with an accurately-fitting ground stopper. Before adding the soap solution the free carbonic acid is removed by shaking the water, and then sucking out the air from the bottle through a glass tube. Then add from a burette or pipette graduated into cubic centimetres 1 c. c. of a standard solution of soap,[1] shake the bottle vigorously, and repeat the process after each addition, the quantity of soap test being gradually decreased until it is added only drop by drop as the reaction approaches completion. When a dense, delicate lather is formed which will endure for the space of five minutes, the bottle being laid down on its side, then the operation is finished, and the quantity of soap solution must be accurately noted.

The number of cubic centimetres of soap solution required to produce a lather being known, the degree of hardness can be ascertained from Table No. 1 or 2.

---

[1] *Standard Soap Solution.*—To make a potash soap, which keeps well, 40 parts of dry potassic carbonate and 150 parts of lead plaster (*Emplastrum plumbi*, B. P.) are rubbed together in a mortar until thoroughly mixed. Methylated spirit is then added and triturated to a cream, and after allowing to rest for a few hours, transfer to a filter and wash repeatedly with methylated spirit. The strength of this is determined by adding it to 50 c. c. of standard calcic chloride solution (the preparation of which will be explained); proceeding as in determining hardness. Dilute with water and alcohol until exactly 14·25 c. c. are required to form a permanent lather with 50 c. c. of solution of calcic chloride. The water is added in quantities such as to make the proportion of water to spirit as one to two.

Standard Calcic Chloride Solution.—This may be prepared by weighing 0·2 gram of any pure form of calcic carbonate, such as Iceland spar, into a platinum dish and gradually adding dilute hydrochloric acid until it is dissolved; loss may be prevented by covering the dish with a clock glass. Excess of HCl is driven off by successive evaporations to dryness, with distilled water, then re-dissolve in distilled water, and make up to one litre.

1. *Clark's Table of Hardness*—1000 *grains of Water used.*

Degree of hardness.	Measures of soap solution.	Difference for the next degree of hardness.	Degree of hardness.	Measures of soap solution.	Difference for the next degree of hardness.
Distilled water = 0...	1.4	1.8	9	19.4	1.9
1	3.2	2.2	10	21.3	1.8
2	5.4	2.2	11	23.1	1.8
3	7.6	2.0	12	24.9	1.8
4	9.6	2 0	13	26.7	1.8
5	11.6	2.0	14	28.5	1.8
6	13.6	2.0	15	30.3	1.8
7	15.6	1.9	16	32.0	1.7
8	17.5	1.9			

2. *Table of Hardness in Parts per* 100,000, 50 *c. c. of Water used.*

C. C. of soap solution.	CaCO₃ per 100,000.	C. C. of soap solution.	CaCO₃ per 100,000.	C. C. of soap solution.	CaCO₃ per 100,000.	C. C. of soap solution.	CaCO₃ per 100,000.	C. C. of soap solution.	CaCO₃ per 100,000.
.7	.00	3.8	4.29	6.9	8.71	10.0	13.31	13.1	18.17
.8	.16	.9	.43	7.0	.86	.1	.46	.2	.33
.9	.32	4.0	.57	.1	9.00	.2	.61	.3	.49
1.0	.48	.1	.71	.2	.14	.3	.76	.4	.65
.1	.63	.2	.86	.3	.29	.4	.91	.5	.81
.2	.79	.3	5.00	.4	.43	.5	14.06	.6	.97
.3	.95	.4	.14	.5	.57	.6	.21	.7	19.13
.4	1.11	.5	.29	.6	.71	.7	.37	.8	.29
.5	.27	.6	.43	.7	.86	.8	.52	.9	.44
.6	.43	.7	.57	.8	10.00	.9	.68	14.0	.60
.7	.56	.8	.71	.9	.15	11.0	.84	.1	.76
.8	.69	.9	.86	8.0	.30	.1	15.00	.2	.92
.9	.82	5.0	6.00	.1	.45	.2	.16	.3	20.08
2.0	.95	.1	.14	.2	.60	.3	.32	.4	.24
.1	2.08	.2	.29	.3	.75	.4	.48	.5	.40
.2	.21	.3	.43	.4	.90	.5	.63	.6	.56
.3	.34	.4	.57	.5	11.05	.6	.79	.7	.71
.4	.47	.5	.71	.6	.20	.7	.95	.8	.87
.5	.60	.6	.86	.7	.35	.8	16.11	.9	21.03
.6	.73	.7	7.00	.8	.50	.9	.27	15.0	.19
.7	.86	.8	.14	.9	.65	12.0	.43	.1	.35
.8	.99	.9	.29	9.0	.80	.1	.59	.2	.51·
.9	3.12	6.0	.43	.1	.95	.2	.75	.3	.68
3.0	.25	.1	.57	.2	12.11	.3	.90	.4	.85
.1	.38	.2	.71	.3	.26	.4	17.06	.5	22.02
.2	.51	.3	.86	.4	.41	.5	.22	.6	.18
.3	.64	.4	8.00	.5	.56	.6	.38	.7	.35
.4	.77	.5	.14	.6	.71	.7	.54	.8	.52
.5	.90	.6	.29	.7	86	.8	.70	.9	.69
.6	4.03	.7	.43	.8	13.01	.9	.86	16.0	.86
.7	.16	.8	.57	.9	.16	13.0	18.02		

Clark was the first to introduce the term "degree of hardness," and in Table No. 1 each measure of soap solution = 10 grains, and each degree of hardness = 1 grain of carbonate of lime or its equivalent of another calcium salt, or equivalent quantities of magnesia or magnesium salts in 70,000 parts (= 1 gallon).

At the present time one degree of hardness is suitably estimated as equal to one part of calcium oxide in 100,000 parts of water.

Should it be found that the quantity of soap solution required to produce a permanent lather exceeds 16 volumes of the solution to 50 of water, a second experiment would be necessary. In such a case a smaller quantity of the sample of water—even as low as 10 c. c. if the water appears to be very hard—to which a sufficient quantity of recently-boiled distilled water has been added to raise the bulk to the required 50 c. c. The same process is then performed as above described, but the number expressive of hardness must be multiplied by 2 or some other figure, according to the degree of dilution of the sample.

For the determination of the permanent hardness, 500 c. c. of water are gently boiled in a sufficiently large matrass for at least one and one-half hour, a part of the evaporated water being replaced by distilled water.

While the water is boiling the steam should be allowed to escape freely, and precaution must be observed to prevent the steam from the matrass from condensing and flowing back into the boiling water, because the escaping carbonic anhydride would be dissolved by the condensed water, which would thus be continually returned to the contents of the matrass in sufficient quantity to interfere with the complete precipitation of the carbonate of lime. The boiled water, when cold, is poured into a flask having a capacity of 500 c. c., and the matrass rinsed out with distilled water, the rinsing being added to the water in the flask. The latter is then filled with distilled water up to the mark, and the entire contents filtered through a dry filter into a dry glass.

The degree of hardness of a definite number of cubic centimetres is then determined in the manner above described.

The English degrees of hardness are reduced to German by multiplying the degrees found by 4 and dividing by 5, the reduction of German to English degrees being *vice versa* accomplished by multiplying by 5 and dividing by 4.

Waters possessing the properties of hardness are unsuitable without purification to be used in mills where the best qualities of white papers of any class are manufactured.

Various methods for purifying water for use in paper-mills have been proposed, some mills using surface streams run their water first into large settling ponds into which it is admitted only when it is comparatively clear.

Other manufacturers use filters especially manufactured for this purpose; but as these filters are made in a great variety of ways and their virtues fully set forth in the advertising columns of various trade papers, we will not devote space to a description of them.

In some mills separate filters are attached to each washing and beating engine.

22

List of Patents for Pulp-washing and Straining, issued by the Government of the United States of America, from 1790 *to* 1885 *inclusive.*

No.	Date.	Inventor.
	Dec. 31, 1833.	S. A. Sweet.
615	Feb. 22, 1838.	R. Carter.
1,441	Dec. 27, 1839.	N. Hebbard.
1,753	Sept. 2, 1840.	
Reissue		
171	June 11, 1850.	} G. Spafford.
Extended 7 years from Sept. 2, 1854.		
1,760	Sept. 3, 1840.	W. Dickinson.
3,354	Nov. 24, 1843.	
Reissue		} J. Phelps.
196	March 25, 1851.	
4,341	Dec. 31, 1845.	W. Bishop.
8,306	Aug. 19, 1851.	G. West.
12,283	Jan. 23, 1855.	
Reissues		
340	Jan. 8, 1856.	} H. W. Peaslee.
2,515	March 19, 1867.	
28,062	May 1, 1860.	C. S. Buchanan.
34,214	Jan. 21, 1862.	J. Piercy.
34,945	April 15, 1862.	S. S. Crocker.
44,059	Sept. 6, 1864.	A. Anderson.
46,030	Jan. 24, 1865.	G. E. Sellers.
46,915	March 21, 1865.	S. Lenher and H. H Spencer.
54,993	May 22, 1866.	L. M. Wright.
62,517	March 5, 1867.	W. Adamson.
62,942	March 19, 1867.	S. Curtis.
66,258	July 2, 1867.	G. E. Sellers.
79,935	July 14, 1868.	J. E. Andrews.
84,850	Dec. 8, 1868.	G. L. Witsil.
87,385	March 2, 1869.	A. S. Winchester.
90,472	May 25, 1869.	R. R. Sylands.
96,515	Nov. 2, 1869.	H. Voelter.
99,735	Feb. 8, 1870.	S. W. Wilder.
103,506	May 24, 1870.	C. G. Sargent.
105,354	July 12, 1870.	W. H. Merrick.
105,755	July 26, 1870.	A. St. C. Winchester.
125,810	April 16, 1872.	G. W. Hammond and T. J. Foster.
128,625	July 2, 1872.	L. Hollingsworth.
136,002	Feb. 18, 1873.	H. H. Olds.
137,696	April 8, 1873.	G. L. Lovett.

No.	Date.	Inventor.
140,166	June 24, 1873.	} J. Robertson.
Reissue	Jan. 14, 1879.	
8,542		
145,159	Dec. 2, 1873.	S. and J. Deacon.
147,595	Feb. 17, 1874.	C. J. Bradbury.
147,717	Feb. 17, 1874.	J. S. Warren.
148,643	March 17, 1874.	A. Annandale, Jr.
154,733	Sept. 1, 1874.	J. S. Warren.
156,885	Nov. 17, 1874.	G. Gavit.
165,192	July 6, 1875.	J. S. Warren.
170,471	Nov. 30, 1875.	S. E. Crocker.
175,286	March 18, 1876.	K. Hollingsworth.
188,474	March 20, 1877.	G. L. Lovett.
190,390	May 1, 1877.	W. C. Tuttle.
192,107	June 19, 1877.	W. Blizzard and E. Mather.
193,344	July 24, 1877.	R. A. Morton.
194,960	Sept. 11, 1877.	W. H. Elliot and L. F. Clark.
197,764	Dec. 4, 1877.	F. A. Cloudman.
206,187	July 23, 1878.	E. Mather.
206,632	July 30, 1878.	S. Snell.
206,877	Aug. 13, 1878.	H. Hollingsworth.
209,326	Oct. 29, 1878.	G. Campbell and W. Lidgett.
210,521	Dec. 3, 1878.	L. L. Could.
210,612	Dec. 10, 1878.	J. W. Hyatt and J. G. Jarvis.
210,853	Dec. 17, 1878.	H. Hollingsworth.
216,243	June 3, 1879.	J. S. Warren.
216,565	June 17, 1879.	} J. Tyler.
Reissue		
10,042	Feb. 21, 1882.	
221,221	Nov 4, 1879.	M. S. Drake.
221,330	Nov. 4, 1879.	W. L. Longley.
223,969	Jan. 27, 1880.	B. F. Warren.
225,545	April 13, 1880.	C. Pinder and W. A. Hardy.
226,819	April 20, 1880.	L. Zeyen.
230,029	July 13, 1880.	A. McDermid.
230,287	July 20, 1880.	B. Klary.
232,383	Sept. 21, 1880.	G. A. Whiting.
234,559	Nov. 16, 1880.	S. L. Gould.
234,719	Nov. 23, 1880.	C. Pindar and W. A. Hardy.
238,126	Feb. 22, 1881.	H. Judson.
235,213	Dec. 7, 1881.	J. Cornell.
239,276	March 22, 1881.	J. M. Shew.
235,976	Dec. 28, 1880.	L. Zeyen.

No.	Date.	Inventor.
239,837	April 5, 1881.	C. Pindar and W. A. Hardy.
242,428	June 7, 1881.	C. Bremaker.
258,209	May 23, 1882.	C. Anderson and T. Patten.
262,877	Aug. 15, 1882.	J. and R. Wood.
276,250	April 24, 1883.	N. Kaiser.
276,596	May 1, 1883.	G. Kaffenberger.
276,989	May 1, 1883.	S. Wrigley and J. Robertson.
277,239	May 8, 1883.	P. H. Cragin.
284,232	Sept. 4, 1883.	E. J. F. Quirin.
287,164	Oct. 23, 1883.	H. Reinicke.
310,469	Jan. 6, 1885.	H. Schlatter.
313,037	Feb. 24, 1885.	F. Williams.
315,420	April 7, 1885.	R. Kron.
316,938	May 5, 1885.	F. K. Black.
318,180	May 19, 1885.	W. Gray.
325,206	Aug. 25, 1885.	W. Gray.
331,304	Dec. 1, 1885.	R. Kron.

CHAPTER XI.

BLEACHING POWDER—ESTIMATION OF CHLORINE IN BLEACHING
POWDER—PREPARING AND USING THE BLEACHING SOLUTION—
ZINC BLEACH-LIQUOR—ALUMINA BLEACH-LIQUOR—DRAINING
—SOUR BLEACHING—BLEACHING WITH GAS—BLEACHING PULP
MADE FROM OLD PAPERS OR IMPERFECTIONS — BLEACHING
STRAW—BLEACHING WOOD FIBRE—METHOD FOR BLEACHING
WOOD, STRAW, ETC.—BLEACHING JUTE—BLEACHING MATE-
RIALS COMPOSED OF HEMP, FLAX, ETC.—BLEACHING VEGETABLE
TISSUES WITH PERMANGANATE OF POTASH—BLEACHING PAPER
PULP BY APPLYING THE BLEACHING AGENT IN A SPRAYED
CONDITION—BLEACHING IN ROTARIES—LIST OF PATENTS FOR
BLEACHING PULP.

BLEACHING POWDER.

BLEACHING powder or chloride of lime is the chemical
which is the active agent in the bleaching processes employed for paper pulp.

Numerous investigations have been made of late years to
determine the constitution of this substance.

Gopner,[1] Richter,[2] and Juncker,[3] support the old view that
bleaching powder is a direct compound of chlorine with lime,
$CaO.Cl_2$.

On the other hand Schorlemmer states[4] " that hypochlo-

---

[1] J. pr. Chem. [2], viii. 441.   [2] Dingl. pol. J. ccx. 21.
[3] Ding. pol. J. ccxii. 339.
[4] Deut. Chem. Ges. Ber. vi. 1509; Chem. Soc. Journ. [2], xii. 335.

rous acid is very easily obtained by distilling bleaching powder with the requisite quantity of nitric acid, a colorless
distillate being thereby produced, which bleaches much more
strongly than recently prepared chlorine water, and when
shaken up with mercury yields a considerable quantity of
brown mercuric oxychloride.[1]

" The fact that when bleaching powder is exhausted with
successive small quantities of water, the last extracts still
contain calcium and chlorine in the proportions required by
the formula, $CaOCl_2$, merely shows that the product of the
action of chlorine on lime is not a mixture of calcium chloride and hypochlorite $(CaCl_2 + CaCl_2O_2)$, but a compound
constituted according to the formula, $Ca \begin{cases} Cl \\ OCl \end{cases}$, as first suggested by Olding.

" In the preparation of aqueous hypochlorous acid by the
action of chlorine on water containing calcium carbonate in
suspension, the compound just mentioned is first formed and
then decomposed according to the equation :—

$$Ca \begin{cases} Cl \\ OCl \end{cases} + Cl_2 = CaCl_2 + Cl_2O."$$

The experiments of Kingzett[2] and of Kopfer[3] corroborate
this view of the constitution of bleaching powder.

The results of the experiments of these two chemists show
that bleaching powder contains either a mixture of calcium
chloride and hypochlorite, or the compound $CaCl(OCl)$.

The production of hypochlorous acid is explained equally

[1] Gmelin's Handbook, vi. 60.   [2] Chem. Soc. Journ. [2], xiii. 404.
[3] Chem. Soc. Journ. [2], xiii. 713.

well by both Kingzett and Kopfer and satisfactorily accounts
for the formation of bleaching powder by the action of chlorine upon calcium hydroxide. " One atom of chlorine first
replaces the group OH, which combines in the nascent state
with the hydrogen atom of another hydroxyl to form water,
whilst the second atom of chlorine goes into the place of the
hydrogen atom thus removed."

Sometimes bleaching powder becomes injured by packing
it too quickly after it has been manufactured, and in such
cases, especially on hot summer days, it is liable to decomposition, and sometimes so quickly as to become worthless in a
few hours.

Bleaching powder should be used as fresh as it is possible
to obtain it, as it undergoes alteration by keeping, the loss
of active chlorine being greater in summer than in winter.
The rooms in which bleaching powder and the bleach solution are stored should be kept dark and moderately cool.

" Bleaching powder ought to be a pure white powder,
which in the case of a strong article is mixed with lumps ;
but these on crushing ought to show just the same properties
as the powder ; they ought to be completely transformed,
and not to contain a core of lime. These lumps are sometimes removed by riddling. In the air, bleach gradually
attracts moisture and carbonic acid, and finally deliquesces
to a pasty mass. It has a peculiar smell, different from that
of chlorine, and usually ascribed to hypochlorous acid set
free by the carbonic acid of the air ; but this cannot be so,
as bleach solutions to which an excess of alkali has been
added exhale the same smell, even after boiling and cooling

in an atmosphere free from carbonic acid (Winckler, Dingl. Journ., cxcviii. 149).[1] Mixed with a little water, bleach forms a stiff paste, with a perceptible rise of temperature; if ground up with more water, most of it enters into solution (according to Fresenius first the calcium chloride), but there always remains a considerable residue, consisting chiefly of calcium hydrate, containing some bleaching chlorine, which can only be washed out by a very large amount of water. The aqueous solution has a faintly alkaline reaction, the smell of bleaching powder, and a peculiar astringent taste. This solution is almost exclusively employed in bleaching, as the residue would contaminate the paper-pulp, the fabric, etc., and even locally destroy them. M. F. Hodges has proved that after complete washing the insoluble residue of bleaching powder is quite harmless.

" Bleaching powder decomposes gradually, even in the absence of air, as is proved by the instance communicated by Hofmann of the explosion of a tightly stoppered bottle, also in well-packed and protected casks—but especially under the influence of air and light. Sometimes the decomposition takes place quite suddenly, but only when warm bleach has been packed in hot summer weather. The shaking in a railway truck or a wagon also injures it more than quiet lying in a dark dry place. Hence the strength of bleach is nearly always guaranteed only at the place of shipment;

---

[1] According to Phipson (' Compt. Rend.' lxxxvi. p. 1196), sulphuretted hydrogen passed over bleaching powder causes the production of a smell of free chlorine: first hypochlorous acid is formed; and this with H_2S decomposes into H_2O, S, and Cl.

but bleach shipped with 35 per cent. in England ought still to show at Hamburgh or New York 33 or at least 32 per cent. Pattinson ('Chem. News,' xxix. p. 143) examined the speed at which bleaching powder loses its available chlorine. In the course of twelve months the strength of the following examples of bleach was lowered—

	A_1	A_2	A_3	B_1	B_2	B_3	C_1	C_2	C_3
from	28.7	37.4	37.1	32.9	35.2	36.7	31.8	37.6	37.6
to	20.8	31.2	30.2	22.2	27.9	28.0	26.4	28.2	32.3

"The samples A, B, and C were taken from different works, but the three numbers of each letter from the same chamber in different stages of saturation. The average loss of chlorine in the first three months, from February to April, was 0.33 per cent. per month; from June to September, 0.86 per cent. per month; from November to January, 0.28 per cent. per month. The greatest loss occurred in August, viz., on the average 1.4 per cent. per month. The monthly loss of chlorine on an average of the whole year was *in maximo* 0.90, *in minimo* 0.50, average 0.63 per cent. It is very noteworthy that weak (28.7 per cent.) bleach lost strength quite as rapidly as the strong (37 per cent.), which contradicts the formerly general assumption. Pattinson's observations were made with samples kept in loosely corked bottles sheltered from direct sunlight; possibly bleach packed in good casks may behave somewhat differently. Dullo ('Wagner's Jahresb.,' 1865, p. 253) showed that bleaching powder continually gives off oxygen. At a lower temperature slowly and gradually, at a higher one quickly; but his suggestion (impracticable in any case) that no bleach should be made above 30 per cent. is shown to be useless by Pattinson's experiments."[1]

ESTIMATION OF CHLORINE IN BLEACHING POWDER.

Bleaching powder, or chloride of lime, is an important chemical in a paper-mill, and while it is possible to have the powder tested by analytical chemists still it is desirable that the manager of a mill should be able to make the tests and to quickly discover whether the material with which the proprietor is supplied is of the desired quality.

Bleaching powders are usually quoted as on the spot or to arrive, and 32 per cent. is considered the standard strength by the trade, and powders of less strength should not be accepted as good delivery.

There are various methods for the volumetric estimation of chlorine in bleaching powder, which contains hypochlorite of lime, chloride of calcium, and hydrate of lime.

The latter two ingredients are for the most part combined with one another to basic chloride of calcium. Fresenius[2] states: In freshly prepared and perfectly normal chloride of lime the quantities of hypochlorite of lime and chloride of calcium present stand to each other in the proportion of their equivalents. When such chloride of lime is brought into contact with dilute sulphuric acid the whole of the chlorine it contains is liberated in the elementary form, in accordance with the following equation:—

$$CaO,ClO + CaCl + 2(HO,SO_3) = 2(CaO,SO_3) + 2HO + 2Cl.$$

[1] Lunge, Sulphuric Acid and Alkali, vol. iii. p. 172 *et seq.*
[2] Pages 504 *et seq.*

On keeping chloride of lime, however, the proportion between hypochlorite of lime and chloride of calcium gradually changes—the former decreases, the latter increases.
Hence from this cause alone, to say nothing of original
difference, the commercial article is not of uniform quality,
and on treatment with acid gives sometimes more and sometimes less chlorine.

As the value of bleaching powder depends entirely upon the
amount of chlorine set free on treatment with acids, chemists
have devised various simple methods of determining the
available amount of chlorine in any given sample. These
methods have collectively received the name of Chlorimetry.
We describe from the authority above quoted the best in use.

The preparation of the solution of the bleaching powder
to be tested is prepared best in the following manner:—

Weigh 10 grams, triturate finely with a little water,
add gradually more water, pour the liquid into a litre flask,
triturate the residue again with water, and rinse the contents of the mortar carefully into the flask; fill the latter to
the mark, shake the milky fluid, and examine it at once in
that state, *i. e.*, without allowing it to deposit; and every
time before measuring off a fresh portion shake again. The
results obtained with this turbid solution are much more
constant and correct than when, as is usually recommended,
the fluid is allowed to deposit, and the experiment is made
with the supernatant clear portion alone. The truth of this
may readily be proved by making two separate experiments,
one with the decanted clear fluid, and the other with the
residuary turbid mixture. Thus, for instance, in an experi-

ment made in my own (Fresenius's) laboratory, the decanted clear fluid gave 22.6 of chlorine, the residuary mixture 25.0, the uniformly mixed turbid solution 24.5. 1 cubic centimetre of the solution of chloride of lime so prepared corresponds to 0.01 gram chloride of lime.

A. *Penot's Method.*[1]

This method is based upon the conversion of arsenious acid into arsenic acid; the conversion is effected in an alkaline solution. Iodide of potassium-starch paper is employed to ascertain the exact point when the reaction is completed.

a. *Preparation of the Iodide of Potassium-Starch Paper.*—The following method is preferable to the original one given by PENOT :—

Stir 3 grms. of potato starch in 250 c. c. of cold water, boil with stirring, add a solution of 1 grm. iodide of potassium and 1 grm. crystallized carbonate of soda, and dilute to 500 c. c. Moisten strips of fine white unsized paper with this fluid and dry. Keep in a closed bottle.

b. *Preparation of the Solution of Arsenious Acid.*—Dissolve 4.436 grms. of pure arsenious acid and 13 grm. pure crystallized carbonate of soda in 600–700 c. c. water with the aid of heat, let the solution cool, and then dilute to 1 litre. Each c. c. of this solution contains 0.004436 grm. arsenious acid, which corresponds to 1 c. c. chlorine gas of 0° and 760 mm. atmospheric pressure.[2]

---

[1] Bulletin de la Société Industrielle de Mulhouse, 1852, No. 118; Ding. Poly. Jour. 127, 134.

[2] Penot gives the quantity of arsenious acid as 4.44 ; but Fresenius has cor-

As arsenite of soda in alkaline solution is liable, when exposed to access of air, to be gradually converted into arseniate of soda, PENOT'S solution should be kept in small bottles with glass stoppers, filled to the top, and a fresh bottle used for every new series of experiments.

According to Mohr the solution keeps unchanged, if the arsenious acid and the carbonate of soda are both *absolutely* free from oxidizable matters (sulphide of arsenic, sulphide of sodium, sulphite of soda).

c. *The Process.*—Measure off, with a pipette, 50 c. c. of the solution of chloride of lime prepared according to the directions already given, transfer to a beaker, and from a 50 c. c. burette, add, slowly, and at last drop by drop, the solution of arsenious acid, with constant stirring, until a drop of the mixture produces no longer a blue-colored spot on the iodized paper; it is very easy to exactly hit the point, as the gradually increasing faintness of the blue spots made on the paper by the fluid dropped on it, indicates the

---

rected this number to 4.436, in accordance with the now received equivalents of the substances and specific gravity of chlorine gas after the following proportion : 70.92 (2 eq. chlorine) : 99 (1 eq. AsO_3) : : 3.17763 (weight of 1 litre of chlorine gas) : x ; $x = 4.436$, *i. e.*, the quantity of arsenious acid which 1 litre of chlorine gas converts into arsenic acid.

This solution is arranged to suit the foreign method of designating the strength of chloride of lime, viz., in chlorimetrical degrees (each degree represents 1 litre chlorine gas at 0° and 760 mm. pressure in a kilogramme of the substance). This method was proposed by Gay-Lussac. The degree may readily be converted into per cents., and *vice versá*, thus : A sample of chloride of lime of 90° contains $90 \times 3.17763 = 285.986$ grm. chlorine in 1000 grams or 28.59 in 100 ; and a sample containing 342 per cent. chlorine, is of 107.6° for 100 grm. of the substance contain 34.2 grm. chlorine. : 1000 grm. of the substance contain 342 grm. chlorine, but 342 grm. chlorine $= \dfrac{342}{3.17763}$ litres $= 107.6$ litres. : 1000 grm. of the substance contain 107.6 litres chlorine.

approaching termination of the reaction, and warns the operator to confine the further addition of the solution of arsenious acid to a single drop at a time. The number of $\frac{1}{2}$ c. c. used indicates directly the number of chlorimetrical degrees (see note), as the following calculation shows: Suppose you have used 40 c. c. of solution of arsenious acid, then the quantity of chloride of lime used in the experiment contains 40 c. c. of chlorine gas. Now the 50 c. c. of solution employed correspond to 0.5 grm. of chloride of lime; therefore 0.5 grm. of chloride of lime contains 50 c. c. chlorine gas, therefore 1000 grms. contain 80,000 c. c. = 80 litres. This method gives very constant and accurate results.

Preparing and Using the Bleaching Solution.

The bleaching solution is best prepared in large wooden vats lined with lead, or in cisterns constructed of brick and cement, and furnished with suitable agitators driven by power.

There should be at least two vats or cisterns in each mill; but if there is sufficient room it is desirable to employ three cisterns in rotation, thus allowing three extracts to be made from the powders and insure their thorough exhaustion.

The residuum, consisting of sand, carbonate of lime, etc., remaining on the bottom of the cisterns is washed out.

The bleaching solutions are run from the cisterns into a suitable receiver where the liquor is kept at the required strength by diluting each solution until the prescribed specific gravity is indicated on the hydrometer, it being necessary that liquor in the receiver should always show degree of strength in order that the same quantity may be expected to reproduce the same results.

After the rags, etc., have been washed and reduced to half-stuff in the engine, the desired quantity of liquor is drawn from the receiver and added to the material to be bleached.

The quantity of chloride used varies according to the nature of the rags: 2 to 2.5 per cent. for white rags, while others, on the contrary, require 7 to 10 per cent.; chlorine gas being preferable for the latter class of rags on account of its greater economy.

Gas bleaching for half-stuff is seldom resorted to in the United States or Great Britain, but in Russia it is almost indispensable for bleaching the coarse linen rags so plentiful in that country.

Sometimes a fresh solution is prepared for every engine of pulp by placing, in a suitable receptacle, the quantity of powders required to bleach one engine of rags (5 to 10 pounds per 100 pounds of paper), and after filling the receptacle with water the contents are stirred with a paddle, and after sufficient time has been allowed for the liquor to rest it is drawn off when clear into the engine through a faucet placed a few inches above the bottom of the receptacle.

A box or barrel lined with lead and having a capacity of from 25 to 30 gallons, and arranged on a suitable platform above the engine, makes a good vessel in which to prepare the solution; such a vessel containing 15 to 20 pounds of good bleaching powder will produce a solution testing 6° to 8° B. The sediment, after drawing off the clear solution, is removed from the vessel, dissolved in another receptacle, and the weak solution thus obtained is used with the pulp in the next engine.

If sufficient time is allowed the chlorine will bleach the pulp without the aid of an acid; but to facilitate the disengagement of the chlorine, sulphuric acid is generally employed, thereby shortening the time and greatly decreasing the number and capacity of the pulp receivers which would otherwise be required. The proportion of sulphuric acid employed may vary from nothing to one pound for every four pounds of bleaching powder, and depends upon the class of stock to be treated, the available draining-room, and time. Instead of adding the sulphuric acid of the ordinary strength, it is preferable to dilute with ten to twenty times its own weight of water. The dilution of the sulphuric acid should be made very carefully, as many a workman has lost his eyesight or been otherwise injured by carelessly adding sulphuric acid to water. When the acid has been thoroughly incorporated with the pulp it is emptied from the engine into a drainer, *i. e.*, if the pulp is not to be at once used.

ZINC BLEACH LIQUOR.

Strong acids are often objectionable for liberating chlorine from bleaching powder, and especially in bleaching some classes of paper pulp. If a solution of zinc sulphate be added to a solution of bleaching powder, calcium sulphate is precipitated, and the zinc hypochlorite formed at once splits up into zinc oxide and a solution of free hypochlorous acid. Zinc chloride acts similarly; for a saturated solution of zinc in strong muriatic acid decomposes as much bleaching powder as half its weight of concentrated oil of vitriol (Varrentrapp). The reaction must be—

$$CaOCl_2 + ZnCl_2 = CaCl_2 + ZnO + 2Cl.$$

Accordingly these zinc salts can be employed in place of sulphuric acid and thus bleach the paper pulp very quickly. When this mixture is employed in bleaching paper pulp, the precipitated calcium sulphate and zinc oxide remain in the pulp. This solution was introduced by Sace ('Wagner's Jahresb.,' 1859, p. 548), and has been recommended by Varrentrapp (Ib., 1860, p. 189). Lunge, 'Sulphuric Acid and Alkali,' iii. 281.

ALUMINA BLEACH-LIQUOR.

Orioli ('Wagner's Jahresb.,' 1860, p. 188) recommended, especially for paper-mills, a bleach liquor made by decomposing equivalent quantities of a solution of bleaching powder and aluminum sulphate; this had been known for several years as Wilson's bleach liquor. Gypsum is thrown down and aluminum hypochlorite remains dissolved. This is very unstable, and hence can be employed for bleaching without adding an acid, splitting up into aluminium chloride and active oxygen. Consequently the liquid always remains neutral, and the difficulty caused by the obstinate retention

23

of free acid in the fibre by which it is strongly acted upon on drying, in this case does not exist. The aluminium chloride also acts as an antiseptic, so that the paper stock can be kept for many months without any fermentation or other decomposition. The solution is allowed to act for about ten minutes in the engine. Lunge, 'Sulphuric Acid and Alkali,' iii. 281.

DRAINING.

The drainers are best constructed of brick and cement, and they can be built so as to allow the pulp to be removed from the open top or through a door near the bottom.

The open-top drainers are usually about five or six feet deep, and are arranged in rows conveniently near the beaters; the pulp being shovelled out by a workman who throws it on a platform which should be on about a level with the engine-room, whence it is carried on trucks to the beaters.

When the pulp is removed from the bottom of the drainers their height is not restricted, and they can be built at any convenient point below the floor of the engine-room, and in some mills they are placed directly below the engines. The doors at the bottom of such drainers should be of sufficient size to allow the workmen to pass in and out, and so arranged as to be conveniently opened and closed. Sufficient room must also be allowed at the bottom for running the trucks on which the pulp is carried to the hoists. The walls of the drainers should be carefully built and then plastered with cement; the thickness of the walls being sufficient to resist the pressure of a body of liquid as large as the drainers will hold.

The bottoms of the drainers are perforated to permit the escape of the water. The perforated bottoms of these screens or drainers have hitherto been made in many different ways, all of which have been found subject to serious objection in practical use. In some instances difficulty was experienced from the liability of the holes or perforations to become clogged or closed by the wet material. In other cases difficulty arose from the fact that the strainers were made of metal, the oxidation of which caused a discoloration of the stock. These difficulties are largely overcome by making the strainer of tile and providing it with perforations of a flaring or conical form larger at the bottom than at the top, and in supporting the tile thus made by beams or bars, of tile or concrete, which, like the tile, are free from liability of oxidation.

Fig. 114 represents respectively a plan view of an improved tile strainer, a cross section of the same, and a top-plan view, illustrating the manner of constructing the floor or bottom of the drainer. It is the invention of Mr. Samuel Snell, of Springfield, Mass.

The tiles *A* are made of any suitable clay or composition, of a flat and regular form, and provided with numerous vertical openings or perforations, *c*, the lower ends of which are larger than the upper.

In constructing the drainer it is provided at the bottom with a series of longitudinal bars or supports, *B*, made of tile or concrete, and arranged parallel with each other at suitable distances apart. Upon these beams the tiles A are arranged in such manner as to form a continuous unbroken floor to support the paper-stock. The perforations c permit

Fig 114.

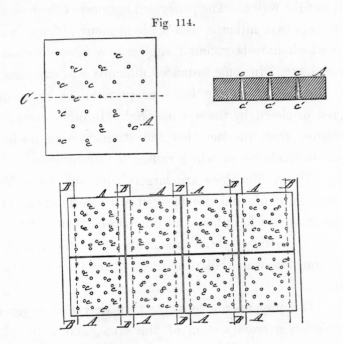

the water to flow from the stock, and, owing to their conical form, avoid the danger of the material which may enter their upper ends lodging within them.

If it is desired to save the bleach liquor, which is a questionable economy where only rags are worked, it is allowed to run into a "junk" or reservoir placed below the bottom of the drainers, whence it is pumped into an upper receiver to be strengthened, or to be employed for preliminary bleaching. In order, however, to save the fibres which escape with the liquor from the drainers it is desirable to run the waste liquor into a cistern where the fibres can be deposited. Scrupulous cleanliness should be the rule in this as in every department of the mill, and stuff which may fall upon the floor and become soiled should be thrown in a box provided for that purpose; all trucks, boxes, etc., used in the drainers should be periodically washed and kept in a cleanly condition, and the floor should be washed each day. In order to economize in time and capital, various methods have been tried to quickly drain the pulp by mechanical means, and one of the most acceptable methods for extracting the water from the stuff is by the centrifugal drainer, which is similar in construction to that used for drying cloth.

Sour Bleaching.

This method of bleaching may be conducted partly or wholly in the drainers, those of good height and large capacity being best suited for this process.

The stuff should be emptied from the engines into the drainers where it should be kept packed closely around the sides, as the stuff shrinks away as the water drains off.

The bleach solution is admitted at the top of the drainers in sufficient quantities to saturate the entire body of stuff, and then the largely-diluted vitriol solution is admitted in a short time after the bleach solution.

The strength and quantity of each of the solutions necessary for different classes and quantities of pulp are soon ascertained by practical experience.

This method of bleaching may be modified in a great variety of ways; some manufacturers add the acid solution first and the bleach solution afterwards; others add one-half of the bleach solution in the engine and then empty the stuff in the drainers, and after running in the acid solution they then admit to the stuff in the drainers the remaining one-half of the bleach solution.

This process is manifestly not suited for bleaching fine stuffs, as its operation cannot be watched in its different stages, and some portions of the material are liable to come from the drainers in an unbleached condition.

BLEACHING WITH GAS.

The process of bleaching with gas can be conducted in the same kind of drainers in which the process of sour bleaching is carried on.

When the half-stuff is emptied from the washing engine into the drainers the water immediately commences to leave it; but as it would require a long time for the stuff to become sufficiently dry in the drainers to be bleached with gas, a centrifugal drainer is commonly employed, or the pulp is run over the wet machine.

In order to achieve satisfactory results with gas-bleached half-stuff, it is necessary that it should be neither too dry nor too moist. A good method of testing the condition of the half-stuff for gas bleaching is to squeeze samples of it between the hands; if the pressure should cause no water to escape and the samples still possess a damp appearance, the material may be considered to be in a suitable condition to be bleached with gas.

"The method of bleaching is as follows: Put 1600 pounds of half-stuff, in the condition mentioned above, loosely into a stone chamber, and seal it in such a manner that it will be perfectly air tight. Into the lead retort connected with this chamber by leaden pipes, pour 3 pails of water and 66 pounds of common salt; stir thoroughly, and add 65 pounds of manganese; stir again, and close the retort. Next charge a leaden vessel with 119 pounds of vitriol and allow the acid to drop through a bell-mouthed bent siphon into the retort containing the mixture of manganese, salt, and water, three hours being allowed for the vitriol to drop into the retort.

"The retort is then heated to 212° F. with steam for several hours, and two hours are allowed for the gas to escape up the mill chimney. For fine stuff, such as willow-rope, one hour extra must be allowed for the escape of the gas."

Chlorine gas may be prepared either with sulphuric or hydrochloric acid, the relative cost of these two acids determining which should be employed.

The proportions of the ingredients to be employed are not absolute, as they must vary with the composition of the manganese and the strength of the acid used. When sulphuric acid is employed the proportions may be about as follows:—

Manganese	1 part.
Common salt	0.5 to 2 parts.
Sulphuric acid	2 parts.
Water	2 "

When hydrochloric acid is employed for preparing the gas, 3 parts of the acid and 1 part of manganese are simply used. Lump manganese is preferable to the powdered manganese, on account of the slower action of the acid on the latter, which also requires agitation to effect a mixture of the two substances.

The chlorine gas in escaping from the retort carries with it some hydrochloric acid, which, if allowed to escape into the half-stuff, would injure the strength of the fibres. There are various methods of freeing the gas from this acid; one of the simplest means adopted for attaining this object being to lead the pipe which conveys the gas from the retort into a receiver containing a small quantity of water. "The pipe should be allowed to enter only a few lines below the surface of the water, in order that the height of the column of liquid may not exercise a sensible pressure in opposition to the escape of the gas. The bubbles of gas give up their particles of hydrochloric acid at the contact of the water, which mechanically washes them away."

In order to impart greater brilliancy to their stuffs, some manufacturers bleach with chlorine gas, and then with liquid chlorine, which method by the employment of a smaller quantity of gas diminishes the danger of excessive action upon the fibres.

The waste from bleaching varies, of course, with the different classes of rags; for fine whites it ranges from 1.5 to 3.5 per cent., and for coarse rags and thirds the loss is from 3 to 8 per cent.

Bleaching Pulp made from old Papers or "Imperfections."

The pulp made from old papers or imperfections is bleached in the same manner as rags. The manipulation of the pulp in the engine during the bleaching process depends upon the class of paper which is to be produced; the whole bleaching process can be finished in one engine if only one class of "imperfections" are to be made into a low grade of paper; but if the pulps of different kinds of old papers are to be mixed, and the best quality of paper produced, the bleached pulp should be emptied from the engines into the drainers.

Bleaching Straw.

The bleaching of straw pulp is done in about the same manner as rags are bleached, the only difference being that a larger proportion of chemicals is required. Some manufacturers commence the bleaching with gaseous chlorine, but the process must always be terminated with the chloride of lime; otherwise the pulp would be reddish.

The proportion of bleaching powder required for bleaching straw varies from 12 to 25 pounds of bleaching powder and a corresponding proportion of vitriol to every 100 pounds of paper made entirely from straw.

In order to obtain a satisfactory result from straw pulp it is imperative that the boiling should be properly done, as it is false economy to curtail the quantity of soda used in the digester, and then be compelled to force the bleaching by the employment of an excess of chemicals which weaken the fibres.

Thorough washing out and neutralization of the alkali is also necessary after boiling straw, and if these points receive proper attention much of the trouble usually experienced in bleaching straw pulp may be obviated.

Care must be observed to thoroughly wash out the drainers with clear water in order to carry away all the hydrochloric acid, which imparts a disagreeable yellowish-gray color to the pulp.

The bleach solution should not, if sulphuric acid has been used, be allowed to remain for a protracted period in contact with the pulp; it should not only be drained as quickly as possible, but it has been recommended to also empty the bleacher with a larger quantity of water, and thus to soak and wash the white pulp immediately in the drainers.

The waste bleach liquor quickly parts with its chlorine when exposed to the action of the air, and consequently if the liquor which escapes from the drainers is to be again used, it is best to add it to the gray pulp in the washing-engine, and employ it as a preliminary bleacher, and then wash it out again, for if the waste bleach liquor is allowed to become transformed into hydrochloric acid through too long contact with the air the injury which it will inflict upon the color and strength of the fibres will be irreparable.

Burns's Bleaching Process for Straw, etc.

On page 231 we gave a description of Burns's process for boiling and disintegrating straw, and in Fig. 115 is shown a vertical section of the apparatus in which the bleaching of the straw is conducted.

Fig. 115.

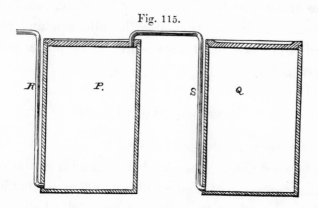

P represents a tank or tub, preferably of wood, in which is placed a quantity of water and chloride of lime. It is provided with a steam-pipe, R, entering the tub at the bottom, so that the entire contents may be impregnated. By this means the chlorine gas is eliminated from the solid particles, and it passes (together with a quantity of vapor and water charged with the gas) into tub Q, by means of pipe S, where it enters at the bottom. In this tub Q the paper stock to be bleached is placed, and the chlorine, entering the stock or pulp from the bottom, filters through to the top, where it passes off. It will be seen that by this construction no solid particles or bleaching-matter touches the pulp. Where the bleaching material itself is placed in the tub with the pulp it rots it and makes it brittle when made into paper.

Straw-board heretofore made by the existing processes is usually extremely brittle; but by Burns's process it is claimed that the fibre of the straw is not destroyed, nor is the pulp made brittle by the introduction of solid particles of lime, caustic alkali, and the like into the bleaching tub, thereby eliminating and destroying also the albumen and gluten properties of straw, which are essential to the proper making of straw paper.

Bleaching Wood Fibre.

There are numerous methods of bleaching the various kinds of wood fibre. The bleaching engine is similar to the common rag-washing engine with the difference that it has no plate or sand catcher. The engine is filled about two-thirds full with warm water and the necessary proportion of bleach liquor. The rolls of fibre, which, having been run over the wet machine, resemble paper, are supplied to the engine in sufficient quantities to nearly fill it, after which the steam is turned on and the contents of the trough are heated to about 210° F. If the fibre has not been sufficiently bleached more bleach liquor is added, and when the desired color is obtained the cylinder washer is lowered and the cold water is turned on. After it is cooled a little the whole is emptied into the drainers and the liquid is allowed to drain off.

Another method of bleaching is to treat the fibre substantially the same as just indicated, to neutralize the excess of chlorine with antichlore, and run the fibre over a wet machine, thus dispensing with drainers. This is an improvement, but it is nevertheless crude and imperfect. ('The Paper Trade Journal,' xiv. p. 43.)

Wood fibre, excepting chemically prepared wood-pulp, will not bleach whiter than a light-yellow or cream color by the chlorine process without decomposition, and partial or entire destruction of fibre taking place.

Mr. Goldsbury H. Pond, of Glenn's Falls, N. Y., has patented a process in which he uses metallic oxides with chlorine or any other substance or solution that will yield oxygen by the action of these metallic oxides, depending entirely in this process upon the generation of oxygen in contact with the material to be bleached, either in air or in a bath of any substance which may be compatible with a bleaching operation, and which is capable of yielding up its oxygen through the agency of a metallic oxide.

In this solution wood-pulp or any other fibrous material to be bleached is thoroughly wet or mixed therein, thereby bringing into a close and positive contact the material to be bleached and the innumerable points of generation of the oxygen.

To accomplish these results there is prepared a bath of a weak solution or mixture of any metallic oxide mixed with water, such as the oxide of iron, copper, zinc, lead, nickel, or cobalt. The inventor prefers the oxide of iron and zinc to all others.

In using the oxides they are mixed with water, and are then thoroughly mixed with the material being bleached, so that the fibres thereof are completely covered with it, the metallic particles being deoxidized and oxidized to an un-limited extent, thereby developing a large quantity of oxygen.

In bleaching wood pulp and any other material of a fibrous nature—such as hemp, jute, flax, cotton, or the waste of any of these—for the manufacture of paper, take the common bleaching tank or engine now in general use in paper-making, fill it partly full with water and with the metallic oxides mixed therein in the proportions of one pound of the oxide to one hundred pounds of the material to be bleached, this proportion being varied according to the amount of oxygen required ; then add a quantity of the solution of either chloride of lime or chlorine water, or any other solution capable of yielding oxygen by the action of the metallic oxides. Then fill the tank or the engine to its working capacity with the wood pulp, heating with steam to nearly the boiling point. The beating-roll by its revolution thoroughly mixes the pulp with the solutions of chlorine or chloride of lime and the metallic oxides therein, when by the action of the heat large volumes of oxygen are produced within the mass of the pulp and in contact with each fibre thereof, and as the oxygen is generated the bleaching is immediately effected. After the bleaching is completed the whole mass of the wood pulp is acidulated with dilute acid, decomposing the oxides, which are then washed out in combination with the acid, leaving the pulp clean, and it is claimed perfectly white.

Chlorine gas or water saturated with chlorine, or any solution susceptible of yielding oxygen by the action of the metallic oxides, may be used and mixed in the same manner as before described, with the pulp, water, and oxides in the bleaching engine, and treated in the same manner, and when heated the same results it is claimed will follow, and a large volume of oxygen will be generated, and when in contact with various fibrous materials and the wood pulp bleaching it to a permanent white in a few minutes.

To accelerate the process of bleaching, more metallic oxide can be added at any time during the operation.

It is a well-known fact that in the use of chlorine when it is heated it leaves its solution and goes off into the air. In this process the oxygen is not formed and the bleaching is not accomplished unless heat is applied and the chlorine or other solution or mixture containing oxygen brought into contact with the metallic oxide. The nearer to, but under, the boiling-point this process is operated the more voluminous will be the generation of the oxygen, and the more efficient and immediate will be its bleaching properties.

METHOD FOR BLEACHING WOOD, STRAW, ETC.

In the manufacture of white paper from wood, straw, etc., it is of great importance, after such stock has been boiled in an alkaline solution, that the pulp thus produced should be thoroughly cleansed from the alkali and saccharine or glutinous matters remaining in it before the pulp is subjected to the action of chlorine for bleaching it for white paper. After the stock has been bleached with chlorine, it is equally important that all traces of the chlorine be removed from the pulp before it is run out on the paper-machine into any white paper.

It is a fact well known to the best paper-makers that the more thoroughly the pulp is cleansed from alkali and glutinous matter remaining in it after it has been boiled, the less chlorine it takes to produce a given shade of white, and it is equally well known that the action of the chlorine tends to weaken and destroy the fibre, and that the less chlorine used, the stronger and more pliable is the paper produced from straw and like fibrous material. Hence the necessity of a thorough washing or cleansing of the pulp before it is bleached, as well as afterward.

Another fact is, that the longer the stock is allowed to remain in the pulpy state after being bleached, before the chlorine is washed out, the more hard, crisp, and brittle is the paper produced; hence it is of importance that such fibrous material be bleached quickly, and the chlorine immediately washed out.

Still another fact, but one not generally known, is that, if a certain amount of chlorine is to be used for bleaching a certain amount of stock, if, instead of using the whole quantity of chlorine at one time, a portion only is used to bleach the stock, and the stock then dried sufficiently to be handled, and the chlorine washed out, and then the remaining chlorine used to rebleach the stock, a much whiter and cleaner paper is produced.

For example, if twenty pounds of chlorine are to be used to bleach one hundred pounds of pulp, instead of using all the chlorine in one operation, first use, say, fifteen pounds to bleach the one hundred pounds of pulp, and then, after thoroughly washing the pulp, use the remaining five pounds of chlorine to rebleach the whole quantity of pulp. But, to perform these operations in mills as now constructed would require a great amount of extra time and labor.

As usually practised in the best paper-mills in the country, the pulp, as it comes from the boilers, is placed in an ordinary rag engine, and partially washed. It is then emptied into a tank or stuff-chest, and drawn off, by means of a pump, to the mixing-box of an ordinary wet-paper machine, where it is mixed with the water extracted from pulp previously passed through the machine.

The pulp thus mixed is then passed through the screen or pulp-dresser and onward to the wire-gauze cylinder, if a cylinder machine is used, where the water is partially extracted and carried off through the cylinder and the pulp thereby formed into wet paper on the surface of the cylinder, and from whence it is taken off by the " wet felt."

The wet machine has been in use for the purpose of running out straw and wood-pulp, and removing uncooked portions of the pulp as it passes through the screen or pulp-dresser, a small portion of the alkali and saccharine matter being also incidentally removed as the wet paper passes through between the press rolls.

In carrying out a process used by one of our most successful paper manufacturers there is made use of, among other devices, the ordinary wet machine, with certain improvements.

These improvements consist, first, in allowing the water which passes through into the perforated cylinder to run off to waste, carrying with it, of course, the alkali and saccharine matter held in solution, and, second, in keeping

24

up the supply of water necessary to work the pulp through the screen or dresser, and to form the paper on the cylinder, by discharging fresh, clean water into the receiving vat or mixing trough, and an additional supply of clean water into the pulp after it has passed the screen, but before reaching the cylinder, as hereafter more fully described.

As a much larger amount of water can be passed through the cylinder than through the screen, the amount of water added to the pulp after passing the screen may be very considerable, so that, by these improvements, a much greater quantity of water can be washed through the pulp than by the old method, and consequently the pulp is more thoroughly cleansed.

Fig. 116 is a top plan view of a common cylinder wet-paper machine, having the improvements mentioned applied.

Fig. 117 illustrates a side view of the same, with portions broken away to show the interior.

Fig. 118 shows a side elevation of a train of three wet machines, with the accompanying bleaching tanks, stuff chests, etc., as arranged for carrying out the process.

A, Figs. 116 and 117, represents a "wet machine" of the ordinary construction, so far as regards the general features of screens, vats, perforated cylinder, endless felt, press-rolls, etc., and in which *a* represents the receiving vat for the pulp; *b* the screen or pulp dresser; *c* the vat which receives the pulp passing through the screen; *d* the trough in which the wire-gauze cylinder is mounted, and which is kept filled with pulp and water; *e* the gauze cylinder; and *f* a pipe communicating with the interior of the cylinder, and with an exhaust pump for causing a suction inward through the cylinder.

Fig. 116.

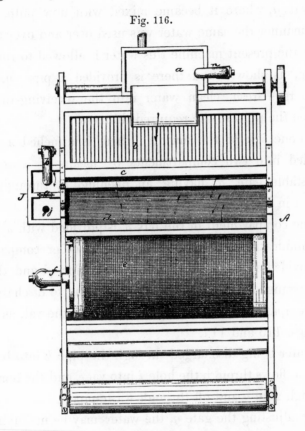

Fig. 117.

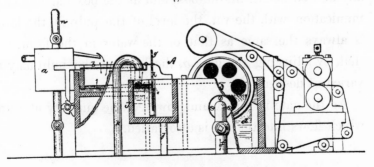

Heretofore it has been customary to have the pipe f discharge the water exhausted through it back into the receiving-vat a, where it became mixed with new pulp, and in this manner the same water was used over and over again.

In the present machine this water is allowed to run off to waste, as shown, and there is provided a pipe, n, which discharges fresh, clean water into the receiving-vat a, to reduce the pulp to the required thinness.

To one side of the machine there is attached a box, J, divided into two compartments by means of a vertically adjustable gate, g, having an opening, h, through it, as shown in Figs. 116 and 117.

One compartment of the box J is provided with a hole, i, communicating with the vat c, and the other compartment is provided at its bottom with a waste-pipe, j, and there is also arranged a water-supply pipe, k, so as to discharge into the compartment which is connected with the vat, as shown in Figs. 116 and 117.

Water being discharged through the pipe k into box J, a portion flows through the hole i into vat c, and the remainder through gate g and off through waste-pipe j.

By adjusting the gate g, the water may be maintained at any desired height in the box, and as the box has free communication with the vat, the level of the pulp in the latter is always the same as that of the water in the former, so that, by adjusting the gate g, the level of the pulp may be varied as desired.

As the gate is raised and lowered, the quantity of water which flows into the vat is also varied.

By this arrangement of parts it is possible to work a very large quantity of water through the machine with the pulp, so as to thoroughly dissolve and wash out all alkali and saccharine matters, ink, chlorine, and other foreign matters, which are carried off with the water through the cylinder *e*.

In the old styles of machine, where less water could be worked through, the alkali and saccharine matter were only partially dissolved, and the only portions that were removed were those that were squeezed out from the wet paper in passing through the press-rolls, as all that was held in solution by the water passing through the gauze cylinder was returned with the water to the receiving-vat, and remixed with fresh pulp or stock.

In carrying out this process there are provided two or more of the wet machines A, and double the number of bleaching-tanks *B*, all arranged as shown in Fig. 118, placing first a machine and then two tanks, side by side, and so on.

Fig. 118.

Near the first machine there is located a tank or stuff-chest, *C*, and connected by a pump, *D*, with the receiving vat *A*, of the first machine.

At the back end of the machine there is arranged an elevator, *E*, which may be shifted so as to communicate with either one of the two tanks, *B*, there located.

Each of the tanks B is provided with a discharge-pipe, emptying into the receiving-vat of the second machine A', and at the back end of this second machine is an elevator discharging into the second pair of tanks, and so on continuously, each machine connecting with a pair of tanks, and these tanks with the next machine. The last tank is connected with either a rag engine, as shown, or with a tank or stuff-chest, according to circumstances.

In operating the train of machines, as shown in Fig. 117, the pulp, after being boiled and then washed in an ordinary rag engine for about an hour, is discharged into the tank C as unwashed pulp; or, if properly boiled, the pulp may be taken directly from the boiler to the tank C without being passed through the rag engine.

From the tank C the pulp is carried by the pump D to the first wet machine A, where the pulp is thoroughly washed and cleansed, and freed from alkali, saccharine matter, ink, chlorine, and other impurities contained in it.

The pulp, in passing through the machine, is converted into wet paper, and in this form it is delivered to the elevator E, which carries the wet paper over into the first bleaching tank B, which is charged with a solution of chlorine, and provided with stirrers or agitators for breaking up the paper and reducing it to the form of pulp again.

When the first of the tanks, B, is sufficiently charged with the wet paper, the elevator E is shifted so as to discharge into the adjoining tank, which is, like the first one, charged with chlorine, and provided with agitators.

While the second of the tanks B is being filled, steam is

admitted into the one which is full, and the contents heated to about 100° above the temperature of the surrounding air, which causes the chlorine to act with great rapidity.

This is continued for about two hours, the agitators being kept in motion all the while, when the pulp is discharged in the receiving-vat of the second wet machine A^1, through which it is passed to wash out the chlorine and coloring matter dissolved by it while in the tank B.

This second machine again forms the pulp into wet paper, and then discharges it on to the second elevator E', which delivers it into one of the second pair of bleaching tanks B^1, where it undergoes the same treatment as in the first, except that the chlorine solution is of only about one-third the strength of the first.

The pulp is then discharged from this tank B^1 to the third machine A^2, where it is washed from the chlorine, formed into wet paper, and then, by the elevator E'' delivered into the third tank B^2, where it is again broken up and reduced to pulp.

From this third tank the pulp is discharged into an ordinary rag engine, and the coloring and sizing matter added to it, and then discharged into a tank or stuff-chest, from whence it is delivered on to the usual paper machine, and by it formed into finished paper.

It is obvious that the bleaching tanks may be located in the basement of the building, immediately under the machines, and pumps used to draw the pulp up from each tank to the next machine, and also that the last tank could be dispensed with, and the wet paper put into the rag engine by hand, but it is preferable to locate them as shown, for the reason that thereby hand labor is entirely saved, and also the cost of extra pumps.

There are numerous advantages claimed for this process over the old methods, among which we mention :—

First, a much more rapid and perfect cleansing of the pulp from the alkali and saccharine matter, as in the same length of time there is washed about ten times the quantity of water through it that could be done by the old method.

Second, it requires no drainers, and therefore the labor required under the old method, to remove the stock from the drainers to the rag engine, is dispensed with, and, besides, in the old method, more or less of the stock was lost in the many handlings from drainers to cars, cars to rag engines, etc., and still more of it injured by dirt of various kinds getting into it while being thus handled.

By the present arrangement it will be seen that the only handling required is in passing the stock, in the first instance, from the boiler to the washing engine.

This process it is claimed produces the finest kinds of book paper from straw, wood, and like fibrous materials, with less labor, trouble, and expense than attended the production of common news printing paper by the old method, and the quality of paper that can be produced is said to be far superior to that made upon the old plan, and hence will conduce to the use of straw, wood, etc., for the finest paper, instead of the far more expensive stock formerly used.

Bleaching Esparto.

The usual method employed in Great Britain for bleaching esparto has already been described on p. 325 *et seq.*

BLEACHING JUTE.

It is already some years since the chemists, Messrs. Cross and Bevan, began to study the jute fibre, yet in spite of the importance of their discoveries they seem not to attract the attention they deserve. According to the investigations of Messrs. Cross and Bevan, jute does not contain cellulose under the ordinary conditions, but in the form of one of several ethers of the cellulose, which they comprise under the denomination of bastose. While cellulose belongs to the class of hydrates of carbon, bastose constitutes the link between the latter and the aromatic compounds, and consequently jute has properties very different from those of the other vegetable fibres. When treated with chlorine this bastose is transformed into a chlorinated compound which offers two characteristic reactions. The treatment with sulphite of soda gives a bright magenta red coloring matter, which, by the action of alkalies, is decomposed in soluble substances belonging to the group of tannic acid. This latter fact is of great importance practically, since, while in the coloring of cellulose, it is necessary to mordant it previously, jute already possesses this mordant from the beginning; it is, in a certain sense a mordanted cellulose, and therefore takes more easily the aniline colors.

Messrs. Cross and Bevan have also discovered two other

remarkable properties of jute, which are technically of very great importance. When large quantities of fibre are kept for some time in a damp condition, principally in presence of sea-water, the fibrous material is decomposed in substances analogous to tannin, and in acid belonging to the group of pectic acid. The fibre itself is rendered more or less rotten, and in some cases reduced to a powder. It is very likely that the larger quantity of jute brought over to Europe is already more or less injured, either from long exposure on the voyage or from circumstances which have affected it before it is shipped.

Another singular property of jute is its decomposition by means of acids, principally mineral acids. At a comparatively low temperature jute is transformed by acids into solid combinations, which at a higher temperature are changed in their turn into brown substances and into volatile products of disagreeable odor, such as furfurol and others. This is the origin of the color and smell of the majority of the jute fabrics treated by acid. Dyers never employ the same receipts for jute that they would for cellulose, as the latter is a great deal more resisting towards acids. It has been found that a small amount of acetate of soda prevents the destructive qualities of the mineral acids, one pound of acetate of soda being sufficient for about five to six gallons of the water used.

Jute can be very easily bleached by means of permanganate of potash; the loss experienced by this process is about 3 to 4 per cent. Of course before treating with permaganate the jute must be thoroughly cleansed either by means of alkali or soap. This method, which would give good results, is unfortunately too expensive, since $3\frac{1}{2}$ to $4\frac{3}{4}$ pounds of permanganate of potash would be required for the bleaching of 100 pounds of jute. According to Messrs. Cross and Bevan the hypochlorites are the only available materials for using on a commercial scale, but great care must be taken in their employment on account of the action asserted by free chlorine on the fibre. Chloride of lime cannot be used as such in the bleaching of jute, as it transforms the same into chlorinated compounds; this latter is easily distinguished by the magenta-red coloration it takes when wetted with sulphite of soda. When such a chlorinated jute is steamed a decomposition takes place, muriatic acid is liberated, a brown coloration is formed, and the tissue is rendered completely rotten.

The hypochlorites oxidize the jute into combinations, which form with insoluble compounds, and which are very difficult to eliminate.

It is well known that in the laboratory jute can be easily bleached by being suspended over phosphorus in a damp atmosphere, and also by means of oxygenated water.

According to a process recently patented the jute stock is first thoroughly washed in the usual manner, after which a composition of 10 pounds of alum, 4 pounds of South Carolina clay, and 2 gallons of water is boiled and added to 500 pounds of the stock, which is then run in the mixing engine for fifteen minutes, after which 50 pounds of bleaching powder are added.

In like proportions the composition and bleaching powder

are used with larger quantities ·of the stock, so that it is claimed that one-half of the quantity of bleaching powder used in the processes heretofore employed is saved, as it commonly requires 100 pounds of bleaching powder to bleach 500 pounds of this stock.

In addition to the claimed saving in the quantity of bleaching powder employed, it is also claimed for the present process that the annoyance of "foam" and the use of coal-oil to kill the same are obviated, and that the wire-cloth jacketing and felts used on the paper-machines are saved.

We here refer the reader to the description of Conley's process for bleaching jute.

Bleaching Materials composed of Hemp, Flax, etc.

The following process, which is the invention of Mr. Auguste Demeurs, of Huyssinghen, Belgium, relates to an improved process of bleaching applicable to materials composed of hemp, flax, or other products containing stalks, straws, or the like (which generally resist bleaching by chlorine alone), and permitting the materials named to be utilized in the manufacture of the finest white paper.

The process is carried into effect in the following manner: The materials having previously been subjected to steeping or boiling in lye, more or less strong according to their quality, and then suitably bleached with chlorine gas, are introduced into the chlorous vat, or subjected to a primary washing of the pulp in the ordinary manner. When this first operation has been suitably effected—that is to say, when the pulp has completely lost its yellow color, produced by the action of the bleaching with gas, and when the washing-water, at first acid and cloudy, has become clear and neutral—the supply of water is stopped, the washing drums or rollers being still allowed to operate until the half-stuff has attained in the vat the degree of concentration desired.   It is at this moment that the straws are attacked, the color of which has become almost completely black by reason of the washing.   For this purpose a caustic lye composed of equal parts of carbonate of soda and lime is introduced into the vat.   This bath, the degree of which is in proportion to the kind of material to be treated, is generally prolonged for two or three hours, at the end of which time it is claimed that it will be impossible to discover in the pulp the least trace of straws.   After a second washing, which destroys the brown color produced by the lye, the bleaching of the pulp is proceeded with by means of a solution of chloride of lime, the quantity of which can be considerably reduced, because the filaments have acquired a certain degree of whiteness by the action of the alkali, and there is, it is claimed, no longer any fear of the presence of straws, which resist the action of the bleaching.

This process, which is claimed to be much more economical than the ordinary methods, offers also the great advantage of bleaching the straws in such a manner that they do not reappear in the finished paper at the end of several months, which enables the manufacturer to keep his pulp in stock or on sale with impunity.

Although the same result can be obtained by the use of

other alkalies, the patentee considers the caustic lye of carbonate of soda and lime the most advantageous. The manufacturer should choose that alkali which appears to be most advantageous for his purpose.

The bleaching with chlorine gas which sometimes follows the reduction of rags into half-stuff can be dispensed with and replaced by a solution of chloride of lime. The result is claimed to be a sensible diminution in the cost of the operation, its duration, and the cost of labor, without taking into account the difficulties and inconveniences heretofore experienced by a number of manufacturers, who can now employ the above-described process without requiring new plant.

BLEACHING VEGETABLE TISSUES WITH PERMANGANATE OF POTASH, AND NEUTRALIZING WITH OXALIC ACID, SULPHITE OF SODIUM, AND CHLORINE.

This process, which is the invention of Mr. John A. Southmayd, of Elizabeth, N. J., is intended as a substitute for the bleaching processes usually practised in pulp-grinding engines, wherein from five to ten hundred pounds of pulp are acted on at once, and a period of eight to twelve hours is consumed in working the pulp through the bleaching liquor. In mills of large capacity from ten to twenty such engines, using from seventy-five to one hundred and fifty horse-power, are employed all the time in bleaching and washing the pulp, it being absolutely necessary to remove every trace of the chlorine to prevent yellowness in the product. In the present invention it is claimed that any required amount of the stock can be bleached at one time in a single vessel, the chemicals acting in a much more rapid manner than does chlorine, and a charge of five tons, it is claimed, may be washed and bleached in about five hours.

The material to be bleached may first be treated with caustic potash to soften and prepare the tissues for the action of the permanganate of potash, and such treatment it is stated is best effected in a closed boiler, where the heat and pressure of steam may be used to facilitate the mechanical and chemical action of opening the fibres. The permanganate attacks and decomposes the coloring matter, but does not remove it, such matter being subsequently removed by the use of the oxalic acid and sulphite of sodium.

We will first describe the application of the process to the bleaching of hard spruce, and then state the modifications employed with other fibres. The first stage of the process consists in boiling the fibrous matter or tissue with a solution of permanganate of potash, in the proportion of ten to fifteen pounds of the agent to one ton of the fibre, until the same appears to be thoroughly oxidized, the operation requiring from one to two hours if performed in a closed vessel under steam pressure, but a rather longer time if boiled in an open vessel. The application of heat in this stage of the process is essential to produce the required effect; but the subsequent treatment may be performed in a closed or open vessel without heat, as may be most convenient. When the fibre is properly affected by the permanganate, treat it with a solution of oxalic acid and sulphite of sodium, which effects the bleaching in about two hours by the decomposition of the permanganate and coloring matter. With certain kinds of tissues this treatment suffices to discharge all the color; but in cases where the fibre, owing to its place of growth and the presence of certain salt in its texture, is not wholly bleached by such treatment, it is desirable to prepare this acid and sulphite with the addition of a small amount of chlorine, thus securing a totally different action with the chlorine from that produced by either the acid or chlorine alone. For a ton of such fibre, the acid solution is prepared by dissolving from forty to sixty pounds of the acid and fifteen to twenty-five pounds of sulphite of sodium in two hundred gallons of water, and the fibre is preferably boiled in such solution to produce the desired effect, although the same results can be obtained by using the acid without heat, if a longer time be allowed for its action. As the spruce fibre is very difficult to bleach, other tissues can be whitened in a shorter time and with a smaller proportion of the agents employed.

In treating manufactured fibres, as rags, the patentee states that he has found that, owing to the twisting of the threads and the variety of thicknesses, textures, and colors, which he subjects to treatment at one time, he is compelled to use nearly the same amount of chemicals as for hard crude fibres; but with soft grasses and soft woods like poplar he uses only from seven to twelve pounds of permanganate, twenty-five to thirty-five pounds of acid, and seven to twelve pounds of the sulphite of sodium for each ton of the tissues. He also finds in practice that it requires about sixteen hundred gallons of water to soak a ton of spruce fibre, and the preparatory boiling with potash, which prepares the fibres so peculiarly for the action of the permanganate, therefore requires that amount of water. When the alkali is drawn off before the bleaching operation, the bleaching agents are then applied with the water in which they are dissolved, and enough water is added to thoroughly boil the charge.

BLEACHING PAPER PULP BY APPLYING THE BLEACHING AGENT IN A PULVERIZED OR SPRAYED CONDITION.

Mr. Jean B. Fessy, of Saint Etienne, Loire, France, has lately patented an invention having for its object to effect economy in the cost of bleaching by reason of the small quantity of bleaching agents required, and to save time, owing to the rapidity with which the process can be performed, by effecting the thorough utilization of the action which develops from the decolorizing agents when in a nascent condition.

To attain these objects the bleaching is effected by submitting the materials to be bleached to the action of a solution or solutions of the decolorizing agents when in a state of pulverization, spray, or fine division, which may be effected by causing the agents, while under pressure, to come in contact with a resisting medium, or to be dispersed by steam or compressed air, and the materials to be bleached are submitted to the action of this finely-divided agent or agents, whereby it is claimed the bleaching can be effected readily and to any required degree. The bleaching agents may be of the usual kind; but for decolorizing paper pulp chlorous acid is preferred.

25

Fig. 119 is an end elevation, and Fig. 120 a side elevation of the apparatus employed by Mr. Fessy for spraying the bleaching agent.

Fig. 119.

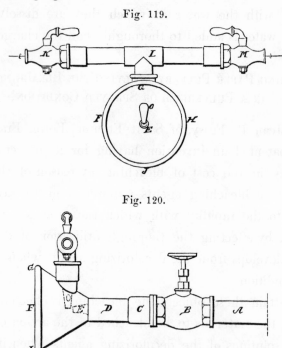

Fig. 120.

The apparatus employed is analogous to the pulverizer of Koerting Brothers, and is constructed as follows: The steam or other fluid passes through a tube, *A D*, and stopcock *B* into a nozzle, *E*, from whence it issues under pressure into a funnel, *F G H J*, into which opens in front of the nozzle *E* a nozzle, *O*, through which the bleaching liquids to be pulverized pass by separate branches, *L*, provided with stopcocks *M K*, so that the liquids are instantly combined in any required proportions capable of being regulated by the stopcocks $M K$. The two nozzles constitute an arrangement resembling a spray-producer inclosed in the funnel or trumpet mouth $F G H J$, and produce a thorough pulverization and mixture of the bleaching agents, which issue therefrom in a fine state of division and act instantaneously upon the pulp or other materials to be bleached or decolorized. The stopcock B on the pipe for the steam or other fluid under pressure enables the supply and pressure to be regulated to give the proper pulverization of the agents for their due action on the materials to be bleached or decolorized. The steam or other fluid under pressure, escaping by the one nozzle, draws the liquids or agents from their respective supply-pipes and nozzles in regulated quantities according to the adjustment of the stopcocks, and instantaneously sprays and thoroughly combines the same, and this finely pulverized combination acts upon the matter to be bleached, and effects the bleaching it is claimed by instantaneous reaction.

The pulverizing apparatus described is the best with which the patentee is acquainted for the purposes of this invention, the essence of which is the instantaneous pulverizing and combination of the bleaching agents which will give the reaction necessary for bleaching. The bleaching thus effected it is claimed does not deteriorate, and by its aid paper pulp can be bleached with rapidity, and it is claimed that it is possible to bleach pulp which hitherto could not be bleached in a practically available manner, and facility is also given for obtaining an absolutely regular and uniform bleaching to any desired tint, as the operation can be arrested at any stage. There is economy in the use of the bleaching agents, as any surplus is not wasted.

BLEACHING IN ROTARIES.

Rotary boilers are used in some mills instead of bleaching engines. It is true that in these rotary bleachers the chlorine gas has no means of escaping into the atmosphere; but for some classes of pulp the friction produced by the rotary motion of the boilers is objectionable, and then again the progress of the bleaching operation cannot be watched.

The construction of rotary bleaching boilers is often defective and dangerous. Mr. Harrison Loring, of Boston, Massachusetts, has patented an invention which consists in so constructing and arranging the gudgeons and induction pipes for rotary bleaching boilers of all the known forms that the induction pipes pass into the boiler separate and independent of the trunnions, and the latter made hollow, in the form of an annular ring, with flanges attached to the head of the boiler, thus obtaining a larger bearing surface of the journal, greatly strengthening the head, and avoiding cutting a large hole in the centre of the same, as is necessary in the old form of solid gudgeon. The metal is thus distributed in a uniform manner, thereby avoiding strain of the casting in cooling, and by making the induction pipes separate, heating and expanding of the gudgeons are prevented. The gudgeons are so made that there shall be no communication between the outer casting and the inside of the boiler, thereby removing all liability of explosion by reason of defects in or accidental breaking of the casting.

Fig. 121 represents a longitudinal section of Loring's invention. Fig. 122 is an end elevation. *A*, represents the end or head of the boiler; *b*, a section of the gudgeon

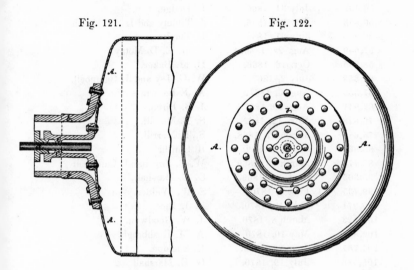

Fig. 121.   Fig. 122.

attached to the same; *c*, the stuffing-box, also attached to the boiler-head within the gudgeon; and *d*, the induction pipe passing through the stuffing-box.

List of Patents for Bleaching Pulp, issued by the Government of the United States of America, from 1790 to 1885 inclusive.

No.	Date.	Inventor.
4,616	July 2, 1846.	J. G. Kendall and J. H. Kendall.
13,008	June 5, 1855.	⎫
Reissue		⎬ H. Loring.
2,320	July 24, 1866.	⎭
16,100	Nov. 18, 1856.	J. A. Roth.
25,975	Nov. 1, 1859.	
44,250	Sept. 13, 1864.	J. B. Meldrum.
46,774	March 14, 1865.	G. W. Billings.
51,569	Dec. 19, 1865.	J. W. Dixon.
52,250	Jan. 23, 1866.	J. Short.

No.	Date.	Inventor.
53,152	March 13, 1866.	
Reissue		H. L Jones and D. S. Farquharson.
2,384	Oct. 25, 1866.	
55,834	June 26, 1866.	J. W. Dixon.
56,732	July 31, 1866.	L. Dodge.
56,833	July 31, 1866.	J. Tiffany and H. B. Meech.
56,860	July 31, 1866.	F. Perrin.
57,649	Aug. 28, 1866.	C. M. E. DuMotay.
58,935	Oct. 16, 1866.	H. M. Baker.
66,353	July 2, 1867.	W. C. Joy and J. Campbell.
67,559	Aug. 6, 1867.	A. J. Loisean.
67,941	Aug. 20, 1867.	J. B. Biron.
70,878	Nov. 12, 1867.	S. T. Merrill.
75,691	March 17, 1868.	S. T. Merrill.
85,860	Jan. 12, 1869.	B. Smith.
87,779	March 16, 1869.	W. C. Joy and J. Campbell
95,365	Sept. 28, 1869.	G. E. Marshall.
99,735	Feb. 8, 1870.	S. W. Widder.
100,071	Feb. 10, 1870.	E. J. Rice.
100,523	March 8, 1870.	J. W. Goodwyn.
102,868	May 10, 1870.	A. M. Koshbrugh.
104,781	June 28, 1870.	E. Sheldon.
105,585	July 19, 1870.	G. E. Marshall
106.711	Aug. 23, 1870.	H. B. Meech.
108,509	Oct. 18, 1870.	C. E. O. Hara.
116,020	June 20, 1871.	J. Campbell.
116,338	June 27, 1871.	H. Monroe.
122,783	Jan. 16, 1872.	J. W. Rossman.
125,658	April 16, 1872.	J. Campbell.
153,775	Aug. 4, 1874.	H. J. Lahousse.
162,043	April 13, 1875.	G. W. Dubuisson.
165,307	July 6, 1875.	E. Conley.
166,117	July 27, 1875.	H. Loring.
266,782	Sept. 11, 1882.	A. Demeurs.
294,619	March 4, 1884.	E. Mermite.
302,055	Aug. 5, 1884.	J. A. Southmayd.
307,390	Oct. 28, 1884.	E. A. D. Guichard.
311,425	Jan. 27, 1885.	J. B. Fessy.
312,525	Feb. 17, 1885.	P. Souders, C. Smith, H. C. Craighead, and N. Souders.
321,452	July 7, 1885.	G. H. Pond.
322,655	July 21, 1885.	A. W. Wilson.

CHAPTER XII.

BEATING—BEATING ENGINES—LIST OF PATENTS FOR PULP
ENGINES AND BED-PLATES.

BEATING.

THE paper may not be made in the beaters as some manufacturers claim it is, but the beating-engine department in the paper-mill is a very important one, and in addition to being roomy and well lighted it should be kept in an orderly and cleanly condition.

The foreman of the beating department should be a thoroughly practical and trustworthy man, upon whom the superintendent or the owner of the mill can rely implicitly that all orders will be carried out exactly as to the quantities and as to the order and time as given.

It is one of the most delicate and important operations in paper-making to determine the composition of the pulp, or the relative proportion of each grade, the combination of which is to supply the beating engine, and the responsibility should rest solely with the superintendent of the mill, who alone should decide the matter.

Papers are usually made upon orders for a certain pattern, and it is important to impart to them the qualities required, although it must be admitted that they are not always compatible. It is for the manufacturer to appreciate the practical value of these conditions and then to regulate the general work of the mill accordingly.

Taking a certain theoretical composition for his pulp he must see whether the cost to which the paper will come is not too high.

The superintendent ought, therefore, to have in his mind, or near at hand, everything necessary for his information; the cost of the raw materials, and the expenses of cutting, boiling, bleaching, loading, sizing, and coloring. He should also see whether the supply will allow him to employ a certain grade in preference to another. Some of these items of cost can, of course, be varied when the pulp, such as wood, straw, etc., is purchased already prepared.

When these various points have been settled he must foresee the difficulties that may be met in making the paper by machinery. If the paper is to be glazed or colored, will such and such a material not be apt to introduce too many lumps, etc., into the pulp, and will this not result in wearing out the wires and felts too rapidly?

It is in the matter of the composition of pulps that the knowledge of the successful manufacturer is displayed. Before settling the question he must indeed have gone over all those involved in the art of paper-making. It generally requires but a few minutes' reflection for one who thoroughly understands the capabilities of his mill. The foreman of the beating department ought to pay attention so as to see that each lot of pulp contains the grades required for his working lists for the day, and a sufficient quantity for the number of lots of stuff he is ordered to prepare. The assistants bring in the fixed quantity of pulp to be beaten and throw it into the tank of the beating engine, and when this is filled washing is commenced and continued for a sufficient length of time.

When the pulp has arrived at such a degree of tenuity that it may be in danger of passing between the wires of the strainer this is closed and the washing ceases.

When animal-sized papers are being made, a sufficient quantity of hyposulphite of sodium or "antichlorine" to neutralize the chlorine is introduced just as soon as the washing is completed.

But with engine-sized paper the loading material should first be added after the washing is finished; then the size is introduced, then the alum, and finally the coloring matter.

When the washing is completed the washing drum is raised and the beating roll is then gradually lowered upon the plate; but before the roll is lowered the engineer must be satisfied that the chlorine in all its combinations has been thoroughly eliminated.

There are numerous methods of testing for chlorine. Small slips of blue litmus paper are sometimes used for this purpose, the washing being continued as long as the blue litmus color of the slips is changed to red after being immersed for a moment or so in the contents of the engine trough.

A more sensitive method by which the presence of chlorine can be established with greater certainty is based on the characteristic color which iodine produces in contact with starch.

In order to test the contents of an engine by the latter method a handful of the pulp is taken out, pressed so that the excess of liquor runs off, while leaving the pulp still moist, when a few drops of a solution prepared as follows are poured on it: $\frac{1}{4}$ oz. of starch is mixed with sufficient cold water to form it into a paste, enough boiling water being then added to make the mixture up to one pint when two drachms of iodide of potassium are added and thoroughly incorporated; the test is ready for use when cold. If chlorine is present in the pulp a few drops of this bleach test will color the stuff blue-black; but if the stuff is free from chlorine no change in color will take place.

As soon as the wash-water is cut off from the engine and the washing drum is raised the engineer commences the beating by lowering the beating roll sufficiently to begin the operation, the space between the knives being gradually curtailed as the operation proceeds. The theory of the beating process is that the fibres are not to be cut but are to be drawn out to their utmost extent by the action of the knives and the friction among the mass of material itself while in the trough of the beating engine.

This theory cannot be fully carried out for ordinary grades of papers; but where the number of engines and the capacity of the mill and margin of profit will allow it the principle should be worked up to as closely as possible.

Long pulp is produced by blunt knives and slow working; short pulp by sharp knives and quick work.

The beating of the stuff into pulp is usually timed according to the thickness of the paper to be made from it, and in direct proportion to the uniformity of time consumed in the preparation of a lot of pulp to be made at a specified weight, will the regularity in quality and weight run on the wire of the machine.

Three to five hours may often be sufficient time for running off a beater of pulp to be used for thick paper; but twenty-four hours or even longer will be required when the pulp is to be used for the thinnest sheets.

The touch is sometimes relied upon by those having considerable experience, to determine the fineness of the pulp; but this fact is best determined by the "proof." A cylindrical vessel made of copper, zinc, or gutta percha is generally employed for this purpose, a small quantity of the pulp being placed in the vessel and diluted with a large quantity of water; as the thin mixture is slowly poured off and flows over the rim as a very thin sheet, the fibres of the pulp will take a direction parallel to that of the current of water, at the point where it flows over the rim, thus allowing the length of the fibre to be determined.

If the pulp looks cloudy and quite a quantity of little white points or lumps remain visible, it will be necessary to make these imperfections disappear, and for this purpose the engineer lowers the roll so that the extremity of its blades almost impinge those of the plate, but they must not actually touch. In fifteen or twenty minutes the lumps can usually be brushed out under the action of the roll.

Many manufacturers are handicapped in their efforts to produce the best qualities of the different classes of papers by having too small a number of beating engines in their mills, as the pulp in such cases must be worked off too hastily.

Much, however, depends upon the workmen in the beating department, whether the pulp is of the desired quality or not. It is possible to work even a comparatively weak material into a reasonably strong paper, if care is exercised to properly handle it in the beating engine; but, as has been previously intimated, "if the stuff is not correctly treated, such as by sending out stuff for laid paper too fast and long, or too soft and carrying too much water, the weight will vary, and the paper crush at the couchers and stick at the press rolls, causing all sorts of trouble and confusion to the machine man, and a considerable amount of waste."

When the beating is about three-quarters completed the loading materials, the sizing, and coloring matters are thrown into the trough of the engine; and while the pulp is finishing, the mixture of the contents takes place, this, however, may often be facilitated by stirring.

When the lot of pulp is finished the engineer lifts the plug and the pulp is run off through larger pipes made of copper or other suitable material into the supply vat of the paper machine. The trough of the engine is then carefully rinsed and the plug replaced; after which the engine is in readiness for the commencement of another operation.

The different materials used for the manufacture of paper, of course, require to be treated in the beating engine according to their nature; pulp made from old paper requiring only a thorough brushing so as to prevent any small pieces of paper from passing through without being reduced to fibre.

If the waste papers or "imperfections" are to be mixed with rags, the latter must be thoroughly reduced before the paper stuff is added.

Bleached straw and wood pulps, as has been previously stated, are already reduced to fibres, and are generally only mixed in beating engines, some of which in leading mills have only a smooth bed plate.

"Antichlorine:" Its Preparation.

Hyposulphite of sodium, or so-called "antichlorine" is used in paper-making to discharge the bleach from the pulp. There are several ways of procuring hyposulphite of soda. Very fine crystals may be obtained by passing sulphurous acid gas, well washed, into a strong solution of sodium carbonate, forming neutral sulphite of soda, and then digesting the solution with sulphur at a gentle heat.

The following simple method of preparing antichlorine will answer for most paper-mills: Have a large square wooden box or cask constructed, and place it upon a platform about 3 feet and 6 inches high. On the inside of the box or cask nail a sufficient number of blocks to support two movable frames, which are to be covered with old close fishing seine or netting made of twine; upon each frame there are to be placed 250 pounds of the common crystal soda of commerce, care being observed to have the meshes of the screens sufficiently close to prevent the soda from falling through. A tight cover is then put on the box or cask, and "daubed" or luted around with soft clay so as to make the cover air-tight. It is preferable to attach an air-cock to the cover so as to allow the air to escape from the interior of the receptacle; but two or three small holes made in the clay luting will answer every purpose. The receptacle containing the soda is connected by means of a suitable pipe with a retort into which are placed eleven pounds of sulphur.

In a short time after fire is started under the retort the sulphur will commence to melt, at which point a piece of red-hot metal should be thrust into the sulphur, the fumes from the burning of which will pass through the connecting pipe into the receptacle containing the soda which will be thus converted into an antichlorine. When the first eleven pounds of sulphur are consumed, another eleven pounds should be placed in the retort, as the five hundred pounds of soda will require twenty-two pounds of sulphur for its complete conversion into hyposulphite of sodium.

The antichlorine is then dissolved in the box or cask, and drawn off into carboys, which are then taken to the beating-engine department of the mill.

Upon a fairly large scale hyposulphite of sodium is now prepared by treating tank waste liquor, or red liquors, with sulphurous acid obtained by the combustion of pyrites. The sulphurous gas is passed up a wrought-iron tower packed with coke, down which the liquors are run. This process yields a cheap product and is preferable to the old method of treating tank waste.

BEATING ENGINES.

The Kingsland Pulp Engine.

The Kingsland pulp engine made by Messrs. Cyrus Currier & Sons, Newark, New Jersey, is shown in Figs. 123 to 126.

Fig. 123 is a perspective view, Fig. 124 a front view,

Fig. 123.

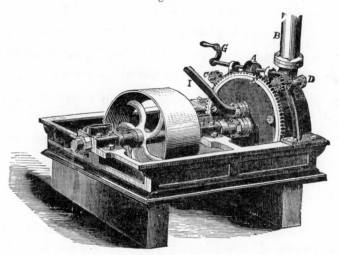

Fig. 124.   Fig. 125.   Fig. 126.

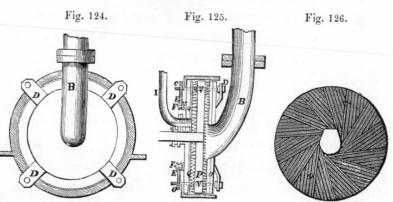

Fig. 125 a vertical cross section, and Fig. 126 a front view of the plate O of the Kingsland engine.

The half-stuff descends through the pipe B, Fig. 124, and passes into a circular chamber, the sides of which are formed of two plates, O, Q, provided with steel teeth; these are stationary, and can be brought closer together, or placed further apart, by the handle and gearing, G, A, C, E, Fig. 122, so as to grind the half-stuff in pulp of the desired length of fibre. The threaded bolts V, passed through lugs D, bring up the back plate O, while F forms guides for E. Between O and Q a plate, P, is placed; it has steel teeth, and is rotated rapidly between them by a shaft and belt. This shaft works in journals, and has no collars, so that it can adjust itself to the varying distances between the outer plates. The pulp, when ground, passes out through pipe I in a continuous stream.

Usual Construction of Beating Engines.

Beating engines are very similar in construction to washing engines. A beater-roller set with knives around its periphery is used in combination with a bed-plate, also set with knives, the parts being operated in a vat or trough, in which a constant circulation of the material to be pulped is maintained.

Heretofore, ordinarily, the material has been circulated horizontally around in an upright partition termed a " mid-fellow," and the beater-roll and bed-plate have been placed in the alley or channel between this mid-fellow and one side of the tank. The beater-roll lifted the material over a sort

of dam (termed a " back-fall"), and the material then flowed
by the action of gravity around the mid-fellow, and entered
again between the beater-roll and the bed-plate. It has,
however, been proposed to dispense with the mid-fellow, and
have the material returned under the back-fall and bed-plate.
In either case, however, the circulating force is that of
gravity due to the piling up of the liquid or semi-liquid on
the side of the back-fall opposite to the beater-roll. Consequently, the flow is comparatively feeble, and it is necessary to use a large quantity of water in order to prevent
the fibre in suspension from depositing. In the invention
patented by Mr. John Hoyt, of Manchester, N. H., and
shown in Figs. 127 to 130, a much more rapid and vigorous circulation is claimed to be maintained. The beater-
roll in this invention is placed at one end of the vat,
which is of a depth sufficient to contain it, and the other
part of the vat is divided by a horizontal partition or division, which extends from the beater-roll nearly to the
other end. The material to be pulped is carried around
by the beater-roll, and is delivered into the upper section
above the partition. It flows over the partition, then
passes down around the end of the same, and returns
through the lower section of the vat to the beater-roll. The
bed-plate is placed at the bottom of the vat under the
beater-roll. The beater-roll not only draws in the material,
creating a partial vacuum in the lower section of the vat,
but delivers it into the upper section with considerable force,
impelling it forward very rapidly. By the aid of this more
rapid as well as more vigorous circulation not only is the

26

material returned more quickly, and, therefore, acted upon more often by the beater-roll in the same time, but it may be worked with a much less quantity of water, and thereby very important advantages may be secured. These advantages are stated to be, first, in the improved quality of the product, for when a considerable body of the fibrous material is drawn between the knives the different pieces are rubbed together, and thus disintegrated without destroying the length and felting quality of the fibre, whereas when the pulp is thin, the pieces are ground individually, as it were, between the knives, and the integrity of the fibre in large measure destroyed; secondly, in the greater quantity of pulp which can be prepared in a medium of given size, owing to the larger proportion of fibrous material in the charge; and thirdly, in avoiding the liability of the fibrous material depositing out of the liquid, and lodging in the channels.

Hoyt's Beating Engine.

Figs. 127 to 130 represent a beating engine constructed in accordance with Hoyt's invention.

Fig. 127 is a plan with part of the casing or vat removed; Fig. 128, a vertical longitudinal section; Fig. 129, a plan of the bed-plate; Fig. 130, a partial view in cross section.

The cylindrical roll A, provided with knives, B, set radially in the periphery, is mounted concentrically on the shaft C, which is journalled in the sides of the vat D. The vat is of any suitable length, and in depth about equals the diameter of the roll, which is set in the vat close to one end. The ends of the vat are rounded. The beater-roll is slightly eccentric to the curvature of the end of the vat, in order to

Fig. 127.

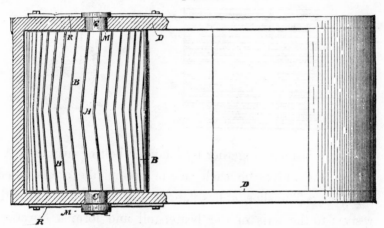

Fig. 128.

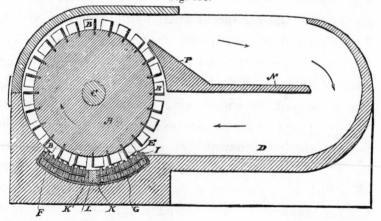

give a clearance (see Fig. 128) and allow the crude pulp to be lifted with less difficulty.

The bed-plate knives E are set in the shoe F, which is fixed in or to the bottom of the vat under the beater-roll.

The knives are separated by strips or blocks, *G*, of wood or other suitable material, and a number of these knives and

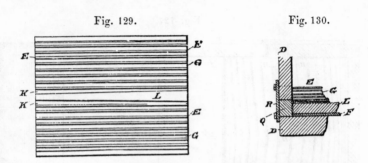

Fig. 129.   Fig. 130.

strips are fastened together by a curved bolt or rivet, *H*. As shown, the knives on each side of the middle are fastened together. The shoe *F* has flanges *I*, which are radial with respect to the axis of the beater-roll and form a dovetail. The two sets of knives with their spacing-strips are placed in the dovetail and are spread apart by the wedges *K* and *L*. Those marked *K* are of wood, the wedge *L* of iron. The wedges *K* are first inserted and the wedge *L* is driven between them. When the bed-plate knives become worn they can be set out by withdrawing the wedges, and placing strips or pieces under the knives. As they are set out the two sets are drawn toward each other, owing to the inclination of the flanges *I*, and it is necessary therefore to plane off a little of the wooden wedges *K* before replacing them and the iron wedge. The bed-plate knives are placed radially with respect to the axis of the beater-roll, and are adjusted in nearly radial planes. The shaft of the beater-roll turns in close boxes, which are further provided with collars, *M*, in order to make the joint liquid-tight. Any ordinary or suitable means can be used to adjust the shaft of the roll.   Between the beater-roll and the opposite end of the vat is the horizontal partition N, which extends to within a short distance of the end of the vat.   There is an upright inclined plate, P, which is brought at the upper edge into close proximity to the beater-knives, but does not touch them.   The vat is provided with the usual valve for withdrawing the pulp and also with the pipe for supplying water.   In the sides of the vat opposite the ends of the bedplate knives are curved slots, through which the knives and wedges can be inserted and withdrawn.   In operation these slots are closed by blocks, Q, of corresponding shape, so as to fit the hole.   The blocks are held in place by the plates R, which are bolted over the slots after the blocks have been put in place.

The operation of the engine is as follows: The beater-roll and bed-plate knives being properly adjusted, the vat is filled with the rags or fibrous material to be pulped and the proper quantity of water.   The beater-roll being revolved at the proper speed—say, for a roll four feet in diameter, at the speed of one hundred and twenty revolutions per minute—the rags and liquid are drawn between the knives, are carried up by the beater-roll, and thrown over the edge of the plate P.   They flow around the partition N with considerable velocity and return again and again to be acted upon by the knives.   The roll is revolved until the pulp is properly reduced.

Umpherston's Beating Engine.

As we have previously explained, pulp engines generally consist of a trough having straight sides and semicircular ends, an operating roll, a co-operating bottom plate and back-fall, the trough being partly divided by a longitudinal partition, called the "mid-fellow" or "mid-feather," around which the pulp flows from the back of the roll to its front, passing between the roll and bottom plate over the back-fall, and again around the "mid-fellow" to the front of the roll, from whence the operation is repeated. Such a construction and arrangement of parts are found in practice to be inefficient, the pulp nearest the circumference of the trough having a greater distance to travel than that portion near the mid-fellow, that in its repeated revolutions is not so often acted upon, and the mass is therefore unequally treated.

The invention patented by Mr. William Umpherston, of Leith, Scotland, is designed to overcome this difficulty; and it consists in providing a longitudinal and direct passage beneath the back-fall, whereby the pulp, in its delivery from the back of the roll and movement through the passage to the front of the roll, is directed as through an inverted siphon and pressed through the passage by its superincumbent weight at the terminus of the back-fall.

As we have already several times described the minor details of construction common to this class of machines, we shall now simply outline a machine containing such essential parts as are necessary to understand Umpherston's improvements.

In Fig. 131 the rotating roll A has a surface adapted for grating, rasping, or filing, and the fixed bottom plate B is also provided with a similar surface co-operating with that upon the roll A, the distance between such parts being

Fig. 131.

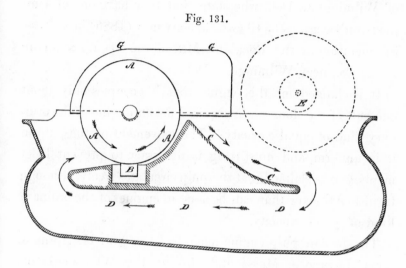

regulated by a vertical adjustment of the latter. The form of the back-fall C is similar to that of ordinary pulping-machines; but in the present invention a return-passage, D, is provided beneath the fall, so that the semi-fluid contents that pass over the back-fall are directed by the return-passage to the front of the roll A, the superincumbent weight of the mass of pulp as it is delivered from the back-fall pressing the mass along this return-passage. The relative position of the drum-washer or cleaning-cylinder is shown at E, and a hood, G, is also provided to prevent the pulp from being thrown out of the machine.

It will be seen that Umpherston's construction provides

for an equal distribution and treatment of every portion of the material acted upon, which insures rapidity and uniformity of treatment.

This engine is manufactured by the J. Morton Poole Co., of Wilmington, Del., who state that they have one of Umpherston's engines of 10 cwt. capacity now (1886) in successful operation at the Rockland Mill of the Jessup & Moore Paper Co., near Wilmington, Del.

It is claimed for this engine that it occupies only about one-half of the floor space required for an engine of the ordinary kind of equal capacity. The movement of the pulp in it is uniform, and no stirring is necessary to prevent lodgments. It is claimed that the pulp circulates freely, although furnished thicker than can be done in engines of the ordinary kind of equal capacity.

The Jordan pulp engine made by Messrs. J. H. Horne & Sons, Lawrence, Mass., and also by the Windsor Locks Machine Co., Windsor Locks, Conn., the Brightman Engine, made by the Cleveland Paper Company, Cleveland, O., the Jeffers Refining Engine, built by the Pusey & Jones Company, Wilmington, Del., as well as a large number of other engines which will be enumerated in the list of patents at the close of this section, are so well known to the trade that it is scarcely necessary to describe them in detail in this volume.

List of Patents for Pulp Engines and Bed Plates, issued by the Government of the United States of America, from 1790 to 1885 inclusive.

No.	Date.	Inventor.
1,760	Sept. 5, 1840.	W. Dickenson.
1,813	Oct. 10, 1840.	R. Daniels.
6,784	Oct. 9, 1849.	W. Clarke.

No.	Date.	Inventor.
8,261	July 29, 1861.	J. C. Fonda.
22,707	Jan. 25, 1859.	F. Stiles, Jr., and J. N. Crehore.
34,214	Jan. 21, 1862.	J. Percy.
26,387	Dec. 6, 1859.	F. Vandeventer.
43,707	June 7, 1864.	G. A. Corser.
46,893	March 21, 1865.	J. G. Fuller.
47,739	May 16, 1865.	T. Lindsay.
47,849	May 23, 1865.	O. Morse.
52,941	Feb. 27, 1866.	J. Easton, Jr., and F. Thiry.
57,355	Aug. 21, 1866.	J. McCracken.
60,645	Dec. 18, 1866.	J. M. Shew.
70,878	Nov. 12, 1867.	S. F. Merrill.
76,270	March 31, 1868.	J. Taggart.
85,386	Dec. 29, 1868.	D. Hunter.
86,858	Feb. 9, 1869.	W. Parkinson.
94,816	Sept. 14, 1869.	P. Frost.
94,843	Sept. 14, 1869.	P. Rose.
98,691	Jan. 11, 1870.	E. Hawkins.
101,008	March 22, 1870.	A. Hankey.
105,728	July 26, 1870.	T. Rose and R. Gibson.
115,274	May 30, 1871.	J. Bridge.
116,039	June 20, 1871.	R. M. Fletcher.
116,045	June 20, 1871.	P. Frost.
116,978	July 11, 1871.	H. B. Meech.
117,122	July 18, 1871.	J. Taylor.
118,092	Aug. 15, 1871.	G. Ames.
118,767	Sept. 5, 1871.	E. Wilkinson.
119,107	Sept. 19, 1871.	B. F. Barker.
120,265	Oct. 24, 1871.	} S. L. Gould.
Reissue		
4,976	July 16, 1872.	
120,787	Nov. 7, 1872.	Wm. R. Smith.
120,837	Nov. 14, 1871.	N. W. Taylor and J. H. Brightman.
121,780	Dec. 12, 1871.	J. Hatch.
121,970	Dec. 19, 1871.	C. Smith.
124,612	March 12, 1872.	T. Nugent.
128,788	July 9, 1872.	J. M. Burghardt and F. Burghardt.
130,067	July 30, 1872.	T. Nugent.
135,631	Feb. 11, 1873.	G. A. Corser.
144,557	Nov. 11, 1873.	S. Moore and R. H. Hurlburt.
150,147	April 28, 1874.	W. B. Fowler.
151,992	June 16, 1874.	A. S. Lyman.
153,774	Aug. 4, 1874.	W. Kennedy.

No.	Date.	Inventor.
155,152	Sept. 22, 1874.	F. Genin.
157,625	Dec. 8, 1874.	M. Meyer.
160,746	March 16, 1875.	M. R. Bonjin.
160,996	March 23, 1875.	B. F. Barker.
163,638	May 25, 1875.	A. Cushman.
166,519	Aug. 10, 1875.	A. Gardner.
163,638	May 25, 1875.	F. A. Cushman.
174,805	March 14, 1876.	S. S. Gould.
178,205	May 30, 1876.	W. E. Taylor.
182,891	Oct. 3, 1876.	J. Chase.
183,349	Oct. 17, 1876.	J. S. Warren.
189,671	April 7, 1876.	J. S. Warren.
190,373	May 1, 1877.	J. H. Robinson.
191,898	June 12, 1877.	E. Sumner.
194,824	Sept. 4, 1877.	} E. D. G. Jones.
Reissue		
8,609	March 4, 1879.	
199,940	Feb. 5, 1878.	A. A. Simonds.
200,828	March 5, 1878.	C. L. Hamilton.
208,292	Sept. 24, 1878.	J. Carroll.
210,937	Dec. 17, 1878.	J. H. Horne.
213,640	March 25, 1879.	P. P. Emory.
216,349	June 10, 1879.	W. H. Russell.
216,505	June 17, 1879.	C. Bremaker.
221,812	Nov. 18, 1879.	A. Hankey.
224,079	Feb. 3, 1880.	G. A. Corser.
225,976	March 30, 1880.	G. H. Ennis.
226,098	March 30, 1880.	O. Morse.
229,201	June 22, 1880.	J. Taylor.
232,460	Sept. 21, 1880.	C. E. B. Cooke, J. Cooke and G. Hibbert.
239,350	March 29, 1881.	A. J. Shipton.
244,220	July 12, 1881.	A. Forbes.
246,528	Aug. 30, 1881.	E. Mather.
248,707	Oct. 25, 1881.	H. P. Case and E. L. Granger.
249,257	Nov. 8, 1881.	A. C. Rice.
253,447	Feb. 7, 1882.	W. E. Taylor.
253,606	Feb. 14, 1882.	J. H. Horne.
254,251	Feb. 28, 1882.	J. R. Abbe.
256,352	April 11, 1882.	G. Miller.
273,801	March 13, 1883.	C. S. Barton.
277,268	May 8, 1883.	S. L. Gould,
282,818	Aug. 7, 1883.	W. Whitely.

No.	Date.	Inventor.
286,216	Oct. 9, 1883.	O. Morse.
288,234	Nov. 13, 1883.	A. Hankey.
289,235	Nov. 27, 1883.	G. W. Cressman.
297,037	April 15, 1884.	⎫
Reissue		⎬ W. Umpherston.
10,658	Nov. 3, 1885.	⎭
299,307	May 27, 1884.	W. Whitely.
302,399	July 22, 1884.	A. Hankey.
303,374	Aug. 12, 1884.	J. Hoyt.
307,237	Oct. 28, 1884.	C. F. Taylor.
308,255	Nov. 18, 1884.	G. F. Harlan.
310,230	Jan. 6, 1885.	A. A. Simonds.
312,390	Feb. 17, 1885.	J. F. Seiberling.
320,612	June 23, 1885.	H. Allen and L. S. Mason.
320,721	June 23, 1885.	F. S. Taylor.

CHAPTER XIII.

SIZING——ENGINE SIZING——BLEACHING RESIN AND PREPARING
SIZE THEREFROM——SURFACE SIZING——HARD SIZING PAPER IN
PROCESS OF MANUFACTURE——"DOUBLE-SIZED" PAPER——TUB
SIZING WITH BENZINE AND RESIN——SIZING THE SURFACE OF
PRINTING PAPER——MATERIALS USED IN SIZING PAPER——WATER-
PROOF SIZINGS FOR PAPER.

Sizing.

Prior to the introduction of paper-making machinery the
sheets were sized only with gelatine or animal size; but resin
or vegetable size is now commonly used; it is added to the
pulp in the beating engine, and greatly facilitates the
manufacture of paper in continuous sheets.

The present century has been rich in great mechanical
and chemical achievements, and the adaptation of inventions
to industrial pursuits; one improvement has engendered
many others, consequently we should not be surprised at
the large number of inventions relating to paper manufacture which were a natural sequence to the invention and
perfection of our modern paper-making machines.

The first attempts at sizing pulp in the beating engine
commenced at the beginning of this century; the experiments of Braconnot, d'Arcet, and others leading to the
preparation of resin or vegetable size.

Papers may be roughly divided into two classes, viz., "tub sized" and "engine sized;" but as most papers, even tub sized, excepting blotting- or water-leaf paper, are more or less sized in the engine we shall first speak of the latter method.

ENGINE SIZING.

The intimate mixture with the pulp and precipitation upon the fibres of a substance which, when desiccated, will virtually fill the interstices between the fibres, and at the same time be comparatively water-proof, is the theory upon which engine sizing is based, and the substance commonly employed for this purpose is a mixture of resin soap treated with alum.

The thorough incorporation of this body with the fibre is best produced by first adding an aqueous solution of resin soap to the pulp in the trough of the engine, and then, after an intimate mixture of the pulp and soap has been made, a solution of alum is run in. A combination of resin and alumina and of sulphate of sodium is formed by a double decomposition in the pulp. The resinate of alumina thus intimately incorporated with the pulp undergoes a fusion when the paper passes over the drying cylinders and communicates to the paper its hydrofuge property.

When there is added to the sizing a small proportion of starch the latter in swelling draws together and unites the fibres of the paper and renders it less spongy.

The resin soap is usually prepared in a wooden tub or

iron-jacketed boiler having a capacity of about 250 or 300 gallons, which is a convenient size for dissolving two barrels of resin; steam is admitted near the bottom of the tub or boiler through a suitable pipe. The desired quantity of water having been run into the receptacle, carbonate or caustic soda dissolved and previously strained is next added to the water, and the contents of the tub or boiler raised to the boiling point, when the finely powdered resin is gradually thrown in, and the contents constantly stirred with a paddle for two hours, or until the resin is entirely dissolved. It is desirable not to use too small a proportion of water in the preparation of the resin soap, as the impurities of both the resin and the soda in such a case will be mixed with the soap.

Some manufacturers of paper prefer to dissolve the resin in a solution of soda-ash of such concentration that its specific gravity is greater than that of the resin soap. The quantity of water required to dissolve the resin and produce the desired concentration is a matter for experiment, and is readily discovered after a few trials.

The impurities of the soda and resin fall to the bottom of the boiler after about two hours' boiling and stirring; the resin soap remains on top, and can be taken off in a clear condition. Should a large quantity of soda remain on the bottom of the tub or boiler after the soap is removed, more resin or less solution must be used next time; but if the resin is not properly dissolved after the boiling has been continued for the usual length of time, the proportion of soda should be increased.

The proportion of resin used to each pound of soda-ash varies in different mills, three, four, and even five pounds of resin being used to each pound of soda-ash.

The proportions of resin, soda-ash, and water can be best determined by practical experience, as no prescription could be devised which would be suitable to every case.

M. d'Arcet, who modified the proportions recommended by M. Braconnot, recommended for the preparation of the resinous soap:—

Powdered resin	4.80 parts.
Crystals of soda at 80° (Fr. alkalimeter) . .	2.22 "
Water	100 "

Theoretically speaking, only 2.45 parts of alum would be required to precipitate the resin; but the waters, which are almost always calcareous, neutralize a part of the alum.

Crystals of soda are much more expensive than soda-ash, but on account of their greater purity they are sometimes preferred to soda-ash. At the present day the resin soap is preferably made by dissolving ordinary resin with a solution of carbonate of soda under a boiling heat in a steam-jacketed boiler, the class of paper to be made governing the quality of resin to be employed. The boiling usually requires from one to eight hours, according to the relative proportions of soda-ash and resin used—the greater the proportion of soda-ash employed, the less the time required for boiling—the process being completed when a sample of the soap formed is completely soluble in water.

As we have previously intimated, the proportion of resin used to each pound of carbonate of soda differs in almost every mill; but about three pounds of resin to one pound of carbonate of soda is the usual proportion. It is really waste to use a greater quantity of soda than is absolutely necessary to thoroughly dissolve the resin, as it only consumes its equivalent of alum, without yielding any beneficial results.

We have several times mentioned carbonate of soda (washing soda) as the material used for dissolving the resin, but caustic soda is used in some mills, and soda-ash in others, all being about equally suitable.

The resin soap is cooled after boiling by running it off into iron-tanks, where it is allowed to settle, the soap forming as a dense syrup-like mass, and the coloring matters and other admixtures of the resin rising to the top are easily removed.

It is important to run off the mother-liquor containing the excess of alkali, for when the soap is used it consumes alum to neutralize it.

After the impurities have been removed from the tank containing the resin soap, the latter is dissolved in water. If, owing to imperfect boiling, the resin is not thoroughly dissolved, a small quantity of carbonate of soda is added to the water used for dissolving the soap.

In many mills where starch is used for stiffening purposes the soap is mixed with a quantity (about $1\frac{1}{2}$ part of starch to 1 part of resin) of starch paste, which is prepared in a separate vessel by dissolving farina in hot water. Some manufacturers mix the starch paste with the kaolin in lieu of mixing it with the resin.

The mixtures of either resin soap and starch paste, or of starch paste and kaolin after being sifted very carefully, are in readiness to be used.

From 3 to 4 pounds of the mixture of resin soap and starch paste to each 100 pounds of dry pulp are about the proportions in which the size is generally used; but the quantity added to the pulp in the trough of the beating engine, of course, depends upon whether the paper is to be soft-sized or hard-sized.

The mixture of soap and starch after being dissolved in water, and in some cases even without being dissolved, is put into the beating engine in which the pulp is circulating, and after being thoroughly mixed with the pulp the solution of alum or sulphate of alumina is added.

The "crystallized alum" used by paper manufacturers is valuable in the sizing process only on account of the sulphate of alumina which it contains, the other ingredients, sulphate of potash, water, etc., contained in the alum exerting no influence upon the resin soap, are consequently of no value.

The concentrated alum, which always contains the greatest percentage of sulphate of alumina and other sulphates which have a direct action on the resin soap, are the most economical for use, being proportionately cheaper than crystallized alum ; such concentrated alums as " Pearl" alum, etc., being especially employed.

Some paper-makers do not object to the presence of iron, for *ordinary* purposes, provided it is in the state of a proto-salt, or in the ferrous state, and is accompanied by more or less free acid. Many of the concentrated alums and alumi-

27

nous cakes contain the iron as a ferric salt, and this is readily decomposed, depositing the ferric oxide and making more or less discoloration, which is very objectionable when clear, white papers are desired. The acid alums are not altogether objectionable because they are very largely used, especially for all common papers, such as news and low grade books ; and the bulk of paper made is no doubt of this character. They are especially important, not only as being more active in sizing than neutral or basic alum, with the same, or larger, percentage of sulphate of alumina, but are particularly useful in developing the aniline blue employed in common papers for the purpose of correcting the color. Acid alums are also useful in sizing when the water employed is at all hard, as the free acid more or less neutralizes the hardness of the water. The best type of a concentrated acid alum is Harrison's " Lion" alum, which contains from 52 to 55 per cent. of neutral sulphate of alumina and from one to three per cent. of free acid. This alum is used with the greatest success by numbers of manufacturers of common papers, such as news and low grade book papers.

" Aluminous cake" is used in many mills as a substitute for alum. If aluminous cake consisted entirely of sulphate of alumina it would be a most valuable substitute for alum in the sizing process, but the great objection to it for fine and colored papers is that it sometimes contains an excess of free sulphuric acid and soluble iron ; the sulphuric acid not only discharging some colors from the pulp but also destroying the brass wire-cloths of the paper machines.

The solution of either alum or aluminous cake is prepared

in a lead-lined receptacle of suitable size, and furnished with a pipe for heating the contents of the tank by steam.

The quantity of alum or aluminous cake used varies with the sulphates themselves and with the class of pulps to which they are added, and as a large number of vegetable and aniline colors are brightened by sulphate of alumina it is desirable for many kinds of colored papers to add the solution of alum or aluminous cake to the pulps in quantities which will be in excess of those required for precipitating the resin upon the fibres.

Litmus papers may be employed to detect a surplus of either resin soap or sulphates in the pulp: if, after the incorporation of the sizing with the pulp it turns red litmus paper blue, the proportion of alum or aluminous cake is insufficient, and for uncolored papers these materials should be increased until the pulp turns blue litmus paper red. Practical experience is always the best guide in regulating the proportion of alum and other chemicals to be used, for the reason that there is such a difference in the strength of various alums, etc., employed, and in the nature of the wash-water, etc., used in various mills, that directions which would prove effective in one case might result disastrously in another.

We could here again remark that neither the solution of resin soap, alum, nor any other chemical should be run into the engine without being previously strained, either through a wire gauze or flannel cloth, and the size, alum, kaolin, coloring matters, etc., should be accurately weighed or otherwise determined.

Various substances have been employed in special mills as a substitute for resin in the sizing process; but as most of these lack that most important requisite—economy—none of them have as yet come into extended use. Wax dissolved with a concentrated solution of caustic soda and precipitated with alum has been proposed, and makes an excellent size, but its costliness would confine its use to the highest grades of fine papers. The addition of about 12 pounds of gum tragacanth to each 500 pounds of resin has been proposed and used in preparing some kinds of engine-sized papers, and imparts to them, it is claimed, an appearance resembling tub-sized papers.

BLEACHING RESIN AND PREPARING SIZE THEREFROM.

The following method of preparing resin size was patented in 1868, by Mr. Thomas Gray, of London, Eng.

Operating on any quantity of resin, say eight hundred pounds, throw into a copper or sheet-iron boiler forty gallons of water; dissolve into it, in heating, about seventy pounds of salt of soda, or other alkaline salt, agitating and stirring the mixture till perfect dissolution of the alkaline salt occurs.

When this result is attained, gradually add the resin in small quantities, stirring the material, and waiting till all is completely dissolved before throwing in any more.

When the whole is thoroughly mixed turn off the steam, and the first operation being over proceed with the preparation of the size.

Having previously dissolved forty pounds of common salt

in fifty gallons of cold water, take a boiler twice or three times more capacious than that used for preparing the resin, pour in together with the resin, yet hot, one-half part of the salt water prepared, adding thereto thirty gallons of cold water.

Then apply heat, stirring the mixture by means of a spattle, and a homogeneous whitish mass will soon be obtained.

Continue stirring until the mixture assumes a very intense dark color, much like that of wine, which change is produced by the perfect union of the prepared resin and the salt at that moment.

Let the mixture settle, and afterward draw off the liquid portion. Then add some cold water and the remainder of salt water and allow the mixture to again settle, and after decanting the supernatant liquor the mixture is ready for use.

SURFACE SIZING OR SIZING IN THE SHEET AND IN THE WEB.

Papers to be used for writing purposes are commonly coated with animal size, which is as colorless as possible in order not to injure the color of the paper.

Hide and skin trimmings, cartilages, and membranes from animals slaughtered for food or to supply the hides and skins used for the manufacture of leather; hog, hare, and rabbit skins, the hoofs and ears of oxen, sheep, and goats, parchment refuse, eel-skins, etc., form the principal sources of supply from which the materials used for the preparation of animal size used in the manufacture of writing papers are derived.

The treatments which the materials receive in the preparation of size vary with their nature and condition and the variety and grade of papers to which the size is to be applied.

Speaking generally, and having in view the sizing of sheets of paper by the hand method of sizing, the materials named are usually washed, then steeped in lime-water, and afterwards perfectly cleansed from it by washing in acidulated and then in pure water.

In order to convert the fat into an insoluble lime soap, the material is next gently boiled in eight or ten times its weight of water for about six hours, during which time it is sprinkled with a small quantity of very finely powdered lime ; the boiling point being completed as soon as a drop of the liquor placed upon a cold porcelain plate solidifies to a jelly.

There are next added to one hundred parts of the jelly thus produced two or three parts of alum previously dissolved in water, by which the size is coagulated and rendered insoluble, and consequently is in a more suitable form for the sizing of paper.

Sometimes the ready-made size of commerce is used. This is steeped for two or three hours in water and then dissolved in boiling water ; 13 to 18 lbs. of size mixed with 4 to 6 lbs. of alum, dissolved in 22 gallons of water, is usually sufficient for sizing an average-size vatful of paper.

The solution of size is brought to a temperature of 77° F., and then about 100 sheets of paper are dipped into it at one time, and so moved about that each sheet becomes coated with size on both sides; they are next pressed so as to distribute the size in the interior of the separate sheets, and afterwards separated and hung to dry on lines in a drying room.

It is necessary that the drying should proceed slowly; caution, however, must be exercised that desiccation is not continued for a sufficiently long time to permit the decomposition of the moist size.   In summer thunderstorms induce the decomposition of the size, which becomes covered with mould, liquefies, and loses its glutinous properties.   If the sheets of paper are too rapidly dried the size remains distributed throughout the body of the material; but during slow desiccation the size as it dries is drawn, in company with the moisture, to the surface, where it forms an impermeable layer. Thus it happens that strictly animal sized papers if properly dried will blot if there has been an erasure or scraping of the surface.

If the paper is to be sized in the web by the use of machinery, the preparation of the size is conducted on a larger scale than has been just described.

In large paper-mills the size is generally prepared in a room devoted to the purpose, and is commonly located near the machine, adjoining it if the latter is on the ground floor, and usually below the machine-room if it is on the second floor.

The finest grades of light hide and skin clippings are employed for No. 1 letter papers ; but less costly stock is employed for the lower grades of animal-sized papers.

In order to preserve the glue stock tanners and tawers macerate it in milk of lime and afterwards dry it, and as the clippings require to be freed from the lime the first treatment which the glue-stock receives after arriving at the paper-mill is to put it in large wooden tubs filled with water in which it is allowed to remain for several days in soak.

If the paper-mill is situated on the banks of a stream the glue-stock is sometimes packed in large willow baskets and the latter submerged in water by means of a travelling crane, and in this the stock is soaked and freed from lime. But the more desirable way is to soak the glue-stock in wooden tubs and then put it in a larger revolving wash-drum about five feet in diameter and ten feet in length, which drum should be driven by power and so constructed that it will be one-half immersed in a vat. The drum suitably covered with wood is filled with the necessary quantity of stock through a door, which is formed by hinging one of the boards which form the surface of the drum, and clean water being admitted through one of the hollow trunnions the dirty water is allowed to escape through perforations in the periphery of the drum while it is being revolved.

The objection to the glue-stock washing apparatus in common use is that the stock is usually damaged by being broken up too much, and considerable loss results, besides from the fact that the small particles are allowed to escape with the wash water. Mr. W. A. Hoeveler has lately patented a washer by which the defects named are claimed to be remedied and other advantages derived.

Fresh waste, i. e., such as has not been limed and dried, and which is sometimes purchased by paper manufacturers from neighboring tanneries, must be prepared as soon as possible after it arrives at the paper-mill, as otherwise it would taint the air, be attacked by rats and other vermin, and suffer injurious alterations by decomposition..

The fresh waste is first placed in tubs filled with water in which has been dissolved 2 per cent., by weight, of caustic lime. It is best to allow the water to stand for a week or so before using it for the fresh waste. The length of time which the stock remains in the lime bath varies according to the material: trimmings from calf-skins requiring from 10 to 15 days; sheep-skins, 15 to 20 days; and trimmings from heavy hides, such as ox, 25 to 30 days. The milk of lime should be renewed once or twice a week and thoroughly stirred.

The material is washed in the washing drum after being removed from the lime, and is afterward spread out in the yard to drain and dry. When sufficiently dried the materials are ready for boiling to glue, and can be stored until wanted.

The glue stock, after being cleaned or prepared as has been described, is placed in a boiler of cast-iron, sheet-iron, or copper. Its capacity depends upon the quantity of raw material to be boiled at one time. It is best to have boilers holding from 100 to 400 pounds of raw material, and to place two or more of such boilers together. Resting upon the bottom there should be a stopcock for drawing off the gelatinous solution. From 1 to 3 inches above the bottom of the boiler there should be a perforated and movable false bottom supported by flanges, thereby preventing direct contact of the materials with the heated bottom of the boiler, and obviating injury by scorching.

The glue stock having been placed in the boiler, water is poured over it and steam admitted under the false bottom; but the water should at no time be allowed to come to a boil, care being observed not to allow the temperature to exceed 200 F., which should be maintained from 10 to 18 hours, the time depending upon the nature of the raw material. As the gelatinous solution is formed it is drawn off from the boiler into wooden tubs, and is at the same time carefully strained to remove impurities.

Two or three extracts are made from the same lot of glue stock, all the solutions being mixed together in the receiving tubs, where a solution of alum is added in such proportions as to be recognized by tasting the liquor. The object in view in adding the alum solution is to prevent the gelatine from fermenting or decomposing, and as the danger from this cause increases with the higher temperature of the atmosphere, more alum should be used in summer than in winter.

After the solutions cool they are ready for use, the gelatine being removed from the receiving tubs and dissolved in a separate tub as required for use, the tub in which it is dissolved being provided with a steam pipe. The proportion of water (which should be only lukewarm) employed in dissolving the gelatine varies from one-quarter to one-half of the volume of the latter, the nature of the fibre and thickness of the paper regulating the proportion of water to gelatine; the concentration of the solution being greater for thin papers and weak fibres than for thick papers and strong fibres.

The mechanism for supplying the size to the trough, through which the web of paper is passed, and the manner of running it over the Fourdrinier machine are so well understood that it is not necessary to enter upon a detailed description of either.

The best method of drying paper after it is tub-sized is still an unsettled question among manufacturers of paper. In this connection we quote from an editorial in the 'Paper Trade Journal:' "When the paper is passed through the size-tub it is again wet; the fibres expand, and their hold on each other is relaxed. Now it must make a difference to the subsequent strength and quality of this paper whether it is hung up in a loft to dry or run over a drying machine. If it is hung in the loft no strain is put upon it, and the fibres are at liberty to shrink or slowly contract in all directions; whereas if it is run over a drying machine, consisting of from fifty to one hundred reels, the longitudinal strain prevents the fibres from shrinking and resuming their normal position in that direction. Attempts have been made to obviate this defect by regulating the speed of each section of the machine in such manner as to allow for the shrinkage; but this only remedies the evil by preventing the paper from breaking as it travels over the machine. Everything else being equal, it would seem that loft-dried paper must be superior to that dried on the drying machine. Our home manufacturers indorse this view, inasmuch as they continue to prefer the system of loft-drying to the less expensive machine methods."

It is, of course, understood that papers which are surface sized may previously have been sized in the beating engine, and this method is mentioned under the head of " Double Sized" Paper. The methods employed for drying the papers will be found treated in Chapter XV.

HARD-SIZING PAPER IN PROCESS OF MANUFACTURE BY ADMINISTERING VEGETABLE AND ANIMAL SIZES SUCCESSIVELY TO THE WEB BEFORE IT IS DRIED UPON THE HEATED CYLINDERS.

The following composition and method of hard-sizing paper was patented in 1873 by Mr. X. Karcheski, of Belleville, N. J. The invention consists in submitting paper in the web, before it is dried, first to a bath of vegetable size, and then to a bath of animal size, both of peculiar composition, and in removing the superfluous sizes by scraping. Also in drying the product upon the heated cylinders of the paper-machine, so that the complete operation of manufacturing paper which is sized is conducted with only one drying process, irrespective of whether the paper is more or less hard-sized or enamelled. The invention further consists in distributing the earthy matter contained in the vegetable size upon a web composed solely of paper-pulp, filling its pores and cavities, and thus producing an even surface upon which the animal size is subsequently deposited in a thin pellicle. Thus the making of the paper brittle is avoided, which brittleness is one objectionable result of mixing earthy matter with the pulp, and preventing the paper from absorbing animal size to such an extent as to become translucent. These sizes contain some novel ingredients, and their composition is varied according to the various results sought to be accomplished.

Mr. Karcheski's process involves the introduction between the drying-cylinders and the last pair of press-rolls, in a Fourdrinier or "cylinder" machine, of two vats or tanks, for containing vegetable and animal sizes respectively, each provided with a suitable device for scraping off the superfluous size from the material operated upon. The first vat contains the vegetable size, consisting of bleached resin dissolved by heat in the least possible quantity of alkalies and water, with the addition to the solution of a quantity of colorless earth and soap. The web of paper from the press-rolls passes through the size and absorbs the resin and soap, while the colorless earth fills its pores.

When the paper is manufactured from new stock and is free from impurities and foreign matters, colorless earth in the first vat may be omitted, if increase in the weight of paper should be objectionable.

The excess of size is removed from the paper by scrapers, and the web then passes into the second vat, which contains the animal size, consisting of a solution of glue, alum, tallow, or other soap, colorless earth, and a trace of chloride of sodium.

In combining these ingredients the proportions are not arbitrary. The operator will soon learn to vary them according to the quality of surface required, as to hardness, lustre, and enamel, and also according to the purity of the ingredients themselves, and the conditions attending their use.

Care must be taken in all cases that a sufficient quantity of alum is present to neutralize the alkalies used in the process. The superfluous animal size is removed by scrapers, as before, and the web then passes on to the drying cylinders.

The apparatus invented by Mr. Kercheski, by means of which the web of paper is subjected to the process of hard

Fig. 132.

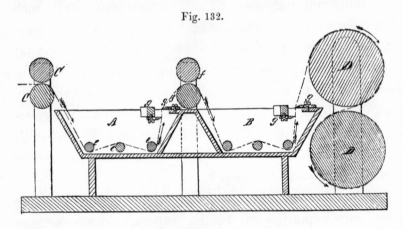

Fig. 133.

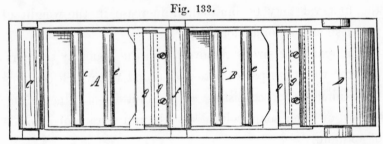

sizing, coloring, or water proofing, is shown in Figs. 132 and 133.

Fig. 132 is a vertical longitudinal section, showing Kercheski's apparatus interposed between the last pair of press-rolls, and the drying cylinders of a Fourdrinier machine, and Fig. 133 is a plan of the same.

The dotted lines indicate the paper or other material under treatment, and the arrows indicate the direction in which it moves.

The two vats, A and B, are interposed between the last pair of press-rolls, $C C$, and the drying cylinders $D D$, of a Fourdrinier or cylinder machine. These vats are provided with guide-rollers, $e e e$, and carrying-rollers, $f f$, for the purpose of conveying the paper or other material operated upon into and from the sizing compositions contained in the vats. Each vat is provided with adjustable scrapers, $g g$, for removing the superfluous size from the surface of the paper.

In sizing paper and other materials which vary in strength of tenacity, and in applying sizes which vary in consistency, it is necessary to regulate the tension to which the web is subjected by the scraping process; hence the importance of the feature of adjustability, as applied to the scraping mechanism.

From what we have previously stated it will be seen that by the present mode of sizing paper the web is in a wet state when introduced into the first or vegetable-size bath, so that the water it contains combines with the vegetable size. The fibre of the web it is claimed absorbs the resin and soap solutions, while the earthy matter fills the pores and cavities in the web, and the scraping operation produces an even surface. The web being then introduced into the animal-size bath, and being incapable of absorbing any more liquid, simply receives upon its surface a thin pellicle of animal size, and the repetition of the scraping operation not only removes the superfluous size, but still further tends to equalize and smooth the surface of the web. The web, which then passes onto the drying-cylinders, is prevented, it is claimed, from adhering thereto by the presence of the chloride of sodium contained in the size, and loss of size by evaporation during the drying process is prevented by the presence of the fatty acids in the size.

Printing-paper manufactured by this process, while it is not increased in cost, it is claimed is greatly improved in quality in respect to superior smoothness, evenness of texture, and capability of bearing writing or receiving even impressions from printer's types. "Hard-sized paper," so called, i. e., writing-paper, it is claimed is not only improved in quality by this process of manufacture, but is greatly reduced in cost.

Heretofore there has been no successful process of hard-sizing paper in the web before drying it, and consequently a second drying operation has been necessary. By the present process the time and labor expended in the second drying operation are saved. In manufacturing writing-paper it is to be observed that the vegetable size of resin, alkaline soap, and colorless earth, which fills the pores of the web, as we have described, prevents the translucency which is a characteristic of paper finished with purely animal size.

To produce the hard surface required in writing-paper the following it is stated will be found to be an effectual composition of animal size, to be applied after the administration of the resin or vegetable size: Dissolve one and a half pounds of white soap in two gallons of water, and two pounds of strong white glue in two gallons of water, mix the two solutions, and add sufficient alum to neutralize the alkalies present, then add colorless earth and a handful of chloride of sodium. This size is to be used warm, say at a temperature of about 120° F.

Paper sized with this succession of resin and animal sizes it is claimed will be perfectly opaque and firm, and will have a hard, even surface.

In manufacturing printing-paper the animal size is made weaker—that is, with a weaker solution of glue, printing-paper not requiring so hard a surface.

" DOUBLE-SIZED" PAPER.

Large quantities of writing paper are sized by the process called "double sizing." The pulp is sized in the beating engine with resin size and alum after the usual method. The pulp thus sized then passes into the stuff chest, and thence on to the paper machine. The paper, as it leaves the dry felt, then passes between two rollers, revolving in a vat containing animal size, thence over the dryers in the usual manner. Instead of drying on the machine the double-sized paper is usually taken wet from the machine and dried in drying lofts.

28

Tub Sizing with Benzine and Resin.

The following process has been employed for sizing paper: One and one-half pounds of resin are added to ten gallons of benzine or naphtha in a close vessel; the resin quickly dissolves, after which the composition is ready for use; but it is stated that the size is much improved in quality if allowed to rest for three or four days before using it.

The quantity of size specified is claimed to be sufficient for about three hundred yards of paper, fifty-eight inches wide, and weighing twenty-five pounds to the ream, when cut into sheets 19 by 28 inches.

The unsized paper, on leaving the drying cylinders, is passed through a sizing trough containing the composition, the rate of speed being about twenty feet per minute, the superfluous composition being wiped or scraped off from both surfaces of the paper and returned to the sizing trough.

The paper thus sized should be air-dried, and it is stated to be usually sufficiently dry after passing through about ten feet of space to be calendered and cut up into sheets or made into rolls.

Sizing the Surface of Printing Paper.

Hover, in 1867, after numerous experiments, proposed the following process for sizing the surface of printing paper: 4 ounces of starch are dissolved in 240 ounces of water, and 12 ounces of commercial carbonate of lime, magnesia, or its equivalent are added to the solution, a small portion of glue being also added if desired. This size is applied to the surface of any of the papers usually employed for printing, and the paper being afterwards dried and calendered will be ready for use.

This sizing may be applied to the paper after it is finished or during the process of its manufacture, while other materials than those alluded to may be used for sizing the paper; the main object of the process being to impart to the surface of the paper a permanent coating of carbonate of lime, or of magnesia, or their equivalents.

In 1869 Hover patented the following composition for treating paper, which compound he claims possesses the property of more thoroughly permeating and becoming incorporated with the "water leaf" than ordinary sizing: Seven gallons of ordinary glue sizing are mixed with one gallon of strong solution of acetate of lime. This sizing may be applied in the same manner as other sizing, and it is claimed for it that it not only renders the paper whiter but that it at the same time improves the surface.

MATERIALS USED IN SIZING PAPER.

Alum.

Alum, in the narrower sense of the word, is such a double combination of two sulphates, which will always contain aluminium as a sesquioxide, when solutions of aluminium sulphate are brought together with sulphates of suitable simple oxides. According to the nature of the sulphate combined with the aluminium sulphate, the following principal distinctions are made in the varieties of alum : Potash-alum, ammonia-alum, and soda-alum. Soda-alum is more readily soluble in cold water than the others; but is not used in paper-making in the United States. It is not a permanent salt, deliquescing rapidly. For coloring purposes potash-alum is sometimes used, or as a substitute for ammonia-alum, or there may be employed a mixture of both in varying proportions. For the purpose of recognizing whether potash-alum is pure, rub a piece of it with caustic lime and moisten the mixture with water. The presence of ammonia will be readily detected by its characteristic odor. It is well for the better grades of paper to subject the alum to a test for iron before using it. This is readily effected. according to Prof. Runge, by throwing a piece of alum to be tested into a solution containing $15\frac{1}{2}$ grains of potassium ferro-cyanide in 7 ounces of water. If the color of the surface of the alum remains unchanged it is free from iron, but in case blue spots make their appearance it contains iron. This test is entirely reliable for alum in pieces; it is claimed to be equally reliable for pulverized alum and alum solution.

Up to about the year 1870 crystal alum was used almost entirely as the sizing agent in paper manufacture. About the date named the Pennsylvania Salt Manufacturing Company introduced a concentrated "porous" alum manufactured under Mr. Henry Pemberton's patent. The company named possessed alumina as a residue from the manufacture of soda from the mineral cryolite. The process of manufacture consisted of simply treating the alumina with sulphuric acid, then running it on to a floor on which there was sprinkled more or less bicarbonate of soda ; the hot sulphate of alumina rapidly decomposed the carbonate of soda, causing the free carbonic acid to rise through the mass, giving the alum its porous properties, and absorbing into its composition the resulting sulphate of soda. This also served to neutralize whatever uncombined acid there may have been present. Because of the large percentage of the sulphate of alumina contained in this alum, paper-makers were enabled to produce the same results by using about one-half the quantity of it that they had previously used of crystal alum. The analysis published by the manufacturers shows some 53 per cent. of sulphate of alumina, whereas crystal alum contains from 36 to 39 per cent. ; and the quantities employed of the respective alums would be in about the ratio of these percentages. This porous alum had been very generally introduced in all the better class of paper-mills in the United States by the year 1880. In the year 1879, Harrison Bros. & Co., of Philadelphia, acquired control of the Laur patent for the manufacture of a concentrated paper-maker's alum from bauxite, and that firm immediately erected a plant for the introduction of this article to the paper trade of the United States. Bauxite is probably the mineral richest in alumina ; the best varieties contain as high as 70 per cent. of anhydrous alumina, and the greater part of it is industrially available. The Laur patent principally covered the use of zinc for reducing and removing the iron, which is always present in bauxite, from the solution of sulphate of alumina made from it. The zinc used in the metallic state is dissolved by the sulphate of alumina solution, reducing and removing the iron and neutralizing all the free acid. Alumina is such a peculiar substance, acting either as a base or an acid, according to the character of the substance opposed to it, that, while in the presence of sulphuric acid, it becomes a base; in the presence of zinc oxide it appears to act as an acid, combining with the zinc as it does with the sulphuric acid. The compound known as " pearl" alum resulting from this treatment consists of sulphuric acid, alumina, and oxide of zinc, all perfectly soluble, and is the most powerful sizing agent yet offered to paper-makers. The aggregate of these materials is stated in the manufacturer's circular to be not less than 63 per cent.; analyses show as high as 70 per cent.; and the claim made by the manufacturers that its sizing power is 15 to 20 per cent. greater than " Natrona" porous alum is generally admitted by careful paper-makers. At the present time this alum divides the honors with the " porous" alum, and " pearl" and "porous" are recognized as the two standard concentrated alums in the paper-making trade. " Pearl" alum is particularly neutral to colors; ultramarine, which is destroyed by the weakest of acids, will remain unaffected by pearl alum for a long time. The inertness of pearl alum to colors like orange mineral is also very marked when compared with other alums.

It has already been stated that it is the sulphate of alumina only which gives to crystal alum its value as a sizing material. Crystal alum contains but 36.31 per cent. of sulphate of alumina, the remaining 63.69 per cent. consisting of water and other inert material. It is on this account, that, for many years, crystal alum has been superseded in engine-sizing by the so-called "concentrated alums," which contain a much larger percentage of sulphate of alumina. The greater the amount of sulphate of alumina which any con-centrated alum contains the greater its sizing power. But it does not necessarily follow that a concentrated alum having the highest sizing power is the purest, and, therefore, the best alum.

The best concentrated alum is that which contains only sulphate of alumina—it should contain no iron. An alum may contain iron and yet be perfectly white in its dry state. But such an alum (or mixture of sulphate of alumina with other substance), if it contains iron (though it may dissolve in water into a colorless solution), will, in time, assume a brown color. This brown color is due to the conversion of the iron it contains into ferric oxide or iron rust.

It requires time to produce this brown color in a colorless solution of alum containing iron, and it also requires time to produce the same effect on paper that has been sized with an alum containing iron. In a perfectly white alum con-taining iron the latter is present as ferrous oxide, which by the action of air is transformed into ferric oxide or iron rust.

For many years there have been, and are still, used by paper-makers enormous quantities of a concentrated alum known in the trade as "Natrona porous alum," made by the Pennsylvania Salt Manufacturing Company of Philadelphia. This alum, made from pure hydrated oxide of alumina, is of uniform strength, and is very rich in sulphate of alumina. It contains neither zinc nor iron, nor any other material that can affect the purest white of the best and finest qualities of high grade paper.

The practical paper-maker can easily distinguish the presence or absence of iron in alum by dissolving half an ounce, more or less, in two or three ounces of water, and adding half as much clear "bleach liquor." Submitted to this test, Natrona porous alum throws down only a white precipitate of sulphate of lime, while the supernatant liquid remains perfectly colorless. The same test (bleach liquor) applied to any alum containing iron (though the solution may be colorless) will at once turn the solution brown and throw down a portion of the iron as a brown precipitate.

Concentrated Alum as a Water Purifier.

Although a paper-mill may have a supply of water which is of excellent quality during dry weather, its water shed may be such that after rain the water is unfit for immediate use. In a few hours a clear brook of crystal purity may be converted into a running stream of dirty water. If it should run into the reservoir of a paper-mill in this condition the water would require many days to clear itself by settling in the natural way. It must, therefore, be " cleared" by artificial means, and the only material that will effect the clearing (or deposition of the impurities held in mechanical suspension) is sulphate of alumina. Crystal alum has been used for this purpose by paper-makers for many years. But, as has been already stated, this material contains, in round numbers, only thirty-six per cent. of sulphate of alumina. Therefore, of late years, crystal alum has been superseded for clearing water by use of sulphate of alumina in a more concentrated form—a " concentrated alum." The quantity of sulphate of alumina required to clear water is very small, and therefore the sulphate of alumina should be in such form that it will dissolve very slowly.   In engine sizing, where it is desirable that an alum should dissolve rapidly, alum should be in a porous or ground condition.   But in such form it would dissolve far too rapidly for water clearing.   It must, therefore, be concentrated and cast into large blocks, which present but little surface to the dissolving action of the water. Such blocks are either thrown into the reservoir of the water to be cleared, or, better, should be placed in a trough through which the muddy water runs into the reservoir.   It must be remembered, that it is the sulphate of alumina only that does the clearing of the water.   A concentrated alum, in blocks or cakes, for water clearing should contain neither iron, zinc, nor free acid.   The Pennsylvania Salt Manufacturing Company of Philadelphia, the makers of the Natrona porous alum already referred to, are also manufacturers of a concentrated alum in blocks, which admirably fulfils all the requirements of a perfect clearing alum.

False Economy in the Use of Alum.

The more advanced paper-makers of the present day have found that there is no economy in the use of too small proportions of alum.   An excess of alum insures a more complete matting of all the fibre, and, in the end, a larger product of paper.   If a pound or two extra of alum per each one hundred pounds of paper produced can be made to yield—

on account of the alum insuring the more complete collection of the stock—from two to five pounds additional weight of paper, the money expended for the extra alum is well invested. The competition of Harrison Bros. & Co. for the alum trade of paper-makers has led, no doubt, to a more thorough understanding of alum requirements on the part of the paper-makers than would have been acquired through their own volition. The firm named first showed that 80 lbs. of the " pearl" alum would do as much sizing as 100 lbs. of the heretofore strongest known alum—the " porous"—would effect, leading to a gradual cutting down of the quantity of alum required. When this quantity was cut down below a certain limit, other troubles would result which were not manifested so long as alum was used in excess. In admitting the advisability always to use an excess of alum in sizing there will still be proportionately less of the more concentrated alums used than there would be of the weaker alum, and of crystal alums, and the strongest alum consequently is always the cheapest. When the alum is carried a long distance from the place of manufacture to the paper-mill, the item of freight becomes of serious importance. At the present time, owing to the great competition between the two leading alum-makers, the manufacture is concentrated almost entirely in and around Philadelphia; no other manufacturing centre supplying, to any extent, the paper-makers' alum; and in the city named the manufacture is narrowed down to four concerns—Harrison Bros. & Co., the Pennsylvania Salt Manufacturing Company, and two others.

Aluminium Sulphate.

Neutral aluminium sulphate (Al_23SO_4) is prepared either by treating clay or bauxite with concentrated sulphuric acid, or from cryolite. In an anhydrous state it contains 30 per cent. of alumina and 70 per cent. of sulphuric acid. With eighteen equivalents of water it crystallizes into octahedrons, or at a temperature of 32° F. into hexagonal rhombohedrons. Aluminium sulphate is soluble in double its weight in water. A solution prepared with the assistance of heat separates, on cooling, crystalline lamina of aluminium sulphate ($Al_23SO_4 + 18H_2O$). Aluminium sulphate is found in commerce in a nearly pure state, the best qualities containing only traces of iron, but from 0.5 to 2 per cent. of free sulphuric acid,[1] which is injurious when the salt is to be used for paper-making purposes. The presence of free sulphuric acid may also be detected by adding to a solution of aluminium sulphate logwood tincture. The solution, if free acid be present, will be colored brown-yellow and deep violet if it is neutral. To make aluminium sulphate containing free sulphuric acid available add to a solution of it 1 to 2 per cent. of zinc chips, the solution of which will be attended by a violent development of hydrogen. By the free sulphuric acid combining with the zinc, zinc sulphate is formed. An excess of zinc is dissolved with formation of

---

[1] To test aluminium sulphate for free sulphuric acid compound, according to Edward Donath, a solution of it at an ordinary temperature with a few drops of potassium iodide and potassium bichromate, and add a little bisulphate of carbon. If free acid is present the iodine is liberated and the bisulphide of carbon, on shaking, assumes a beautiful violet color.

zinc sulphate and separation of basic sulphate of alumina. Instead of zinc chips 1 to 2 per cent. of sodium carbonate may be used.

Aluminium sulphate or alum cake is found in commerce in the United States, generally in the form of powder, excepting when prepared in a very dense form used in the settling ponds of paper-mills. By boiling aluminium sulphate with water, it is gradually and completely dissolved, and yields a colorless fluid with an acid reaction. The solution can be advantageously used in all cases where alum was formerly employed, especially if it contains no excess of sulphuric acid, as it constitutes the only component part of alum which makes the latter valuable to the paper manufacturer.

English cakes have always been imported in the form of lumps or blocks, but American manufacturers reduce theirs by chipping or powdering to a more or less comminuted state, in which state the aluminium sulphate is much more serviceable to the paper-maker. The Harrison " Lion" alum is a fair example of a high grade sulphate of alumina, which is chipped so as to be in a state readily soluble. This alum is soluble without residue. Another class of aluminous cakes is made by boiling together the alumina-giving material and sulphuric acid, and allowing the entire mass to harden into a cake, containing all the insoluble material of the alumina mineral. This insoluble material is generally silica or silicate of alumina, which has not been acted on by the acid. If the alumina mineral contains this silica in a coarse state, it gives a sandy residuum, which may be the cause of a great deal of trouble to the paper-maker, cutting his wires and injuring the machinery generally. When, for very common paper, such as binder's boards, building papers, etc., it is desirable to have a cheap alum, and this sort of a cake is selected on account of its low price, the buyer is cautioned to select such a one as will give this insoluble residue in the finest possible state, for all medium and good classes of papers, the alums which are soluble without residue are, in the end, the best.

Resins.

The resins used by paper-makers vary greatly in quality and in price, and are usually graded commercially as Common to Good Strained, Good No. 2, Low No. 1, Good No. 1, Pale, Extra Pale, etc., the first named being the most inferior in quality, and the other varieties increasing in value in the order named. To prevent injuring the whiteness of the paper, upon which much of the value of the best classes of paper depends, it is necessary that the purest resins be used.

Of the different kinds of resins the American and French have the preference. They are the residues obtained in distilling turpentine with steam in the preparation of oil of turpentine, and are usually known as colophony. This resin is either only slightly yellowish or quite colorless.

Of course, the lighter and cleaner in color the resin is the better will be the quality and appearance of the papers to which it is applied. It may be that the resin at hand is dark, and contains many impurities. If so, it may be necessary to submit it to a purification, which is usually done by boiling it with a solution of common salt, when the impurities and much of its color are precipitated with the salt water. This boiling is often repeated a second or third time to insure a bright color.

The resins are rarely ever approximately pure definite bodies, but are usually mixtures of several analogous oxygenated bodies in various proportions. Their chemical relations are at present but very imperfectly understood.

Starch.

For the paper-manufacturer a specially pure and white starch is made. Of the best two varieties of starch used, corn-starch and potato-starch, the former is the cheaper in price, and the latter the better in quality.

Neither variety should contain more than 18 per cent. of water, which is tested by the loss of a weighed quantity at 230° F. The ash should not amount to more than three or four parts per thousand; a larger quantity would indicate a fraudulent admixture.

Stock for Paper-makers' Sizing.

Adamson's method of preparing stock for paper-makers' sizing is as follows:—

Take ordinary glue-stock, such as the cuttings of raw-hide, bones, etc., which has been treated with lime or alkali in the usual manner, and wash this glue-stock with water until as much of the lime as possible has been removed. Then impregnate the stock with alum by steeping it in a solution of alum and water for from four to ten hours, or as long as the thickness and quantity of lime remaining may demand, after which drain off the solution, leaving the stock in a moist and swollen condition, in which state pack it into barrels, boxes, or bags, ready for transportation to, and use by, the consumers.

The alum acts as a preservative, so that the stock can be kept for a considerable length of time without deterioration.

As the stock thus prepared reaches the consumer while it still retains its moisture, it will readily yield to the dissolving action of heat and water, and can be speedily converted into the desired size, thereby avoiding the long-continued boiling, which has a tendency to discolor the size.

The alum treatment has the further advantage of neutralizing the lime which has remained in the stock, and which would otherwise detract from the quality of the size.

A further important action of the alum is to make the size limpid and transparent, and to render paper and other material to which it is applied partially water-proof.

One of the main advantages of this process is, that the size-stock can be prepared and packed and forwarded directly to the consumer, thereby obviating the usual preliminary drying, which is costly, and frequently injures the stock.

The proportion of alum to the water in the solution will in a great measure depend upon the character of the glue stock, as will also the time during which the stock is steeped in the water; but the desired result may be attained by steeping ordinary raw-hide, clippings, bones, etc., for from four to ten hours in a bath in which the proportions are about three to five pounds of alum to one hundred pounds of wet unprepared glue-stock.

WATER-PROOF SIZINGS FOR PAPER.

Sizing and Water-proofing Paper with a Compound consisting of Water, Soda, Lime, Lard or Tallow, Glue, Bichromate of Potash, and Linseed Oil.

The following compound is designed to be incorporated with the pulp with a view to rendering the paper water-proof. The mixture is prepared as follows: Dissolve twenty-five pounds of soda in thirty-one gallons of water by boiling. To this liquor add gradually twelve pounds of recently burned lime, mixed in a small quantity of water. Let this boil about an hour, and then allow it to settle, and pour off the clear liquor, which should be a caustic lye of 36° Baumé. Melt fifty-six pounds of lard or tallow by a gentle heat, and add fourteen pounds of the above-named lye, stirring well, and not allowing it to boil. When thoroughly mixed, add fourteen pounds more of the lye, stirring constantly, and not allowing it to get to a higher temperature than 148° F. To this add fifty-six pounds of glue, dissolved in twenty-eight pounds of caustic lye, at 18° Baumé, by a gentle heat. Stir well until the whole is a homogeneous paste; then add sixteen ounces of bichromate of potash, dissolved in a small quantity of hot water, and finally add sixteen pounds of linseed oil. Stir continually for about half an hour, and then run it into a box, and keep it covered closely for about twelve hours. This size should be made a few days before using.

By this method of making the size it is claimed that the glycerine of the fat is retained, and forms with the glue a compound very much like India-rubber, which adds greatly to the strength and elasticity of the paper.

To use the size, dissolve three pounds of it in two gallons of water for a two-hundred-and-fifty-pound engine, and when thoroughly incorporated with the pulp add twenty pounds of alum and ten pounds of acetate of lead; or, in place of the alum and acetate of lead, add twenty pounds of sulphate of iron; or ten pounds of chloride of lime and one-fifth of a pound of bichromate of potash, or vary the proportions of these according to the nature of the pulp materials.

When preparing the size for ledger or writing-paper omit the linseed oil, and instead use one ounce of lime, in the form of milk and lime, and two ounces of hyposulphite of soda, dissolved in a small quantity of hot water, to every pound of the size, and mix thoroughly. For straw paper, use the size with the oil, retaining the lime-water used in boiling the straw, using the alum as usual. For photograph-paper, prepare the size the same as for writing-paper, except that the alum is omitted in the engine, and instead two pails of milk of lime are used to the two-hundred-and-fifty-pound engine.

Sizing with a Composition of Soda-Ash (Carbonate of Soda), Resin, Chloride of Sodium, Linseed Oil, and Silicate of Soda.

The following process has for its object to render the paper more or less impervious to ink, so that the ink will

29

lie up clear and distinct, and not be absorbed into the tissue of the paper; also, for making the paper stronger, and of smooth and fine surface finish, at the same time improving the color of the paper, and for acting as a mordant for fixing the colors of the compound. The following is an example of the material and proportion of the same, and the manner of compounding them:—

Soda-ash (carbonate of soda), three hundred pounds; twelve hundred pounds resin; chloride of sodium (common salt), five pounds; raw linseed oil, two gallons, or its equivalent vegetable oil; silicate of soda, thirty-six gallons, gravity 28° Baumé.

The process of compounding these ingredients is as follows, or substantially so: Take the three hundred pounds of soda-ash, and dissolve it in about eighty-five gallons of boiling water, and allow the solution to settle. The solution is then siphoned off into any suitable vessel, the precipitate remaining in the bottom of the vessel in which the solution was effected. This precipitate is utilized for destroying the fatty or animal matter in boiling rags. The twelve hundred pounds of resin is reduced to a powder, which is then added slowly and under constant stirring to the soda-ash solution, after the solution is raised to a boiling point, and the boiling is then continued for about one hour. Then is added the chloride of sodium (common salt). This mixture is continued to be boiled until a chemical union is formed of the said alkaline solution, resin, and common salt. While in this heated state next combine with the mixture the two gallons of linseed or other equivalent vegetable oil, thoroughly incorporating it therewith. Now add the silicate of soda slowly to the mixture, constantly stirring the same all the while, continue the boiling for about thirty minutes after all the ingredients are mixed together. It is then drawn off into suitable vessels for use.

The practical application of this sizing for the purpose specified is as follows: Of this sizing compound take from one to three quarts, and dissolve it in about a pailful of boiling water, and add it to the pulp that will make about one hundred pounds of paper. To this compound and pulp, when thoroughly mixed in the beating-engine, add from three to four pounds of alum dissolved in water. The pulp thus treated and having attained the proper consistency is now run off into receivers, from which it is taken on to the machine when the paper is formed.

Water-proofing Building or Sheathing Paper with a Composition consisting of Resin, Paraffine, and Silicate of Soda.

This composition for water-proofing paper consists of the following ingredients, combined in the proportions stated, viz: Resin, 50 per cent.; paraffine, 45 per cent.; silicate of soda, 5 per cent. These ingredients are thoroughly mingled by heating them together, and by agitation.

In using the above-named composition it is placed in a suitable open tank, to which heat is applied in any convevient manner, so as to keep it hot while being used.

The paper to which the composition is applied is mainly building or sheathing paper. The latter is taken in the condition in which it comes from the paper-machine, being

quite dry. A strip or strips of the paper, from a roll or
other convenient holder, are conducted and drawn through
the tank of hot composition, and upon emerging from the
tank the paper passes between suitable rolls, which press
any surplus composition from it, leaving it hard and smooth.

The proportions of resin, paraffine, and silicate of soda
previously named are employed generally for the purpose
just described; but in some cases, according to the solidity
of the texture of the paper to which they are applied, the
proportions of resin and of paraffine are varied from five to
fifteen per cent. from those stated, but about five per cent.
of silicate of soda being retained.   Thus the proportions of
resin and paraffine may vary, under such conditions, between
fifty and sixty-five per cent. of the former, and between
forty-five and thirty of the latter, making a composition
consisting of said ingredients by which it is claimed the
paper is rendered water-proof and durable when exposed
to the weather, and by the combined effects of such ingredients the proper degree of water-proofing effect is produced
and a surface-finish both smooth and hard is obtained.

Method of Applying Paraffine to Paper and Strawboard.

The following process is intended to facilitate methods
of applying paraffine and similar substances to paper
or strawboard, such, for instance, as paper cartons or
boxes used for packing sensitive articles, as tea, coffee, or
spices, and separating frames or mats used for packing pho-
tographers' " dry-plates," the sensitive film upon which is

liable to injury by moisture or chemical substances which may be contained in the paper or board.

The process constituting this invention, which is that of Mr. Warren B. How, of Chicago, Ill., consists, essentially, in dipping the article to be treated in a bath of melted paraffine or other substance similar in its characteristics when the article is at a temperature lower than the melting-point of the paraffine, and promptly removing it from the bath, whereby the adhering paraffine is prevented from entering the body of the article to any considerable extent, and practically forms by congealment only a thin coating or film upon its surface, and then subjecting the article to heat above the melting-point of the paraffine until the former has been brought to substantially the same temperature throughout, and the superficial paraffine is thereby caused to sink into the paper or board.

It has been proposed heretofore to treat articles of paper or strawboard with paraffine or similar substances by a process in which the article is first heated to a temperature higher than the melting-point of paraffine, and then immersed in such heated state in the bath of melted material. In this older process the article, being when dipped in the bath in condition to readily absorb the melted material, will become fully or substantially saturated, and will also, in addition to the substance absorbed, be covered by a distinctly visible coating of surplus material, forming a considerable external body thereon. The subsequent heating of the article for the purpose of removing this surplus material has been found to leave the article with a greasy surface, and not one of a dry, smooth texture and even color, such as may be obtained by the process herein described, and claimed as new. The inventor of the present method states that he has found by experiment that the quantity of paraffine deposited in the form of a coating upon the article by dipping it when at a temperature lower than the melting-point of the paraffine will be less than will serve to completely saturate paper or board of the usual thickness used in making boxes or other articles intended for packing merchandise, so that when the article is afterwards heated for causing the absorption of the said coating a less quantity is present than will fully saturate the paper, and when absorbed will leave the article with a smooth, dry surface, and one which is not objectionably greasy to the touch or harmful to the contents of the package by reason of an excess of paraffine. The article will be quickly removed from the bath, because if allowed to remain a sufficient time therein to become heated through, the paraffine will soak into and fill the paper or board, with the same objectionable result which follows from previously heating the article, to wit, of applying an excess of paraffine. That portion of the bath substance which strikes and to a slight extent enters the paper in this brief dipping of the article is congealed by the lower temperature of the paper with which it comes in contact, and, the article being promptly removed from the bath, instead of being itself heated by the liquid, it cools the latter, and insures, as stated, that an adhering coat shall be practically confined to the external surface of the article.

The appliances employed in the present process may

obviously be of the ordinary construction, any form of oven or hot-air chamber being all that is required for the heating which follows the dipping. It is to be understood, of course, that the paraffine coating may either be allowed to cool and become solidified after dipping and before the heating of the article, or that the article after being dipped may be immediately heated to cause the absorption of the coating, the final result being the same in either case.

Treating Paper with Ozocerite.

The following process consists in saturating paper with a natural bitumen or wax known as " ozocerite." The object is to supply meat and fish venders with a paper which will be water-proof, and which at the same time will not impart to the article inclosed within the paper discoloration or a disagreeable odor.

Paper has heretofore been rendered water-proof by means of paraffine, which answers the purpose well; but owing to its cost it has not come into general use. Paper has also been rendered water-proof by means of coal-tar and its products; but, owing to its offensive odor, it cannot be used to wrap provisions of any kind, and its use is necessarily confined to wrapping hardware or in wrapping articles to protect them from the ravages of insects.

In carrying out the present process the inventor, Mr. Charles A. Maxfield, of New York, N. Y., utilizes any of the well-known machines in the manufacture of roofing-felt, in which a heating-tank is provided, and devices for removing a surplus of the saturating material. He then places the ozocerite in the melting-tank and reduces it by heat to the desired consistency. The paper, which is in long strips or rolls, is then drawn through the melted or liquid ozocerite, and any surplus of the material removed therefrom. The paper may then be passed through a drying-chamber, so that it can be cut to the desired size and packed into reams and bundles ready for use.

The prime object of this invention is to saturate the ordinary wrapping-paper used by butchers and meat and fish venders with a compound which will be odorless and colorless—that is, will not discolor the article—and at the same time render it water-proof and strong.

CHAPTER XIV.

COLORING.

COLORING is a special branch of paper-making and requires much practical experience, as mistakes made in the coloring of paper are difficult and expensive to remedy. A line of receipts for coloring paper are, of course, useful; but it must be remembered that the ingredients and directions which will result in producing the desired color with one engine of pulp may not be so successful in another case when the pulp may be supposed to be of exactly the same nature.

It is, therefore, well to understand something of the theory of coloring, and of the nature of the different coloring ingredients commonly employed; such knowledge will enable a practical man to quietly confront any problem which he may in the course of his operations encounter, giving at the same time an astonishing accuracy to his calculations for the elementary composition, as well as the production of colors and shades, which he desires to imitate.

Light is the source of all color, and it is the result of the vibrating motions of a very subtle substance, which the natural philosophers term ether.

The ether receives vibrations from self-lighting bodies, such as the sun, and spreads them in the same manner as the air spreads sound, with this difference, that the oscillating motions called light are brought forth many million times quicker than those of sound, because the ether is many million times finer than the air; consequently, the vibrations are more rapid and intensive.

The light entering our eye excites the optic nerve, producing a sensation called vision, and thus light renders objects visible.

Light itself is not a simple body, but is composed of various colors of which we distinguish seven by separate names.   All the colors observed in the organic and inorganic world around us are derived by reflection from the different colors, of which the white light or sunlight is composed.   When a ray of sunlight is admitted into a dark room, and there split by a prism, there is observed upon the screen placed opposite to the hole a series of bright colors, consisting of violet at the top, indigo, blue, green, yellow, orange, and red.   This phenomenon is termed a spectrum. Apparently there are seven colors; really there are twelve; but there are only three primitive colors, namely, red, yellow, and blue, from which all others are derived.   If we look attentively at a spectrum, we soon realize the fact that violet, indigo, green, and orange are the products of amalgamation of either two of the three primitive colors, namely, violet and indigo from red and blue, green from blue and orange, orange from yellow and red.   When speaking of primary colors, hereafter, it must be borne in mind that we refer to the colors of the spectrum.   There exists no primary color substance, that is to say, a color representing nothing but itself, in the true sense of the word. In fact, if we compare all known dye-stuffs, we find that they always contain, besides the principal ones, more or less of some other color. This is not of course by bodily mixture, but their intimate atomic construction is such as to reflect more or less of the others, too, which are the component parts of the white light. In composing our shades for paper there are, in fact, only three primary colors at our disposal, that is, red, yellow, and blue. The modifications which these three colors are capable of undergoing, and the limitless combinations into which they can enter with each other, enable us to reproduce any required color or shade. It should be borne in mind, however, that in all cases, the dye-stuffs which are used for producing the various combinations of colors must be of such a nature as to allow their combination, that is, of their perfect embodying one into the other, if mixed in solution ; so that, after the paper pulp is colored, no separate colors can be distinguished upon the finished material. If we take three very bright artificial dye-stuffs, red, yellow, and blue, products of coal tar, better known as aniline colors, whose chemical composition allows their perfect union by mixing their solutions with one another, we can produce the twelve colors which a close examination and dissection of a spectrum discloses are con: tained in it. These colors are red at the bottom, and following in successive order upwards, red-orange, orange, yellow-orange, yellow, yellow-green, green, blue-green, blue, blue-violet, violet, and red-violet. Nine of these are binary colors, so-called from being composed of two of the three primary colors. If, in the transition from one prismatic color to another we were at each step to exchange only one hundred parts of the one for an equal quantity of the other, it is evident that infinitely more colors might be produced; but these slight modifications would be hardly distinguishable. To our eye the whole color-scale would appear as an amalgam like the spectrum of a white light. But if it is considered that either of the two colors can be graduated, that is, progressively changed from very dark, nearly dark, to light and very light, nearly white, it may readily be imagined that thousands of colors can be produced by this very simple means. The slightest alteration of white is at once perceptible; while a considerable proportion of any other color can be added to black, before the modification becomes apparent to even the trained eye.

All colors growing out of primary colors must be classed under that denomination, until they reach the compound of the primary with the nearly binary standard color. For instance, all colors originating in red are to be classed as reds, until the red-orange is reached, with this difference, however, that they are distinguished as first, second, third, etc., reds, according to the degree of modification the red has undergone. Thus, if red is denominated as No. 15 on the side toward blue, it would signify that this red contains 15 per cent. of blue, as only with the addition of 25 per cent. blue the red becomes red-violet. The same nomenclature should be used for blues and yellows. From the above it is easy to see that a precise knowledge of this circle is sufficient to determine a color at once, thus enabling the paper colorer to reproduce it. The series of modifications of which any

primary or binary color is susceptible, is collectively called
the category of that color, and it should be understood that
the category is not identical with the shade, though the
erroneous application of this term is quite general and popu-
lar. We know what a simple (or primary) color is, of
which there are only three: red, yellow, blue; we know
also, that either two of these simple colors combined with
one another form a binary (or secondary) color. But a
shade is the result of the combinations of one binary color
with one or two other binary colors, not belonging to the
category of the former. Brown, for instance, is a shade.
All the browns are oranges, shaded more or less with blue
or violet. When a paper colorer is called upon to reproduce a certain brown, the first thing he has to do is to ascertain what kind of orange is the base of it, whether red, pure,
or yellow-orange. By contrasting various browns with one
another, it is easy to determine whether the particular brown
in question belongs to the reddish or yellowish category;
the category once defined there is no further difficulty in
deciding upon the base of the coloring, whether reddish or
yellowish-orange. This base is then to be composed, then
to be shaded, and the mixture is ready for coloring the paper;
if the diagnosis of the orange is correct, and the shading
carefully done, there should be no difficulty in producing
the exact brown. From this it may be seen how indispensable it is for the paper colorer to thoroughly understand his
binary colors; upon this knowledge depends, in fact, the
whole art of producing the various shades; the whole
variety of binary colors being so many bases for all imagi-

nable shades. The shades, in their turn, are again susceptible of infinite modification, that is, of being rendered light by the addition of water, or darkened by the chemical action of various salts.

This brings us to the subject of "mordants," which, in addition to being bodies possessing the power of fixing certain dyes upon materials to be colored, are also certain bodies which possess the properties of changing the natural color characteristic to coloring matters, thus producing different shades with one and the same substance. The acids, chlorine, resin, and the alkalies contained in the pulp may act as mordants, and the paper-colorer has consequently to take their effect into consideration.

The mordants commonly employed in paper-coloring are alum, green vitriol, nitrate of lead, and sugar of lead. *Sodium carbonate* (soda) is used by the paper-colorer chiefly for neutralizing purposes and for dissolving coloring-matters.

Alum is used as a mordant with nearly all the colors employed for coloring paper. For a description of the various alums see p. 435 *et seq.*

There are two chromates of potash known, viz., the chromate and the bichromate. Bichromate or acid chromate of potash occurs in commerce in beautiful red crystals, very easily soluble in water, and is employed in the preparation of red, yellow, and green colors.

Green vitriol (*ferrous sulphate*) is a combination of ferrous oxide with sulphuric acid. It is used as a mordant, especially for black, gray, and violet, and also in the production of Berlin blue.

Lead nitrate (nitrate of lead) is prepared by dissolving litharge in nitric acid. It forms white crystals of a nauseously metallic taste, which are difficult to dissolve in cold water, but readily so in boiling water.

Plumbic acetate (sugar of lead) is prepared by dissolving litharge in vinegar, and evaporating the resulting solution. Both nitrate of lead and sugar of lead are poisonous, and when even handled carelessly produce symptoms of colic.

As a rule it may be stated that acid mordants should be employed for aniline colors. Yellow prussiate of potash, *potassium ferrocyanide*, is not used as a mordant for fixing other colors, but in combination with different salts of iron produces different shades of the well-known beautiful blue color, Prussian or Berlin blue. *Potassium ferricyanide*, red prussiate of potash, produces with ferrous salts a desirable blue color, Turnbull blue, which is similar to Prussian blue.

We will mention some of the natural and artificial dyestuffs which can be advantageously employed for producing the various colors and shades upon paper, and will enumerate them in their natural order, namely, red, yellow, blue, etc. It should be remembered, however, that there is quite a difference between the colored pulp in the beating-engine and the finished colored paper, and if a sample of paper is to be imitated in color it will be necessary to have them in the same condition as to moisture. The most expeditious way is to reduce a piece of the colored sample paper to pulp by macerating it in water. Some of the pulp can be taken from the trough of the engine and both the sample and the pulp under treatment can be squeezed so as to reduce both to about the same state of moisture. It can then be readily determined whether it will be necessary to add additional coloring matter to the pulp in the engine. If the pulp is too highly colored it will be difficult to remedy the mistake, and it is best to be on the safe side by not employing too much coloring matter at first.

It will be noticed in the following receipts that the proportion of size to be used with certain colors is not specified, and that it is often suggested to first mordant the pulp and color it afterwards. This is the rational way, but for low and medium grades of paper it will usually be found more expeditious to add the mordant to the coloring mixture. Such points are matters for practical consideration; the object of the present chapter is to suggest combinations of materials which can be used to produce various colors and shades, the proportions of the ingredients to be used and the manner of applying them must be determined by experience as the same directions would not apply to any two mills.

A great many of the coloring matters which we shall now enumerate are not now used in practice, the object being to suggest those capable of employment.

Red Shades on Paper.—The natural dye-stuffs capable of producing red colors and shades on paper are cochineal, and the numerous red woods usually comprised under the name of Brazil wood, kermes grains, etc. Cochineal and kirmes grains are derived from the same species of insect, as is also the so-called lac-dye, the coloring matter in all of them being identical. The ordinary Brazil wood is derived from *Cæsalpinia brasiliensis*, a native of Brazil. Other varieties of red wood are Jamaica wood, Nicaragua wood, Pernambuco wood, Santa Martha wood, and Sapan wood, the latter yielding a somewhat lighter coloring material than that derived from the other red woods which we have named. The extracts of these woods, or rather the coloring matter, the so-called Braziline, may be employed for amaranth, crimson, purple rose, and similar shades, but it should be remembered that the red colors produced from these woods alone are not at all fast. The coloring matter is extracted by boiling the woods, and the extract so obtained is soluble in water. Red upon paper may also be produced by ammoniacal solution of carmine, or of cochineal, or of cochineal heightened by a solution of tin, or by means of a decoction of Brazil wood and alum in malt liquor. Lobster red, rose red, scarlet, crimson, and Morocco red, may be produced upon paper by employing cochineal extracted by diluted ammonia ; or carmine, lac-dye, drop lake, and aqua-fortis can be used in case the cochineal does not answer.

Paper when first colored with decoctions of Brazil wood, freshly prepared, has a perceptible touch of orange, but on exposing the colored paper to the action of the air the orange shade disappears leaving only red; it is possible to impart a bluish tone to paper colored with Brazil wood by the use of alkalies. Cochineal red of a magnificent shade may be applied to paper, the color being best prepared by tying the cochineal in small linen bags and then boiling in water containing about two per cent. of spirit of sal-ammoniac, or liquid ammonia can be used in lieu of sal-ammoniac. Scarlet may

30

be produced by applying an extract of carthamus. The extract of carthamus is dissolved in a solution of one part of tartaric acid in sixty parts of water.

Red sanders wood and barwood cannot be extracted, or, rather, they are difficult to extract with boiling water, but with alcohol and sulphuric ether their coloring matter is readily obtained, and it possesses the advantage of being much faster than any of the red-wood decoctions which we have heretofore named, and is especially valuable when such shades as rose are desired.

Hypernic can be employed in preparing red dye-stuffs for coloring paper, it is not, as is commonly supposed, a species of any particular variety of wood, but is composed of a mixture of chips of numerous varieties of red wood; but it is employed the same as any of the single species for coloring red, and can also be employed for mixed colors.

Extracts of the various woods which we have mentioned are regular articles of commerce, and can be employed in lieu of the various woods. A few years since it was a complaint among persons whose business compelled them to use dye-stuffs that they could not always depend on the extracts of dye-woods; there may have been just grounds for their objection at that time, but since the introduction of improved processes and machinery it is now possible to procure highly satisfactory extracts of dye-woods.

In order to produce a strong dye liquor from extract of Brazil wood, place two and a half pounds of extract in ten gallons of clear water, and boil it for ten minutes, a copper kettle being used if the boiling is conducted over an open fire; but if boiled by steam a wooden tub or barrel should be used. When the boiling has proceeded for the length of time specified, add in small quantities, and always under vigorous stirring, one-half ounce of potash and one-half ounce of soda; the boiling should then be continued for about four or five minutes longer, after which the dye liquor so prepared should be drawn off into a separate barrel, and properly covered for use. The liquor can afterwards be diluted to any desired degree with water.

Venetian red (ferric oxide), when thoroughly washed, is used for delicate brown colors, and is employed in its merchantable form for wrapping and other low grades of paper.

Aniline red colors of various kinds come into commerce under the names azaleine, fuchsine, magenta, mauve, roseine, rubine, solferino, tyraline, etc.; but although they may differ very greatly in the manner of their manufacture, they are all produced from the salts of a base termed "rosaniline," and they find their way into trade in the form of greenish, granular crystals having a metallic lustre, but often the product has the form of a red powder. The acetate of rosaniline, which is commonly known in the United States and in England under the name of fuchsine, is readily recognized, as it forms crystals which are especially beautiful; but in Germany the acetate is known under the name of roseine, and the hypochlorate of rosaniline is known as fuchsine. It is necessary to exercise considerable precaution in buying crystallized fuchsine, as it is not uncommon to find it considerably adulterated, sugar crystals being especially used for this purpose, and all aniline colors in the form of powders, are also liable to be greatly adulterated. When unadulterated, fuchsine may be dissolved quite easily in hot water, and very easily in acetic acid, alcohol, wood-spirit, or even in a solution of tartaric acid, but it is only sparingly soluble in water of an ordinary temperature. The solution which pure fuchsine yields is of a beautiful purple-red color, and the additions of alkalies or strong acids discolor them; but when the red color of the solution is discolored by the addition of an alkali, it may be restored by adding an acid. and when discolored by strong acid the color may be restored by adding water.

Azaleine, known chemically as the nitrate of rosaniline, is not often brought into commerce, and it may be readily distinguished from the various other aniline reds by means of the cherry-red color of its solution.

Diamond magenta, or fuchsine, comes into commerce in the form of large crystals having a greenish lustre, and it possesses about the same properties as fuchsine, but an additional advantage is that it is non-poisonous.

Rosaniline colors are not fast, but with the exception of this class the other aniline products have probably found a permanent employment for coloring paper, their brilliancy and freshness especially recommending them. Another series of colors are derived from the creasote in coal tar, and these are known chemically as phenol colors, and one of the most important of this class is phenol red, or corralline, which is a magnificent red coloring matter, and comes into commerce in the form of a red powder which is with difficulty soluble in water, but is readily dissolved in alcohol, and yields a scarlet solution.

Coralline is also soluble in alkalies, but the solutions thus obtained change very readily, but it is not changed by acids. A beautiful color is imparted to pulp previously mordanted with alum by means of coralline. In order to prepare the color it is first dissolved in alcohol, after which some caustic soda is added, and the alkaline solution thus prepared is mixed with water, which should be sufficiently acidulated with acid in order to fully neutralize the soda. A great objection to the employment of coralline is that the color produced with it will not stand exposure to light.

Aniline red occurs but seldom in commerce in the form of a paste or in solution.

Yellow shades on paper can be produced with various vegetable, mineral, and aniline coloring matters.

Barberry-yellow may be produced by treating the pulp with a liquor produced by boiling two pounds of barberry root and six ounces of alum in every seven gallons of water. It has been stated that the fruit of the barberry might be employed, but this, however, is not the case, the color being derived from the root and the bark of the shrub. A decoction of the ground bark and root can be employed for the purpose of producing a lustrous lemon color when the paper-pulp has been previously mordanted with alum, an additional mordant or striker of tin salt being also employed in the production of the latter color.

Quercitron-yellow can be produced in any desired shade by the use of a decoction of quercitron bark.

Rust-yellow can be produced with a liquor made by boiling two pounds of annotto, and four ounces of potash in every nine gallons of water. There are various grades of annotto, such as the French, East and West Indian, Brazilian, etc., but the first named is most valuable and comes into commerce neatly and securely put up in tin cans, and this grade of annotto is distinguished by its bright red color and its peculiar odor, which somewhat resembles carrots.

In addition to the substances which have been named, fustic, alder bark, bablah or babool, sumac, saw-wort, tumeric, dyer's broom, and the American golden rod, can be used for producing various shades of yellow grays, and orange red on pulps previously mordanted with alum. Canary-yellow can be produced with weld.

Lemon yellow can be produced by digesting one part of tumeric in four parts of ordinary spirit of wine.

Mineral Pigments.—Chrome yellow can be produced by first adding to the pulp a solution of two ounces of bichromate of potash or bichromate of soda in one quart of water, and afterwards adding a solution of one ounce of sugar of lead in one quart of water. The basic acetate of lead may be used in lieu of acetate of lead, and is prepared by boiling a solution of plumbic acetate with an excess of litharge, and comes into commerce in solution under the name of vinegar of lead. The poisonous properties of these salts should not be forgotten. Orange mineral may be used to intensify the chrome yellow, or orange. Yellow ochre should also be mentioned among the mineral pigments employed for coloring paper.

Aniline-yellow, ordinary, is now coming into use as a substitute for chrome yellow, and like most aniline colors it is more soluble in hot than in cold water.

Chrysaniline is a yellow powder which is not in the slightest degree soluble in water, but when dissolved in alcohol, it gives a beautiful yellow color.

Aurin is the commercial name usually applied to the hydrochlorate of chrysaniline, which is somewhat soluble in water and produces beautiful golden-yellow colors, and is readily recognized by its red-yellow needles.

Zinaline comes into commerce in the form of a cinnabar-colored powder; it is not soluble in water, but warm solutions of borax, sodium phosphate, or sodium acetate dissolve it. Zinaline is also soluble in alcohol and wood spirit, but water precipitates it from these solutions. The shades which it yields are a reddish-yellow.

Blue Shades on Paper.—The number of blue coloring matters applicable to paper coloring is not large. A durable blue on paper is obtained by the use of mineral pigments, which produce a color offering greater resistance to air and moisture than that obtained by employing aniline colors.

Prussian blue, or Berlin blue, is produced by the employment of yellow prussiate of potash, which, on coming into contact with iron salts, produces a blue color. Prussian blue may also be obtained by somewhat varying the method of coloring, by employing a solution of ferric salts and ammonium oxalate, which is used to saturate the pulp, which is next treated with yellow prussiate of potash, and finally adding to the pulp a weak acid solution. When produced by either of the above methods Berlin or Prussian blue forms a fast color, which is destroyed only by alkalies, and is not affected by acids.

Pulps which have been sized with vegetable (resin) size would injure these blues because of the alkaline nature of the resin soap, provided the coloring ingredients were added to the pulp before the alum solution was run into the engine trough, hence care should be taken not to commence the coloring until the sizing is complete. Any surplus of alum will be beneficial as it will act as a mordant and intensify the blue color.

This blue color can also be prepared through the action of yellow prussiate of potash upon sulphate of iron, green vitriol, but in this case a slow oxidizing process is sometimes employed in the preparation of the coloring liquor. The oxidation, however, may be accelerated by using a solution of about one pound of bleaching powder to each two pounds of the green vitriol. Nitric acid is now much employed for hastening the oxidation, because of the bright shade of blue which it produces. But after using either of these oxidizing agents the blue precipitate should be washed with clean water several times.

In commerce, yellow prussiate of potash is recognized in the form of yellow crystalline masses, the surface of which by exposure to the air loses water and is converted into a greenish-white powder. In water previously heated the salt is readily soluble, and is distinguished by its bitter taste and neutral reaction, and is poisonous.

The so-called Turnbull blue, may be produced with red

prussiate of potash, which acts upon ferrous salts in the same manner as yellow prussiate of potash acts on ferric salts. Like the latter, red prussiate of potash is soluble in water, and it is poisonous. Yellow prussiate of potash is, however, preferable for producing blue colors as it is cheaper.

Berlin or Prussian blue can be prepared by dissolving twenty-five pounds of yellow prussiate of potash and thirty pounds of sulphate of iron in hot water, in separate vessels, and pouring the two solutions into an empty barrel or other receptacle and thoroughly agitating the contents of the barrel and then filling it with water. Nitric acid or a solution of bleaching powder is then added, and the contents of the barrel are allowed to settle, and after several hours the supernatant clear liquor is decanted. The blue sediment in the bottom of the barrel is again agitated while fresh water is added, and the contents of the barrel are again allowed to settle. After being thus two or three times thoroughly washed the blue sediment is taken from the barrel and placed in another receptacle, which is afterwards filled with water and covered to protect from dirt. The blue liquid must be always thoroughly stirred before furnishing the engine in order to produce the same results with each engine of pulp. The coloring can be accomplished in the beating engine by adding directly to the pulp ninety-five parts of sulphate of iron and one hundred parts of yellow prussiate of potash.

Berlin or Prussian blue is usually employed for low and medium grades of paper, such as news, etc., and the blue is not injured even if the bleaching solutions are not thoroughly washed out of the pulps. The greenish tint sometimes objectionable in papers colored with Berlin or Prussian blue can be neutralized by the addition of a small proportion of red color.

Ultramarine is employed for coloring the finer grades of paper. But this coloring material should not be added to the pulp until some time after the resin size and alum have both been added to the pulp in the beating engine. This method prevents the alum from affecting the color of the ultramarine blue.

Cobalt blue, and a mixture of sulphate of copper and red extract, are sometimes used for imparting a blue color to paper pulp.

Sky blue can be obtained by heavily mordanting the pulp with yellow prussiate of potash and afterwards applying a solution of acetate of iron. The mordant is prepared by dissolving eight pounds of yellow prussiate of potash in twenty-four gallons of water, the pulp being thoroughly permeated therewith; the solution of acetate of iron, which should be afterward added to the pulp, should contain four or five ounces of the salt to each gallon of water.

Formerly aniline blues required the use of alcohol or wood spirit to dissolve them. At the present time, however, paper makers use only these aniline blues, which are soluble in water.

The *bleu de lumière*, which shows a pure color by candle light, and *bleu de Parme*, which is a darker blue, having a violet tinge, and showing a different color by candle light, are two principal shades of ordinary aniline blue.

Bleu de Lyon dissolves only with difficulty in water, but

very readily in alcohol, and the blue color which it imparts to paper is very beautiful. In commerce this blue is readily recognized by its lustrous masses of copper-red color.

Bleu de Paris (soluble aniline blue) is a powder having a blue-black color, with a slight copper lustre, and is soluble in water. It may be precipitated from its aqueous solution with acids or common salt. On account of its easy solubility in water, this blue may be especially recommended for coloring some classes of paper.

Phenol blue or azuline, is one of the direct products of creasote in coal tar. This blue coloring matter, with a shade resembling ultramarine, is a coarse-grained powder, having a slightly copper lustre; it is soluble in alcohol, but insoluble in water. Phenol blue may be dissolved in alcohol, and after diluting with water containing tartaric acid be employed for coloring paper.

Blue rags are still in some mills separately sorted out and employed where a very deep blue colored paper is required, and in such cases the rags are neither boiled nor bleached.

Blueing Paper.—The bleaching operation invariably leaves the paper pulp with a yellowish tinge more or less deep according to the nature of the material from which the pulp has been made, the care exercised, and the method of bleaching employed. This yellowish tinge is unpleasant when the finished paper is exposed to the light, and it is necessary to neutralize it. If blue alone were added the paper would present a greenish tint, consequently a little red color is necessary in order to impart to it a pleasant white appearance. Magenta and aniline blue are usually employed for low and medium grades of paper; and carmine and ultra-marine for the finer grades.

Mr. James Hogben, of Cleveland, O., patented in 1869 the following process for the combination of aniline or other suitable red pigment with sulphate of iron, prussiate of potassa, and sulphuric acid, for the purpose of giving the desired tint or color to paper in the process of its manu-facture.

The compound consists of the following ingredients and proportions, viz., sulphate of iron, sixteen pounds; prussiate of potassa, eight pounds; sulphuric acid, eight pounds; red aniline, two ounces; making altogether thirty-two pounds and two ounces.

Pulverize the sulphate of iron (green is preferred) and the prussiate of potash, and add the sulphuric acid. Mix it in a glass or stone vessel, then let it remain until it is digested or assumes a pasty condition.

Dissolve the aniline in sulphuric acid (one ounce) or suffi-cient to liquefy the same. Add this to the compound of sulphate of iron, prussiate of potassa, and sulphuric acid. Then mix the entire mass thoroughly, dry, and grind the same to a powder.

Previous to drying and grinding, in order to guard against any free acid that may remain in the compound, add one gallon of clear solution of caustic lime. An insoluble sulphate of lime will be formed, if any free acid remains, which sulphate of lime will not injure the paper.

The compound or preparation, without the aniline, may be used first with the pulp, and the aniline or red pigment subsequently added, while the pulp is in the engine or beaters, but allowing sufficient time for the preparation to be dissolved so completely as to leave no specks or spots upon the paper.

Green Shades on Paper.—Dark green can be produced with a decoction consisting of eight parts of quercitron, two parts of logwood, one part of alum, and one part of green vitriol.

Sap-green may be produced by digesting two parts of buckthorn sap in eight parts of spirit of wine; this decoction should be applied to pulp previously slightly mordanted with alum.

Olive green can be produced with a decoction of four parts of quercitron bark, two parts of Hungarian fustic, and a small quantity of dog-wood berries; the pulp to which this mixture is to be added should first be strongly mordanted with alum. Olive green may also be produced with a liquor prepared by boiling one part of the quercitron bark by measure, with two parts of water to which there is added a solution of green vitriol. Olive green, light, can be produced upon paper by first treating the pulp so as to obtain a Berlin or Prussian blue, and then adding a liquor produced by boiling five pounds of fustic and one and one-quarter pounds of archil in ten gallons of water.

Picric green can be produced by first treating the pulp with a moderately strong solution of Berlin or Prussian blue, and afterwards adding a solution of picric acid in water.

Greens of various shades can be obtained by varying the

proportions of a mixture of Prussian blue and chromate of lead.

Aniline green colors are now coming into general use for coloring certain grades of paper. Brilliant green, new Victoria green, and Russia green can be used for coloring paper. All of these colors should be applied to pulps previously mordanted with alum.

The various other aniline greens, such as Hoffman's night green, Lowe & Clift's emeraldine, as well as Fritsche's emeraldine, cannot be utilized for coloring paper.

Brown Shades on Paper.—Dark brown can be produced with a decoction prepared by boiling one-half part of quercitron, one part of logwood, one part of sandal wood, two parts of Brazil wood, and eight parts by weight of Hungarian fustic in a sufficient quantity of water to cover the ingredients to the depth of about two inches. The boiling should be continued for about one hour, after which the liquor is strained through linen, and then allowed to cool. By boiling a second time the ingredients above named the resulting product can be used for a similar color. Before applying this decoction to the pulp it should be previously mordanted with a solution of green vitriol.

Light brown can be produced by using the mixture prepared for dark brown. But instead of mordanting the pulp with green vitriol a weak mordant in the form of a bichromate of potash solution should be used.

Catechu brown can be produced with a decoction composed of two and one-fourth pounds of catechu and four ounces of green vitriol in twenty gallons of water. The pulp to which this decoction is applied should first be slightly mordanted with alum.

Olive brown can be produced by boiling one-half part of logwood, two parts quercitron and four parts of Hungarian fustic, all by weight, in sufficient water to cover the ingredients about two inches deep. After boiling for one hour the liquid is strained through linen, and when cold is applied to the pulp which should first be strongly mordanted with a solution of potash. After the coloring liquor is applied the pulp should be treated with a solution of green vitriol.

Coffee-brown is produced by mordanting the pulp with a solution of one pound and two ounces of acetate of copper in seven gallons of water, and then immediately treating it with a solution of yellow prussiate of potash in slightly acidulated water.

Light leather brown can be produced by employing three pints of logwood liquor, five pints of Brazil wood dye liquor and six gallons of fustic dye liquor. After this mixture is applied to the pulp it should be treated with a solution of one pound of alum in thirteen gallons of water.

Mifonce brown is produced by employing one gallon of logwood dye liquor, one gallon of fustic dye liquor, and one and one-half gallons of Brazil wood dye liquor. After this mixture is applied to the pulp it should be treated with a mixture of one pound of sulphate of copper in ten gallons of water to which should be added a solution of one-fourth pound of sulphate of iron in ten gallons of water. The two solutions should be thoroughly stirred before the resulting mixture is added to the pulp.

Aniline brown. Bismarck brown comes into commerce in the form of a tarry, black brown mass which is soluble in spirit of wine and insoluble in water; but the spirituous solution after being mixed with water can be directly used for producing a dye liquor for imparting a brown color to paper. Coloring matters of a character similar to that of Bismarck brown, and also known as aniline brown, not infrequently consist of only by-products obtained by over-heating the composition in the preparation of fuchsine.

Havana brown is soluble in alcohol, acetic acid, and also in water, and is purified by precipitation from its solution from common salt.

Phenol brown is one of the direct products of the creasote in coal tar. In commerce it is found as a delicate brown powder which is readily soluble in alcohol, acetic acid, and alkalies, especially with an addition of some tartaric acid; but in water this coloring matter is only partially soluble. If such an oxidizing agent as potassium chromate be added to the solution of coloring matter it is possible to obtain a variety of shades ranging from dark wood brown to light leather brown. Phenol brown gives an agreeable brown color, and is readily absorbed by the fibre.

Violet Shades on Paper.—Violet may be produced by first coloring the pulp a pale blue with Berlin or Prussian blue, or with ultramarine, and then treating it with a solution of carmine. Violet can also be produced by digesting two parts of dry shavings of logwood in sixteen parts of spirit of wine, and after adding a little alcohol, the mixture should be applied to the pulp, which should first be slightly mordanted with alum.

Violet can also be prepared with an extract of Campeachy wood and alum.

Aniline Violets.—Dahlia is a beautiful coloring matter of rare purity of color, and in hot water it is readily soluble. Dahlia (dahlia imperial) is probably a by-product obtained in the manufacture of aniline red, and it assumes a brownish-red color when treated with concentrated sulphuric acid, thus differing from the ordinary aniline violet, as the latter color is colored blue when treated with concentrated sulphuric acid.

Hoffman's violet comes into commerce in two varieties, one of which is a reddish violet and the other a blue violet; they both occur in bronze-colored grins or crystals, and are readily soluble in alcohol, wood spirit, etc., but only moderately soluble in water. This dye-stuff is pre-eminent for its beauty and purity.

Perkins's violet commonly occurs in commerce as a green crystalline powder, having a metallic lustre, that is, when in a pure state; but it sometimes comes as a dark violet paste, and is readily soluble in hot water, and also, in the presence of an acid, in alcohol, wood spirit, glycerine, acetic acid, etc., but is only moderately soluble in cold water. This coloring matter is precipitated by alkalies and alkaline salts from its solution, but from spirituous solutions it is precipitated by water.

31

Parisian violet is not soluble in water, but by adding an acid it readily dissolves.

Rosaniline violet dissolves readily in alcohol and acetic acid, but it is with difficulty soluble in water. This coloring matter comes into commerce in the form of a brownish-blue powder having a weak lustre.

In addition to the aniline colors and the phenol colors there is another series belonging to this class which are the so-called naphthaline colors, and they are produced from naphthaline, which is a constituent of coal tar, and is a white crystalline body belonging to the hydrocarbons. Naphthaline violet is one of this class of colors, but it has not as yet been so generally introduced as the regular aniline colors.

Gray Shades on Paper.—Dark gray can be produced by employing a decoction of two ounces of concentrated extract of logwood, four ounces of Indian fustic, three pounds of tan liquor, three pounds of alum, and one pound of green vitriol.

Iron gray can be produced by employing a decoction of one ounce of logwood extract, two and one-half pounds of tan liquor, and nine pounds of solution of green vitriol.

Dark gray can also be obtained by neutralizing the grease contained in lampblack by treating it with a warm dilute solution of soda or potash. The lampblack is then washed until all traces of the alkali have been removed, and the material is then run into the beating engine. The gray color can be deepened or lightened by increasing or decreasing the proportion of the lampblack employed. Gray can also be obtained by employing a mixture of one part of gallotannic acid, two parts of green vitriol, and ten parts of water. The material should be filtered through a linen rag, and the liquor freed from the insoluble particles run into the beating engine.

Aniline gray, sometimes called murine, is one of the important aniline colors, and in many respects it approaches aniline violet; it is soluble in boiling water, and yields a pretty gray. By the action of aldehyde upon aniline violet in the presence of sulphuric acid, there is produced another gray; but this, on account of its high price, has not as yet been generally adopted for coloring paper.

Black.—The pulp is seldom colored black; the color, however, can be produced either by the employment of lampblack, cleansed as has been described, or the color can be obtained by the employment of a concentrated decoction of logwood and acetate of iron, the latter usually being of 2° Baumé, the pulp being previously mordanted with alum.

Aniline black does not occur in commerce as an actual coloring matter, but by the action of oxidizing agents upon the paper pulp, treated with an aniline salt (best aniline acetate), the black color can be produced directly upon the material to be treated. The color which results from this treatment is, in fact, a very dark aniline green, and on account of the insolubility of the products of oxidation formed upon the fibre itself, the color so produced is very fast, and the action of the most energetic acids and bases affect it less than any other black color. Various receipts have been given for preparing this color, but the nature of the fibre used for paper does not allow the employment of every oxidizing agent; ammonium ferrocyanide is probably the best oxident that can be used.

Deep, indelible black, such as is used in coloring paper to be employed in lieu of leather for the manufacture of cheap pocketbooks, etc., can be prepared as follows:—

Blue aniline	.94 parts.
Yellow aniline	.26 "
Naphthaline	.48 "
Red aniline	.32 "
Alcohol	74.00 "

The whole is dissolved in a suitable vessel by agitation, and the liquid afterwards filtered.

Bronze Shades on Paper.—To prepare an aniline bronze according to Fiorillo's formula, proceed as follows: Dissolve ten parts of aniline red, five parts of aniline purple, in one hundred parts of alcohol. Aniline red, as we have previously stated, is also sometimes called diamond fuchsine or roseine, and aniline purple is also known in commerce as Hoffman's violet and also as methyl violet. The strength of the alcohol employed measures about 95°, and, in order to promote the dissolution of the ingredients, the vessel containing the mixture is placed in either hot water or a sand bath. When the anilines above named have been dissolved by the alcohol, there should then be added five parts of benzoic acid, and the mixture is then allowed to boil gently, after which there is added 32 parts of gum benzoine, which will impart a cantharide green color to the mixture, but by continuing the boiling for about eight or ten minutes, the color will become changed into a bright golden-bronze hue.

A brilliant bronze may be prepared as follows: Take of aniline blue, violet, or purple, one ounce; aniline red, three ounces; acetic acid, one pint. The above mixture is slightly heated in order to promote the solution of the aniline, and the mixture is then allowed to cool. One pound of gelatine is next dissolved in a separate vessel in two quarts of acetic acid, and the mixture thus prepared is added to the one first described, and the whole is thoroughly incorporated by vigorous stirring.

Surface Coloring.—The bronze colors as well as some of the other colors and shades which we have mentioned are oftentimes applied only to the surface of the paper, and in such cases the apparatus shown in Figures 132 and 133 can be used for the purpose.

Vegetable Substances not always Desirable for Coloring Paper.—Vegetable substances should be cautiously used in paper-making for coloring the pulp, and, as a rule, they should not be employed except under circumstances where the coloring matters derived from mineral substances and from the anilines would not give pure shades. Vegetable colors are often so unstable that they are readily decomposed under the action of air and light, and vegetable substances should not be employed under any circumstances when the paper colored therewith is to be used for covering books, pamphlets, etc.

Stains used for Coloring Paper after it is Manufactured in order to prepare it for use in the Fabrication of Artificial Flowers, etc.

Sap-colors are only used and principally those containing much coloring matter. The following colors are calculated for one ream of paper of medium size and weight:—

The gum-Arabic given in the receipts is dissolved in the sap-liquor.

Blue (Dark).—I. Mix 1 gallon of tincture of Berlin blue and 2 ounces each of wax-soap and gum tragacanth.

II. Mix $\frac{3}{4}$ gallons of tincture of Berlin blue with 2 ounces of wax-soap and $4\frac{1}{4}$ ounces of gum tragacanth.

Crimson.—Mix 1 gallon of liquor of Brazil wood compounded with borax, 2 ounces of wax-soap, and $8\frac{3}{4}$ ounces of gum-Arabic.

Green (Dark).—I. Take $\frac{1}{2}$ gallon of liquor of sap-green (boiled down juice of the berries of *Rhamnus catharticus*), $4\frac{1}{4}$ ounces of indigo rubbed fine, 1 ounce of wax-soap, and $4\frac{1}{2}$ ounces of gum-Arabic.

II. One-half gallon of liquor of sap-green, $4\frac{1}{4}$ ounces of distilled verdigris, 1 ounce of wax-soap, and $4\frac{1}{2}$ ounces of gum-Arabic.

Yellow (Golden).—Mix $6\frac{1}{2}$ pounds of gamboge with 2 ounces of wax-soap.

Yellow (Lemon).—I. Compound 1 gallon of juice of Persian berries with 2 ounces of wax-soap and $8\frac{3}{4}$ ounces of gum-Arabic.

II. Add to 1 gallon of liquor of quercitron compounded with solution of tin 2 ounces of wax-soap and $8\frac{3}{4}$ ounces of gum-Arabic.

Yellow (Pale).—Mix 1 gallon of liquor of fustic, 2 ounces of wax-soap, and $8\frac{3}{4}$ ounces of gum-Arabic.

Yellow (Green).—I. Compound 1 gallon of liquor sap-green with 2 ounces each of distilled verdigris and wax-soap and $8\frac{3}{4}$ ounces of gum-Arabic.

II. Take 1 gallon of liquor of sap-green, 2 ounces each of dissolved indigo and wax-soap, and $8\frac{3}{4}$ ounces of gum-Arabic.

Red (Dark).—Compound 1 gallon of liquor of Brazil wood with 2 ounces of wax-soap and $8\frac{3}{4}$ ounces of gum-Arabic.

Rose Color.—Mix 1 gallon of liquor of cochineal with 2 ounces of wax-soap and $8\frac{3}{4}$ ounces of gum-Arabic.

Scarlet.—I. Mix 1 gallon of liquor of Brazil wood compounded with alum, and a solution of copper with 2 ounces of wax-soap and $8\frac{3}{4}$ ounces of gum-Arabic.

II. Mix 1 gallon of liquor of cochineal compounded with citrate of tin with 2 ounces of wax-soap and $8\frac{3}{4}$ ounces of gum-Arabic.

Stains for Glazed Papers.

On account of the cheapness of these papers a solution of glue is used as an agglutinant. The following proportions are generally used for one ream of paper of medium size and weight: One pound of glue and $1\frac{1}{4}$ gallons of water.

Black.—I. Dissolve one pound of glue in $1\frac{1}{4}$ gallons of water; triturate with this 1 pound of lampblack previously rubbed up in rye whiskey, $2\frac{3}{4}$ pounds of Frankfort black, 2 ounces of Paris blue, 1 ounce of wax-soap, and add $1\frac{1}{2}$ pounds of liquor of logwood.

II. Take $\frac{1}{2}$ gallon of liquor of logwood compounded with sulphate of iron, 1 ounce of wax-soap, and $4\frac{1}{2}$ ounces of gum-Arabic.

Blue (Azure).—Dissolve 1 pound of glue in $1\frac{1}{4}$ gallons of water, and compound the solution with $1\frac{1}{2}$ pounds of Berlin blue, $2\frac{3}{4}$ pounds of pulverized chalk, $2\frac{1}{4}$ ounces of light mineral blue, and 2 ounces of wax-soap.

Blue (*Dark*).—I. Dissolve 1 pound of glue in $1\frac{1}{4}$ gallons of water, and mix with it $4\frac{1}{4}$ pounds of pulverized chalk, $4\frac{1}{4}$ ounces of Paris blue, and 2 ounces of wax-soap.

II. Mix $\frac{1}{2}$ gallon of tincture of Berlin blue and 1 ounce of wax-soap with $2\frac{1}{4}$ ounces of dissolved gum tragacanth.

Blue (*Pale*).—I. Mix $\frac{1}{2}$ gallon of tincture of Berlin blue and 1 ounce of wax-soap with $3\frac{1}{2}$ ounces of dissolved gum tragacanth.

II. Dissolve 1 pound of glue in $1\frac{1}{4}$ gallons of water, and mix with it 4 pounds of pulverized chalk and 2 ounces each of Parisian blue and wax-soap.

Brown (*Dark*).—I. Dissolve 1 pound of glue in $1\frac{1}{4}$ gallons of water, and mix with it 1 pound of colcothar, a like quantity of English pink, $1\frac{1}{2}$ pounds of pulverized chalk, and 2 ounces of wax-soap.

II. Dissolve 1 ounce of wax-soap and $4\frac{1}{2}$ ounces of gum-Arabic in $\frac{1}{2}$ gallon of good liquor of Brazil wood and a like quantity of tincture of gall-nuts.

Green (*Copper*).—Dissolve 1 pound of glue in $1\frac{1}{4}$ gallons of water, and triturate with it 4 pounds of English green, $1\frac{1}{2}$ pounds of pulverized chalk, and 4 ounces of wax-soap.

Green (*Pale*).—Dissolve 1 pound of glue in $1\frac{1}{4}$ gallons of water, and mix with it 1 pound of Bremen Blue, $8\frac{1}{2}$ ounces of whiting, 1 ounce of light chrome-yellow, and 2 ounces of wax-soap.

Lemon Color.—Dissolve 1 pound of glue in $1\frac{1}{4}$ gallons of water, and mix with it 13 ounces of light chrome-yellow, 2 pounds of pulverized chalk, and 2 ounces of wax-soap.

Orange-yellow.—Dissolve 1 pound of glue in $1\frac{1}{4}$ gallons of water, and mix with it 2 pounds of light chrome-yellow, 1 pound of Turkish minium, 2 pounds of white lead, and 2 ounces of wax-soap.

Red (Cherry).—Dissolve 1 pound of glue in $1\frac{1}{4}$ gallons of water, and mix with it $8\frac{1}{2}$ pounds of Turkish minium previously rubbed up with $\frac{1}{4}$ gallon of liquor of Brazil wood, and 2 ounces of wax-soap.

Red (Dark).—Mix $\frac{3}{4}$ gallons of liquor of Brazil wood with 1 ounce of wax-soap and $4\frac{1}{2}$ ounces of gum-Arabic.

Red (Pale).—Dissolve 1 pound of glue in $1\frac{1}{4}$ gallons of water, and mix with it $8\frac{3}{4}$ pounds of Turkish minium previously rubbed up with 2 ounces of wax-soap.

Rose Color.—Dissolve 1 pound of glue in $1\frac{1}{4}$ gallons of liquor of Brazil wood and mix with it 50 pounds of rose madder previously rubbed up with 2 ounces of wax-soap.

Violet.—Mix $4\frac{1}{2}$ ounces of gum-Arabic and 1 ounce of wax-soap with $\frac{1}{2}$ gallon of good liquor of logwood. After the gum has dissolved in the liquor compound it with sufficient potash to form a mordant.

Stains for Morocco Papers.

The following colors are calculated for one ream of paper of medium size and weight.

Black.—Dissolve $8\frac{3}{4}$ ounces of good parchment shavings in $1\frac{1}{2}$ gallons of water and stir in 1 pound of lampblack, 30 pounds of Frankfort black, and $1\frac{3}{4}$ ounces of fine Paris blue.

Blue (Dark).—Dissolve $8\frac{3}{4}$ ounces of good parchment shavings in $1\frac{1}{2}$ gallons of water, and mix with the solution $8\frac{1}{4}$ pounds of white lead and $4\frac{1}{2}$ ounces of fine Paris blue.

Blue (Light).—Dissolve $8\frac{3}{4}$ ounces of parchment shavings in $1\frac{1}{2}$ gallons of water, and mix with it $8\frac{3}{4}$ pounds of white lead and $2\frac{1}{4}$ ounces of fine Paris blue.

Green (Dark).—Dissolve 13 ounces of parchment shavings in $2\frac{1}{2}$ gallons of water, and mix with 10 pounds of Schweinfurth green.

Green (Pale).—Dissolve 13 ounces of parchment shavings in $2\frac{1}{2}$ gallons of water, and mix with $8\frac{3}{4}$ pounds of Schweinfurth green and 1 pound of fine Paris blue.

Orange-yellow.—Dissolve $8\frac{3}{4}$ ounces of parchment shavings in $1\frac{1}{2}$ gallons of water, and mix with $1\frac{1}{2}$ pounds of light chrome yellow, $8\frac{3}{4}$ ounces of orange chrome-yellow, and 1 pound of white lead.

Red (Dark).—Dissolve $8\frac{3}{4}$ ounces of parchment shavings in $1\frac{1}{2}$ gallons of water, and compound this with $7\frac{3}{4}$ pounds of fine cinnabar and 1 pound of Turkish minium.

Red (Pale).—Dissolve $8\frac{3}{4}$ ounces of parchment shavings in $1\frac{1}{2}$ gallons of water, and mix it with $8\frac{3}{4}$ pounds of Turkish minium.

Violet (Dark).—Dissolve $8\frac{3}{4}$ ounces of parchment shavings in $1\frac{1}{2}$ gallons of water, and mix with $3\frac{3}{4}$ pounds of white lead, 1 pound of pale mineral blue, and $8\frac{3}{4}$ ounces of scarlet lake.

Violet (Light).—Dissolve $8\frac{3}{4}$ ounces of parchment shavings in $1\frac{1}{2}$ gallons of water, and mix with $4\frac{1}{4}$ pounds of white lead, 13 ounces of light mineral blue, and $8\frac{3}{4}$ ounces of scarlet lake.

Yellow (*Pale*).—Dissolve $8\frac{3}{4}$ ounces of parchment shavings in $1\frac{1}{2}$ gallons of water, and mix with 2 pounds of light chrome-yellow and $8\frac{3}{4}$ ounces of white lead.

Stains for Satin Papers.

The following colors are calculated for one ream of paper of medium size and weight.

Blue (*Azure*).—Dissolve 13 ounces of parchment shavings in $2\frac{1}{2}$ gallons of water, and mix with 3 pounds of Bremen blue, $1\frac{3}{4}$ pounds of English mineral blue, and $4\frac{1}{2}$ ounces of wax-soap.

Blue (*Light*).—Dissolve $8\frac{3}{4}$ ounces of parchment shavings in $1\frac{1}{2}$ gallons of water, and mix with 1 pound of light mineral blue and $3\frac{1}{2}$ ounces of wax-soap.

Brown (*Light*).—Dissolve $8\frac{3}{4}$ ounces of parchment shavings in $1\frac{1}{2}$ gallons of water, and mix with 13 ounces of light chrome-yellow, $6\frac{1}{2}$ ounces of colcothar, 2 ounces of Frankfort black, 3 pounds of pulverized chalk, and $3\frac{1}{2}$ ounces of wax-soap.

Brown (*Reddish*).—Dissolve $8\frac{3}{4}$ ounces of parchment shavings in $1\frac{1}{2}$ gallons of water, and mix with one pound of yellow ochre, $4\frac{1}{2}$ ounces of light chrome-yellow, 1 pound of white lead, 1 ounce of red ochre, and $3\frac{1}{2}$ ounces of wax-soap.

Gray (*Light*).—Dissolve $8\frac{3}{4}$ ounces of parchment shavings in $1\frac{1}{2}$ gallons of water, and mix with $4\frac{1}{4}$ pounds of pulverized chalk, $8\frac{3}{4}$ ounces of Frankfort black, 1 ounce of Paris blue, and $3\frac{1}{2}$ ounces of wax-soap.

Gray (Bluish).—Dissolve $8\frac{3}{4}$ ounces of parchment shavings in $1\frac{1}{2}$ gallons of water, and mix with $4\frac{1}{4}$ pounds of pulverized chalk, 1 pound of light mineral blue, $4\frac{1}{4}$ ounces of English green, $1\frac{3}{4}$ ounces of Frankfort black, and $3\frac{1}{2}$ ounces of wax-soap.

Green (Brownish).—Dissolve $8\frac{3}{4}$ ounces of parchment shavings in $1\frac{1}{2}$ gallons of water, and mix with 1 pound of Schweinfurth green, $8\frac{3}{4}$ ounces of mineral green, $4\frac{1}{4}$ ounces each of burnt umber and English pink, 1 pound of whiting, and $3\frac{1}{2}$ ounces of wax-soap.

Green (Light).—Dissolve $8\frac{3}{4}$ ounces of parchment shavings in $1\frac{1}{2}$ gallons of water, and mix with $2\frac{3}{4}$ pounds of English green a like quantity of pulverized chalk and $3\frac{1}{2}$ ounces of wax-soap.

Lemon Color.—Dissolve $8\frac{3}{4}$ ounces of parchment shavings in $1\frac{1}{2}$ gallons of water, and mix with $1\frac{1}{2}$ pounds of light chrome-yellow, 1 pound of white lead, and $3\frac{1}{2}$ ounces of wax-soap.

Orange-yellow.—Dissolve $8\frac{3}{4}$ ounces of parchment shavings in $1\frac{1}{2}$ gallons of water, and mix with $1\frac{1}{2}$ pounds of light chrome-yellow, $8\frac{3}{4}$ ounces of orange chrome-yellow, 1 pound of white lead, and $3\frac{1}{2}$ ounces of wax-soap.

Orange-yellow.—Dissolve $8\frac{3}{4}$ ounces of parchment shavings in $1\frac{1}{2}$ gallons of water, and mix with $4\frac{1}{4}$ pounds of light chrome-yellow, $8\frac{3}{4}$ ounces of Turkish minium, 1 pound of white lead, and $3\frac{1}{2}$ ounces of wax-soap.

Rose Color.—Dissolve $8\frac{3}{4}$ ounces of parchment shavings in $1\frac{1}{2}$ gallons of water, and mix with $\frac{3}{4}$ gallon of rose color prepared from liquor of Brazil wood and chalk, and $6\frac{1}{2}$ pounds of wax-soap.

Violet (Light).—Dissolve $8\frac{3}{4}$ ounces of parchment shavings in $1\frac{1}{2}$ gallons of water, and mix with $1\frac{1}{2}$ pounds of light mineral blue, a like quantity of scarlet lake, 1 pound of white lead, and $3\frac{1}{2}$ ounces of wax-soap.

White.—Dissolve $8\frac{3}{4}$ ounces of parchment shavings in $1\frac{1}{2}$ gallons of water, and mix with $8\frac{3}{4}$ pounds of fine Kremnitz white, $4\frac{1}{4}$ ounces of fine Bremen blue, and $3\frac{1}{2}$ ounces of wax-soap.

Silver White.—Dissolve $8\frac{3}{4}$ ounces of parchment shavings in $1\frac{1}{2}$ gallons of water, and mix with $8\frac{3}{4}$ pounds of Kremnitz white, $8\frac{3}{4}$ ounces of Frankfort black, and $3\frac{1}{2}$ ounces of wax-soap.

Pale Yellow.—Dissolve $8\frac{3}{4}$ ounces of parchment shavings in $1\frac{1}{2}$ gallons of water, and mix with $4\frac{1}{4}$ pounds of light chrome-yellow, 1 pound of pulverized chalk, and $3\frac{1}{2}$ ounces of wax-soap.

Fig. 134.

CHAPTER XV.

MAKING AND FINISHING.

THE materials to be made into paper having been subjected to all the preliminary operations which we have described, the pulp is ready for transformation into sheets by means of the paper-machine. Before proceeding further, we shall recapitulate in a few words the operations of which we have treated. The materials arrive at the mill, are sorted and cut by hand or by machinery, dusted, boiled in water and in alkalies, dripped, washed, and reduced to half-stuff in the rag-engine. The half-stuff is drained, pressed or air dried, and submitted to the action of an hypochlorite or of gaseous chlorine, or to the action of both these chemicals. After the bleaching with chlorine or other chemicals, the pulp is washed, and, if necessary, the last traces of chlorine are eliminated by means of antichlorine; then the refining is proceeded with in the beating engine, and the minute fibres or pulp thus obtained is treated with agglutinative, loading, and coloring materials, intended to give weight, body, and finish to the paper. The pulp having been thus prepared is passed into the stuff-chest of the paper-machine.

In Fig. 134 we show an interior view of a machine room in a modern paper-mill, containing a Fourdrinier machine.

The pulp is passed into the stuff-chest of the paper-machine and is kept in suspension by means of an agitator ; it is delivered to the paper-machine in constant quantities, and after passing through several purifying contrivances, intended to free it from the last traces of sand, it is run in a thin and wide sheet upon endless metallic cloth, horizontally disposed at its anterior part. A continuous forward movement is communicated to this metallic cloth, technically termed the " wire," which receives during its forward movement a continuous succession of lateral shakes, in imitation of the " shake" which the vatman gives to the sheet of paper when moulded by hand ; the " shake" is intended to favor the dripping and felting of the pulp upon the " wire." A suction contrivance operated underneath a certain part of the metallic cloth also assists in the abstraction of moisture and renders the sheet more solid. The sheet, having reached the extremity of the horizontal parts of the metallic cloth, passes upon a cylinder which delivers it to two laminating cylinders covered with felt ; thence the sheet passes through the pressing rolls and to the drying cylinders heated by steam ; lastly it may be made to pass between one or more pairs of circular scissors which trim and slit the paper into strips of the desired width ; these strips are then rolled upon a cylinder fixed on a mandrel.

It will not be possible in a volume of the size of the present one to enter upon a detailed description of all the parts and the manner of operating our modern paper-machines. We shall consequently devote space only to such portions of the machines employed for making and finishing the paper as possess features susceptible of improvement, and in this connection illustrate some of the recent inventions of practical paper manufacturers. The practical parts of the work of forming and finishing the paper will be treated only so far as may be necessary for comprehension, as, in the opinion of the writer, a description of the mechanical skill which can be acquired only through actual experience offers no advantage to either the theorist or to the practical paper-maker; the " machine man" can become familiar with the best manner of adjusting and operating the paper-making machine under his charge only by long and practical experience.

It was the writer's first intention to have made the present chapter as well as the various other chapters in this work entirely " practical," and to this end he secured personally in the United States, as well as in the other great paper-making countries in the world—Great Britain, Germany, France, and Belgium—a large number of notes entirely practical in their character. But upon a full consideration of the subject, and by the advice of some of the leading paper makers, the author decided that the present volume could be made more valuable by surrendering the space to a description of some of the leading mechanical inventions of the present day which have done so much to increase the product and lessen the cost of all kinds of paper. In adopting the latter plan the writer is well aware that he renders himself liable to the old complaint of the " practical" man who neglects and condemns books because they do not enter into the minutest details contributing to successful operations; but this criticism does not by any means lessen the value of books. After all, we can only reiterate that

32

" practical information" can be obtained only by actual practice. The province of a book on any of the arts is that of a guide-post which points out the correct road, but leaves the traveller to take it and to encounter and remove for himself all the small obstacles found on the way. Hence, while books cannot take the place of actual experience, their value, as embodying the experiences of others must not be ignored, and nothing can be more useful or valuable to the really practical man than a volume describing and illustrating in detail the state of the art in any manufacture and thereby pointing out the roads that are being pursued by successful manufacturers and inventors of the present time. Such a book, if intelligently used, should prove a never-failing means of suggestion and inspiration to all progressive practical men. Whereas, a description of methods and mechanical appliances which have for a long series of years been employed in various portions of the world, and have already become common property can, in reality, prove of little value to the advanced paper-makers of the day.

Fig. 135 shows an apparatus for purifying the pulp. The machine oscillates, and receives a shaking movement from blades arranged for that purpose. The pulp is received from the supply-box through the tube *H*, and is distributed in compartments provided with stops for the sand and other heavy bodies; then, passing through grooves made in a copper plate, it falls in the reservoir *B*. It then passes through the pressure of its upper level, in the compartment *C*, going through other grooves, passes over rubber sheets, *D*, and falls in the channel *F* to be conveyed to the machine.

In order to insure satisfactory weight and uniformity in the color of the paper, some manufacturers employ two stuff-

Fig. 135.

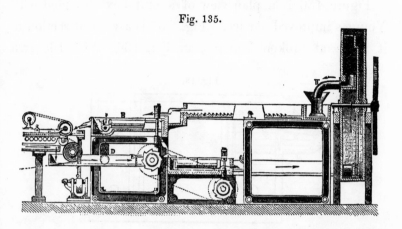

chests with each machine, into one of these chests the pulp is emptied from the beater while the "machine" is being supplied from the other. In some cases it may be found advantageous to work the "wire" taut on the machine, and in other cases it may be found more economical to work the "wire" quite slack; but these matters, as we have previously hinted, are subjects for practical consideration.

Stuff Regulator for Paper-Making Machines.

The invention shown in Figs. 136 to 140, which is that of Mr. Cornelius Young, of Sandy Hill, New York, relates to improvements in stuff-regulators for paper-machines; and it consists in providing an adjustable gate, the movements of which are automatically controlled by the movements of a balanced stuff-chute, and in providing the stuff-box with a vertically-sliding gate.

The object is to regulate the flow of pulp or "stuff" from the stuff-box to the paper-machine proper.

Figure 136 is a plan view of a stuff-box provided with Young's improved device. Fig. 137 is a vertical section of it taken at broken line $x\,y$ in Fig. 136. Fig. 138 is a

Fig. 136.

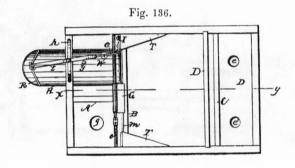

Fig. 137.

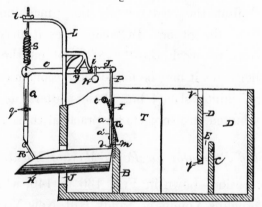

front elevation of it, with part of front wall broken away to show gate I closed. Fig. 139 is same showing gate I open. Fig. 140 is a perspective of gate I and pivoted arms.

The box is divided by partitions A, B, and C into four compartments. One of the compartments is provided with

one or more apertures, e e', in the bottom, through which
the stuff is forced by a pump into the box, until the compartment is filled to the top of partition C, when it flows

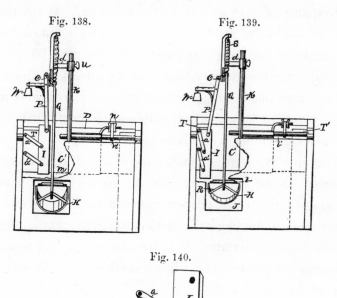

Fig. 138.   Fig. 139.

Fig. 140.

over between C and D into the next compartment, which
retains the stuff until filled to the top of partition B, which
extends across the box from block T to block T', the blocks
serving to narrow the compartment on the side next the
partition B. When this compartment is filled, the stuff
flows over partition B, one portion entering the chute H,
from which it is conveyed to the paper-machine proper, and
the other portion into the compartment provided with out-

let g, through which it flows back to the reservoir from which it was originally pumped into the box. The thickness and weight of paper produced depend upon the thickness and rapidity of flow of stuff to the machine. As the thickness or quality of stuff is not subject to absolute control, uniformity in the weight of paper produced must be secured by subjecting the rapidity of flow to the quality of the stuff. Such control has been heretofore attempted in various ways —by means of valves or adjustable gates controlled by an attendant or by floats and balances.

The present invention makes use of an adjustable gate, I, attached to one side of the box by means of the parallel arms, $a\ a'$, which are pivoted one end to the gate and the other end to the block T, so that when the gate is lifted it travels horizontally away from the side of the box to which the arms are pivoted, and when it is allowed to fall it travels toward the side of the box. There is also used another adjustable gate, G, supported by rod c, on which it is adapted to slide to and fro between the blocks TT'', the lower end of the gate resting upon the upper edge of the partition B, or a metallic strip, m, projecting therefrom. The gate is also provided with a threaded arm passing through stop n', fixed upon rod c, having adjusting-nuts, n, by means of which the gate may be secured in different positions. By sliding gate G toward gate I the opening C is diminished in size, and less stuff will flow to the machine, the opening b' will be increased in size, and more stuff will escape by outlet g. By sliding gate G in the opposite direction toward opening b' more stuff is allowed to flow to the machine and less back

to the stuff-reservoir through outlet g. The gate G is therefore first set at the proper point to make paper of the desired
weight with a constant flow of the stuff to be used, provided
the flow is even in quantity; but experience shows the
impossibility of securing such a flow. The pump forces a
constant quantity into the box; but the thickness—*i. e.*, the
relative quantity of pulp and water—varies continually.
When the stuff flows thick, more of it must be held back
from the machine, and diverted to the outlet g, and when it
flows thin a greater supply must be sent to the machine.
This is accomplished by means of the gate I, which is connected by link P with one end of the sweep O, the other
end of the sweep being connected by link Q and bail R with
the chute H at one end, the other end of the chute being
hinged upon partition B directly beneath the opening C', so
that the stuff which goes to the machine passes through the
chute.   The weight of the stuff in the chute will depress its
projecting end, which raises the gate I, and, as before explained, narrows the opening C'. By means of the weight
W, adapted to slide upon the sweep O and the spring S, the
gravity of the stuff in the hinged chute may be balanced to
secure the desired width of opening C'. After the respective
parts have been once adjusted to produce paper of a given
weight from a constant flow of stuff of known average quality they will thereafter be automatically adjusted to the
varying quality of stuff as the latter passes through and from
the stuff-box. If the stuff suddenly thickens, its progress
upon the chute is slower, and it dams up, as it were, thereon.
The additional weight overcomes the force of the spring S,

and the gate I is elevated and forced toward opening C' to close the latter, which forces a larger proportion of the stuff through opening b', and admits a smaller quantity to the chute and machine. If the stuff is thinner than the average, it flows more readily from the chute, leaving a less lighter quantity thereon, which lowers the gate and widens the opening C' and permits a larger flow of stuff to the chute and machine.

Fig. 138 shows the gate closed to narrow opening C'', and Fig. 139 shows the gate opened to widen opening C'.

In Fig. 137 the link Q is shown in two slotted lapping parts, which permits of its longitudinal adjustment by means of the thumb-screw q. The spring S is also attached to its supporting-arm, L, by a threaded rod, which permits of the adjustment of the spring by the thumb-screw t.

There is shown sweep O, pivoted at the arm y; but it may be pivoted at arm i instead, through another aperture in the sweep, to change the leverage of chute and gate.

In Figs. 138 and 139 the supporting-arm d is shown vertically adjustable upon the upright K', by means of set-screw u.

To prevent a sudden rush of thick pulp upon the chute, which might cause it to overflow, there are provided the partitions, D and C, the latter extending from the bottom of the box upward about half the height of the box, and the former being situated a little one side of the latter and extending both above and below the top of C. It can be slid vertically in grooves, V, in the sides of the box.

When the stuff is of the usual thickness, it flows through

the aperture between D and C, and does not attain a level much above the top of outlet-partition B or m; but if a considerable quantity of thick stuff is suddenly forced into the compartment D' it will not run so freely between D and C, and rises in the compartment D' until the stuff runs thinner, or until its height affords sufficient pressure to force it through the opening.

By raising or lowering the sliding partition D, the narrow opening between it and C is lengthened or shortened, which gives it more resistance when lengthened and less when shortened. The inventor claims he is thus able to secure a perfect adjustment of the flow of stuff to the machine automatically and produce an even quality of paper.

Automatic Wire-Guide for Paper-Making Machines.

The objects of the invention shown in Figs. 141 to 145, which is that of Mr. Thomas P. Barry, of Stillwater, N. Y., are to provide means by which the guide-roll cylinder which supports the wire apron is made to operate mechanism placed intermediate between the guide-roll cylinder and guides at the side edges of the wire apron, so that the guides will be automatically operated to truly and properly guide the wire apron in its forward movement, and also to provide means by which an attendant will be enabled to cause the wire apron to be guided in its forward movement should certain parts of the automatic mechanism become disarranged.

Fig. 141 represents a plan view of a section of a paper-making machine and Mr. Barry's improved wire-guiding device attached. Fig. 142 is a side elevation of the guide-

roll cylinder carrying the wire to be guided.   Fig. 143 is a
front side elevation of the device.   Fig. 144 is a rear side
elevation of it; and Fig. 145 is a cross-sectional view of it,
taken at line No. 1 in Fig. 141.

Fig. 141.

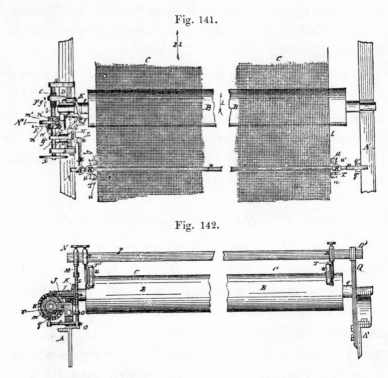

Fig. 142.

A A represent the frame of a paper-making machine, and
B is a guide-roller supporting the wire web or apron *C*,
which the present device is intended to guide as it moves
forward.

Secured to one of the side portions, *A*, of the frame of the
machine is the bed or way *D*, made preferably with a
V-shaped form, as shown in Fig. 145.   Made with the way,
and extending outward and in a lateral direction from the

Fig. 143.

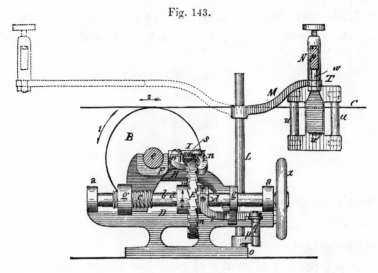

Fig. 144.

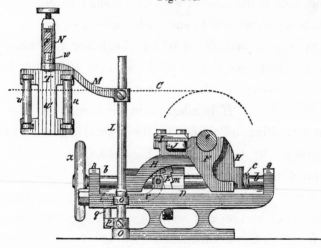

same, are brackets, *a a*, which brackets support shaft *b*, which is free to revolve in bearings made in the rear ends of the brackets. The shaft is held from being moved longitudinally by the shoulders of the journals of the shaft bearing against the sides of the bearings in which they work.

Made with shaft b is a screw-thread section, c. Mounted on the shaft is a duplex ratchet-wheel, E, which is secured

Fig. 145.

from turning on the shaft by a feather or spline, d, fixed in the shaft and working in a seat made in the hub of the ratchet-wheel.

F is a bearing of shaft e of guide-roll cylinder or roller B, which bearing is made solid with or attached to bracket H, which bracket is pivoted to shaft b by arms g g', and is supported by way or bed D, on which the lower foot ends of the bracket rests, as shown in Figs. 141, 143, and 145. Being thus pivoted to shaft b, and supported by way or bed D, the bracket H is adapted to be turned up from a horizontal position, as shown in the several figures, to the position shown by dotted lines in Fig. 145.

The pivoting eye or bearing in arm g of the bracket H is provided with a screw-thread, which corresponds with the screw-thread of section c of shaft b, and the screw-threaded section works in the screw-threaded eye of arm g', as shown in Figs. 141 and 143.

Mounted on the upper side of bracket H, and against or adjacent to bearing F of guide-roll cylinder B, is housing I, in which freely works bar J, which bar is adapted to be moved in either direction transversely to the direction of shaft b. A notch, v, Fig. 141, is made in the end of the reciprocating bar J next to the end of guide-roll cylinder B, and in its side facing. Secured to the shaft of the guide-roll cylinder, so as to revolve with it, is cam-wheel K, formed by an annular flange arranged in one direction slightly oblique in the axis of guide-roll cylinder B, as shown in Fig. 142. The cam-wheel works in notch v of bar J, and moves the bar alternately in opposite directions as guide-roll cylinder B is revolved.

Pivoted to the end of reciprocating bar J, opposite to its notched end, are dogs n n', Figs. 141 and 143, which dogs are held and drawn toward each other by spring s, Fig. 143, and engage respectively with the teeth m m' of wheel E, as shown in Figs. 141 and 143, according as the wheel E is in situation for engagement with the dogs, as will be hereinafter described.

Supported in arms o o, projecting from the way or bed D, is a vertical shaft, L. To the lower end of the shaft is secured crank p, provided with vertical arms q, Figs. 142, 143, and 144. Pivoted to the upper end of the vertical arm q is a forked pitman, r, which pitman is yoked to a loose collar or sleeve, t, on the hub of wheel E.

Secured to the vertical shaft L at its upper end is arm M, carrying a clamping device, consisting of upright stud N, provided with an oblong slot and set-screw. (Shown in Figs. 143 and 144.) Secured in the clamping device at one of its ends is the shifting-bar P, the opposite end of which

works freely in sleeve Q, supported by standard Q', attached
to the frame of the machine, as shown in Figs. 141 and 142.

Secured to the shifting-bar P by clamping devices R R'
are guides T T', which guides are each composed of two
rollers, u u, Figs. 141, 142, 143, and 144, arranged and connected with plate u' by being pivoted to ears made with the
plate, as shown in Fig. 144. The guides thus composed are
each pivoted to their respective clamping devices R R', as
indicated by dotted lines w in Fig. 144, and are each capable
of a swiveling movement.

Secured to one end of shaft b is a hand-wheel, X, which
may be operated in either direction for shifting the bearings
of guide-roll by hand when circumstances require, as will be
hereinafter described.

The manner in which the several parts of Mr. Barry's
device operate is as follows : The guides T T' are set near
to the side edges of the wire apron C, with their rollers u u
at a distance from the same as the machine-tender may
select. The guide-roll cylinder B revolves in direction of
arrow No. 1 in Figs. 141 and 143, and carries the wire
apron to be guided in direction of arrow 2 in same figures.
The cam K, attached to the shaft or journal of guide-roll
cylinder B, revolves with the same, and works in the notch
made in the reciprocating bar J, and causes the bar to be
moved once in each direction at each revolution of the guide-
roll cylinder and operate the bar, so as to carry the dogs n
n' to a full movement back and return at each side of the
ratchet-wheel. The ratchet-wheel E being mounted loosely
on shaft b, and held by its feather so as to turn with the shaft,

is adapted to move longitudinally on the same in either direction and between dogs n n'. When the wire apron is moving forward uniformly and truly with its side edges in straight lines of direction the dogs will be moved back and forth without engaging with the teeth of the ratchet-wheel. When the wire apron C begins to shift or run from side A of the machine the off side edge, 1, of the apron will crowd against the rollers u u of the guide T, when the shifting-bar P will draw arm M in direction of arrow 4 and cause crank-arm p to move in direction of arrow 5, and through pitman r and sleeve t move ratchet-wheel E toward dog n, so that the dog will work in engagement with the teeth m of the same. As the revolution of guide-roll cylinder B is continued cam K will impart to bar J a reciprocating movement, by which the dogs n n' will be moved back and forth, the dog n' being out of engagement with the ratchet-wheel, while the dog n will be in engagement with the teeth m of the same, and at each return movement draw on the teeth and cause the ratchet-wheel to be moved in direction of arrow 6, when the screw c on shaft b will be turned and cause bearing F to be shifted in direction of arrow 7. This shifting of the bearing will be attended by a gradual shifting also of bar J and its attached dogs n n' in the same direction, so that in a short time the dog n will be out of engagement with the ratchet-wheel, while dog n' will be thrown into engagement and will operate to turn the ratchet-wheel in an opposite direction, and also operate the screw of shaft b to shift bearing F, and also the dogs, in direction opposite to arrow 7.

It will be seen that the dog n operates to turn the ratchet-wheel when it is drawn toward the wire apron, and moves the ratchet-wheel in direction of arrow 6 ; also, that dog n' operates to turn the ratchet-wheel when it is pushed back, and moves the ratchet-wheel in an opposite direction to that indicated by arrow 6. When the dogs are operated back and forth, and at the same time free from engagement with the teeth of the ratchet-wheel, the ratchet-wheel will be idle, and there will be no relative shifting of parts. When this idle condition of parts exists the wire apron is running evenly and uniformly straight without its side edges exerting any great pressure on their respective guides T T'', while when the edge at guide T bears against that guide it will tend to throw the ratchet-wheel toward dog n at engagement with the same, when the dog, operating with teeth m, will gradually shift the parts and cause the opposite operating-dog, n', to be carried toward teeth m' of the ratchet-wheel, while the dog n will be thrown out of engagement with teeth m. When the side edge of the wire apron presses against guide T' the pressure will operate to slightly move arm M in opposite direction to arrow 4, and cause the crank-arm p to operate pitman r so as to shift the ratchet-wheel toward dog n' and in engagement with the same, when the screw will be turned in an opposite direction and carry bearing F and bar J and dogs n n' in direction opposite to arrow 7, when dog n' will be released from engagement with teeth m' of the ratchet-wheel. These alternate movements or reversals of action of parts operate to hold the ratchet-wheel at nearly one situation, and consequently it is made to resist the excessive pressure of the respective side edges of the wire apron, and be compelled to run with its edges with comparatively uniform lines of movement.

It should be understood that the wire apron does not run around guide-roll B, but is merely supported by the same, and is affected by the slight shifting of its bearing F, in the manner above described. Should the dogs n n' from any cause whatever become disarranged or out of working order, an attendant may, by moving the hand-wheel X in alternate directions, readily direct the movement of the wire apron. This is claimed to be a great advantage, as the machine need not be purposely stopped, but may be continued to run until the web being made is completed or the usual time for stoppage arrives. The support of the bearing F of the guide-roll cylinder directly over the way D removes all weight from shaft b, so that the shaft is rendered easy to be turned.

In Fig. 145 dotted lines illustrate the manner in which the bearing F and its adjunctive parts may be turned up. This adaptation of the bearing to be turned up, as shown, enables the operator to properly set the bearing in line, as required, as he may, by turning the bearing up without moving the shaft and then turning it down and moving the shaft, gradually adjust the bearing in one direction, while by turning the bearing F up and at the same time moving the shaft in the same direction, and then turning the bearing down while the shaft is held from turning, the bearing will be shifted in an opposite direction. It will therefore be seen that by this means the bearing F will be set properly.

33

It is not new to construct paper-making machines with a rule provided with plates and connected to a pair of levers, in connection with a screw on which a double toothed wheel is secured, a curved lever carried by another lever, and a crank which imparts motion to the last-named lever. Nor is it new to provide the mechanism for guiding the wire cloths or belts of paper-making machines with a crank-action for operating a double pawl to engage by draft or thrust a screw-nut ratchet-wheel to slide laterally a slide or purchase. The combination, with the journal of a roller, of an adjustable crank to operate double-acting pawls, which operate in either direction a ratchet-wheel, is not new; nor is the combination of a bed-plate and a fixed screw-bolt with a screw-nut ratchet-wheel to hold and operate a slide by the action of double pawls, which are operated by a crank. Mr. Barry consequently does not claim such constructions in his patent.

Suction-Box for Paper-Making Machines.

The following invention, which is that of Mr. Isaac Bratton, of Wilmington, Del., relates to improvements in connection with the suction-box of a paper-making machine, the object of the invention being to facilitate the operation of the machine and improve the product:—

Fig. 146 is a longitudinal section of the suction-box of a paper-making machine with Mr. Bratton's improvements; Fig. 147, a transverse section of it.

A represents a suction-box of an ordinary Fourdrinier paper-machine; B, the perforated plate forming the top of the box; D, part of the endless web or apron of wire-gauze

on which the sheet of pulp is deposited; *E E*, the usual
deckle-straps for limiting the width of the sheet, and *F F*

Fig. 146.

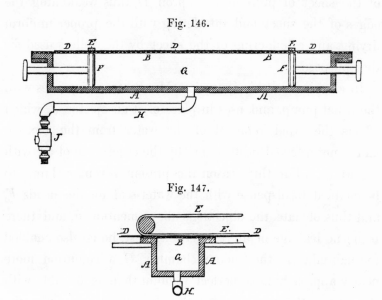

Fig. 147.

the adjustable heads or plungers, which coincide with the
deckle-straps, and are intended to prevent leakage of air
into the suction-chamber *G* of the box.

A partial vacuum being established in the chamber, the
water is drawn from the sheet of pulp as the endless apron
D carries the sheet over the perforated cover of the box *A*,
the water being drawn from the box through a pipe, *H*.
Ordinarily a pump is used for this purpose, and it becomes
necessary to prevent the entrance of air to the suction-chamber *G*, as a mixture of air with the water would interfere
with the operation of the pump; hence it has been usual to
seal the suction-chamber by filling the box *A* with water

outside of the adjustable heads *F*. A portion of this water
finds its way beneath the deckle-straps and wets the edges
of the sheet of pulp on the apron *D*, thus weakening the
edges of the sheet and interfering with the proper uniform
drying and calendering of the sheet by the rollers used for
that purpose.

In carrying out his invention Mr. Bratton dispenses with
the usual pump, and uses in place of it an ejector, *J*, which
effects the rapid removal of the water from the box *A*,
and is not affected in its action by the admixture of air with
the water. For this reason the present inventor claims to
be enabled to dispense with the water-seal for the heads *F*,
and thus obviate the objections above mentioned, and there
being no leakage of air to guard against, he is also enabled
to maintain in the suction-chamber *G* a condition more
nearly approaching a perfect vacuum than is possible with
the pump, the water being thus rapidly drawn from the
sheet of pulp, so that the speed of the apron can be materially increased without risk of carrying off the pulp while
it still retains a surplus of moisture. Another advantage of
using the ejector in place of the pump is the facility with
which the ejector can be cleansed by passing a current of
clean water through it.

It is the common practice to convey the discharge from
the suction-box of the machine back to the mixing-box, in
order to save the particles of pulp carried off with the water.

In making colored paper the particles of colored pulp fill
the interstices in the cylinder, valve-boxes, and other parts
of an exhaust-pump to such an extent that it is difficult to

properly clean the latter when changing from the manufacture of paper of one color to that of another color, or from
colored to white paper, and in consequence the first paper
produced after the change is apt to be streaky or spotted.
This objection is claimed to be effectually overcome by this
invention, there being no parts in the ejector which would
serve to retain the colored particles of pulp, so that the
thorough cleansing of the ejector can be effected in a few
moments.

Incidental to the use of the ejector in place of the pump
are the further advantages of its compactness, its freedom
from liability to get out of order, and the facility with which
it can be used in positions where the use of a pump would
be impossible.

Various forms of ejectors may be used in carrying out
Mr. Bratton's invention; but the form which the inventor
states he has found to answer well in practice, and which he prefers to employ, is
that shown in Fig. 148, a being the steam-
chamber of the ejector, having a branch,
a'; b the inlet and d the outlet branch.
The branch b has a conical end and the
branch d a flaring mouth, and the steam
passes from the chamber a through the
tapering annular passage thus formed, the
water from the suction-box being drawn
through the branch b and forced from the
branch d.   A free and unbroken flow of water through the
branches b and d is thus permitted, whereas in an ordinary

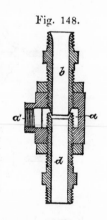

Fig. 148.

ejector the water takes a more or less circuitous course through a contracted passage, and when the water carries with it numerous particles of pulp it has a tendency to clog such a passage and interfere with the proper working of the device.

The ejector is shown in a vertical position at the side of the machine; but it may be located wherever convenience or circumstances may suggest.

Dandy-Roll for Paper-Making Machines.

The invention shown in Figs. 149 to 153 is that of Mr. David McKay, of Holyoke, Mass., and consists of a catch-pan placed inside of and hung upon the centre shaft of dandy-rolls used in paper-machines for the purpose of cleaning the inside of dandy-rolls of froth and other accumulations.

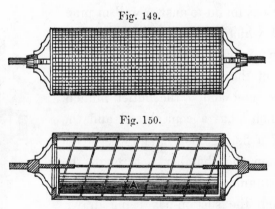

Fig. 149.

Fig. 150.

Figure 149 is an ordinary dandy-roll. Fig. 150 is a long section view of a dandy-roll with the catch-pan *A* attached. Fig. 151 is a transverse section of dandy-roll without the catch-pan. Fig. 152 is a transverse section of dandy-roll with catch-pan *B* attached. Fig. 153 is a view, in perspective, of the catch-pan.

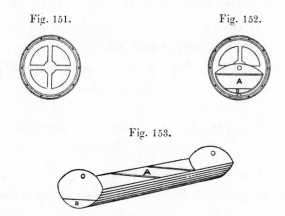

Fig. 151.   Fig. 152.

Fig. 153.

The dandy-roll, the purpose of which is to make both surfaces of the paper alike, is a hollow cylinder of woven wire. As the roll revolves upon the wet paper, particles of the paper-pulp are forced through the wire net-work of the roll until, accumulating in masses inside of the roll, they are again forced out through the wire net-work upon the paper, blotching and marring its surface. Heretofore the manner in which these accumulations have been removed was to take the dandy-roll out of position, causing a stoppage of the machine and a consequent loss of production.

The catch-pan, as seen in Fig. 150, hangs upon the centre shaft of the dandy-roll, and is held by its own weight in an upright position while the roll revolves, catching the drippings from the top of the inside of the roll, and also catching upon the edge of the pan, which may be made of sheet-tin or other material of like thinness, the particles of pulp or other matter forced through the net-work surface of the roll. Through the opening *B*, Figs. 152 and 153, in one end of the catch-pan, the contents of the pan will be constantly carried off at the end of the dandy-roll. If the accumulations in the pan are such as will not run off freely through the opening *B*, Fig. 152, a stream of water is to be poured into the pan at the opposite end, or otherwise.

Regulating the Speed of the various portions of Paper-Making Machines.

That part of the machinery in a paper-mill which drives the " machine," the couch-roll, the presser-rolls, the drying-cylinders, and the calenders, has long been susceptible of improvements; the object being to allow ready access to be had to the machine. This can be accomplished by doing away with a train of shafting, spur and mitre-wheels, belts, pulleys, etc., all arranged on the floor alongside of the machine and its adjuncts, impeding any approach with safety to the machine. A change of speed on any of the parts of the machine singly or together can thereby be made much more easy and convenient. Ordinarily the whole train is driven from the main shaft, which, by means of spur-wheels, mitre-wheels, and pulleys with belts of various widths and lengths (some very long), drive the coucher, press-rolls, driers, and calenders in a long line. The belts are often very close to the train and extend nearly their whole length, making approach to the machine difficult and dangerous, even to the loss of life in more than one case in a single mill, since in order to reach the machine for oiling or any other purpose it is necessary to crawl through and between long wide belts when in motion, and the very first principle in using machinery—safety to the operatives—is quite disregarded.

In running a paper-machine it is absolutely necessary that the harmony between the different parts should be maintained by exact and quick adjustments. The paper, while passing over the different parts of the machine, is always kept under a strong but steady tension, which must necessarily stretch it lengthwise, especially while it is wet. The first press, in drawing the web from the wire while it is in a very soft condition, will stretch it somewhat. Another small addition to its length is made by the second press, while it will shrink on the driers, and again become elongated on the calenders. If all the pulleys for driving the different parts have been fixed for a certain speed, weight of paper, and kind of pulp, so as to adapt themselves to these elongations and contractions, and suddenly a change in the pulp occurs—if it is beaten longer or shorter, if it is thicker, or if " imperfections" have entered into its composition—it will be found that the first press and following parts are pulling the web either too much or too little for its changed character and tenacity, and the paper breaks or is injured. The same experience will be had if the paper be made thinner or thicker; even if the speed of the whole machine only be changed, everything else remaining as before, the paper may be differently formed on the wire. It may leave the coucher with more or less water, and its tenacity will be increased or decreased. Any inequality of tension, too, is liable to make a wrinkle in the calendering, and much more so to break the sheet. Great quantities of paper are destroyed by these causes, which are constantly occurring, often three or four times during a day's work, making it necessary that the relative speeds of the presses, driers, and calenders should be frequently but slightly changed. The common way of accomplishing this is by what is termed " lagging." A number of strips of canvas or felting or thick cloth about as wide as the face of the pulley, called " lagging," are kept on hand. When a change of speed is desired one of these is smeared with " lagging-wax," so called, and laid on the surface of the driving-pulley to increase its diameter and quicken the speed of the pulley or gear at the other end of the shaft. Through the number and length of the strips held on the pulley by the wax any slight change can be produced, and though a very rude and clumsy expedient, this lagging has been in almost universal use up to the present time, as nothing practical has yet been found to supersede it. Expanding pulleys of various forms and other devices have been tried ; but some have been found too complicated, some get out of order too easily, some cannot be adjusted without stopping the machine, and others, still, require occasional lagging. In the usual way, in the train for running the machine, are two or three sets of gearing and from four to six heavy belts from twelve to sixteen inches wide and from eight to twenty-four feet long, all very expensive in their first cost and difficult to keep in repair, strained to their fullest tension, while by the use of the invention shown in Figs. 154 to 156, which is that of Mr. Marshall, of Turner's Falls, Mass., the belts used are only few in number, are short—none over four inches wide—and all run vertically, not interfering with work and ready access to the machine.

Mr. Marshall's invention dispenses entirely with lagging, does away with the long and dangerous train of shafting, wheels, pulleys, and belts on the floor by the side. of the machine, giving free approach to all its parts. Mr. Marshall claims that by his invention danger is avoided, safety increased, regularity of tension secured, speed regulated, power reduced, and production increased.

Fig. 154.

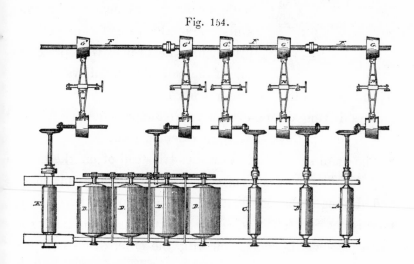

Figure 154 is a plan view of the machinery. Fig. 156 is a side elevation of a set of pulleys, etc. Fig. 155 is a side view of the belt-shipper and the device for actuating it.

Fig. 155.

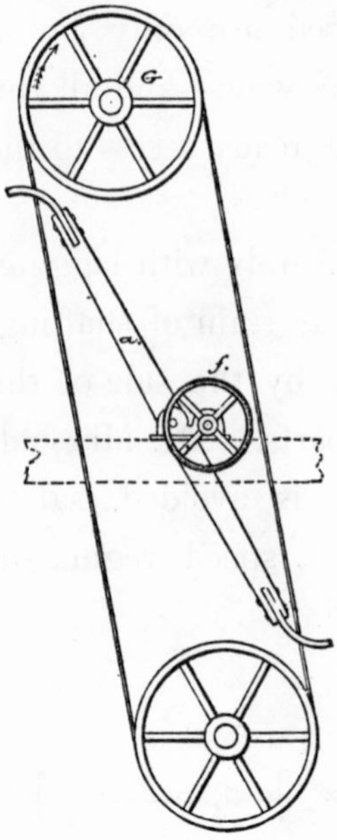

Fig. 156.

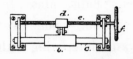

A represents the couch-roll. *B* and *C* represent the first and second press-rolls. *D* represents the drying-cylinders. *E* represents the calenders. *F* represents the main driving-shaft running in hangers overhead. *G* represents one of the pair of cone-pulleys driving the couch. *G'* represents one of the pair of cone-pulleys driving first press-roll. G^2 represents one of the pair of cone-pulleys driving second press-rolls. G^3 represents one of the pair of cone-pulleys driving drying-cylinders. G^4 represents one of the pair of cone-pulleys driving the calender-rolls.

The main shaft *F* runs overhead instead of on the floor. This removes the great obstacle to approaching the machine, and although placing shafting overhead rather than on the floor is not of itself new, yet in running the machinery of a paper-machine, in connection with the other devices of this invention, it has never before been done. On the shaft *F* are fixed conical pulleys or drums, each of which is connected by a belt with another corresponding one below, driving each of the presses, the driers, and the calenders.

To any one at all familiar with mechanics it will be obvious
that when the belt is moved up on the increasing form of
one of these cone-pulleys, it is correspondingly moved down
on the decreasing face of the other pulley, and thereby
increases its speed and consequently that of the roller or
drier, to which it is connected by a mitre-wheel gearing into
a shaft on which the roller or drier is fixed at a right angle
to that running the cone-pulley; and, *vice versa*, when the
belt running over the cone-pulley on the driving-shaft overhead is shifted toward the decreasing face of the pulley, the
one on which it turns below will be diminished in speed.
Now, the capability to increase or decrease the speed of any
one of these rollers—couch, press, calender, or driers—at
will in a moment's time, and to any exent ever so small, has
never been effectually done on any paper-machine ever made,
and it is what Mr. Marshall claims to accomplish by means
of these cone-pulleys and the belt-shippers to be now described.

The belt-shipper *H*—one for each pair of cone pulleys—
is hung at a convenient height midway between the two
pulleys, and of a length suitable to reach from the upper to
the lower pulley. The ends are forked as usual in belt-
shippers of all kinds. To enable the shipper, which is of
some length, to be held firmly in place and yet to move
readily when required, the arms *a* are formed at their
middle into a sleeve or box, *b*—say, fourteen to sixteen
inches long—through which passes a bar, *c*, securely bolted
to supports, and on this bar the shipper moves steadily.
To the upper side of this sleeve *b* is bolted a box, nut, or

female screw, through which passes the male screw e, running in fixed gudgeons on the same supports that hold the bar c, and having on one end a hand-wheel, f, by which the screw is readily turned.

The operation of these devices will be readily seen, and how the speed of any of the rollers can be adjusted instantly to suit circumstances. For instance, if from a slight change in the stock in thickness, or in its being beaten up, it should take more water, and should, in passing over the driers, gain a trifle more on the calenders than it had been running, so as to become a little slack on reaching the calender, the machine-tender in an instant seizes the hand-wheel on the belt-shipper, which moves the belt on G^4, and which runs the calender-rolls, gives it two or three turns to the left, runs the belt up on G^4, and correspondingly down on the opposite pulley on the shaft below, and instantly brings up the speed to the extent required; and so with any one or all.

Cone-pulleys have long been in use for changing speed on various machines, and in various ways, but never in any such combination as Mr. Marshall presents, by the operation of which, as described, he secures greater convenience and certainty in controlling the speed of every member of the machine, more equal tension of the stock and paper, a considerable saving of power in running the machine, a decided saving of labor in avoiding lagging, and also much expense in the care and straining of long heavy belts, by which he gets a large increase in the production of paper, more room in working about the machine, which, most important of all, allows the machine tender easy and comfortable access to every part of the machine with entire safety and unexposed to the constant danger of losing his life among the complication of belts and gearing of machines as at present commonly used.

Mr. Marshall has also made another improvement which relates to that part of the machinery of a paper-mill which drives the " machine" and all the concomitant parts, and it is a continuation of and a combination with the devices shown in Figs. 154 to 156.

In running a paper-machine it is absolutely necessary that the harmony between the different parts should be preserved and maintained by exact and quick adjustments, as has previously been stated; but it is also of equal importance to provide, in the driving apparatus for making paper, for maintaining in certain portions of the machinery a constant and unvarying rate of speed as it is delivered from the controlled action of the steam-engine, water-wheel, or whatever motive power impels the whole machinery. All practical paper-makers know that this applies thoroughly to the pump which returns the water strained from the paper pulp as it is being formed into paper on the Fourdrinier wire, to the screens through which the pulp is strained on its passage to the Fourdrinier wire, to the agitators which keep the pulp in motion in the vat, and to the " shake" which gives the oscillating motion to the " wire." But while it is of such importance to maintain in all the above-mentioned parts of paper machinery a constant and uniform speed, it is equally necessary that the couch-rolls, the press-rolls, the driers, and calenders, including all those parts of a paper-machine on which the paper is formed, pressed, dried, and calendered, must, in order to produce the full amount of paper the machine is capable of making economically, be varied and run at a greater or less speed, according to the character of the pulp and the thickness of the paper manufactured. The provision ordinarily made to accomplish this is to use gears of different sizes at a point on the main line of shafting between the point of transmission to those parts necessary to run at a constant speed and those parts which it is desirable to increase or diminish in speed. This is commonly done by cutting the main line of shafting at the desired point, and employing a counter-shaft running parallel to the main shaft and connected at each end by spur-gears to the two sections of the main line. The employment of different sizes of spur-wheels at these points of connection gives any desired speed to the couch-rolls, press-rolls, driers, and calenders, as the speed at that point of the main shaft transmitting power to these several points is varied by changing the spur-gears at the points of connection with that part of the main line which runs at a constant and unvarying speed. This is an unhandy and expensive way of working, for as the changing of the spur-gear necessitates the stopping of the whole paper-machine while the alteration is being made, considerable time is lost in accomplishing the shifts. Efforts have been made to obviate the necessity of stopping the machine while the changes are being made by combining two cone-pulleys ten to twelve feet long with the speed-gear running from one hundred and fifty to two hundred and twenty-five revolutions per minute; but they have never been so effectual as to be satisfactory. The length of these cones—ten to twelve feet—does not admit of a high speed, on account of the springing of the cones, and it is necessary to use a belt at least ten inches wide, in order to transmit sufficient power to drive the machine, and as a belt of this width can work to advantage only on cones of a moderate pitch, the range of the change of speed as controlled by the pitch and length of the cone is not sufficient to make all the changes of speed required in producing paper at rates of from twenty to two hundred feet per minute, and it has still been found necessary to use three or more sets of spur-gears in connection with these cone-pulleys. The extent of change of speed allowed by these cones from the slowest to the fastest being only about sixty feet per minute, when a change in the thickness of paper being made requires a greater rate of speed than this, the machine must be stopped and a change of the speed-gears must be made. So manufacturers have found that all their efforts to effect the desired changes of speed by the use of the cone-pulleys alone as commonly employed have been but partially successful. Again, the use of two long cone-pulleys in combination with the common back line or driving shafting of a paper-machine necessitated so much additional room on the back side of the machine that it has commonly been found necessary to erect a small building outside the machine-house, but attached to it, in which to operate these unwieldy cones. To accomplish this desirable purpose of changing the speed of the machine at once without the disagreeable necessity of stopping it and changing the gear-

34

ing, and without requiring any additional floor room, Mr. Marshall combines four cone-pulleys in pairs—one of each pair placed above the other—about five or six feet long, or half the length of those commonly used, and of different diameters, as shown in Fig. 157. These cones, being comparatively short, admit of a high rate of speed, which permits the use of a narrow belt of only four inches in width. This gives the required power, and being able to use these narrow belts, it is possible to get a much sharper pitch of the cones, and thus the required range of speed is obtained without the annoyance of stopping the machine to change the gear. By placing these pulleys one above the other, and one pair driving the other, the inventor greatly economizes space and obviates the employment of a long, wide belt, which is always costly and troublesome.

Fig. 157.

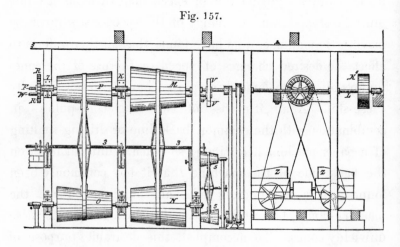

Fig. 157 is an elevation, partly in section, illustrating Mr. Marshall's invention.

L is the main driving-shaft, connected with the motor by pulley and belt, and terminates at *K.*

M is the main driving cone-pulley, connected by a belt with the cone-pulley *N*, placed underneath it. On the same shaft with *N* is the cone-pulley *O*, the larger end of the pulley *O* being of the same size as the small end of the pulley *N*. The cone-pulley *O* is connected with the cone-pulley *P*, placed over it, and driven by a belt passing over both. The cone-pulley *P* is on a short shaft overhead, extending to and stopping at its bearing *K* on the right, and on the left extending far enough beyond the bearing 7 to carry a spur-wheel, *R*. The belts connecting all the cone-pulleys are controlled by the shipper, as shown in Figs. 154 to 156. On the shaft *F*, which is stopped and has its bearing in the rear of bearing 7, is a spur-wheel engaging in the wheel *R*, which carries the shaft *F*, with all its dependencies. So it will be seen that the main shaft *L*, with all its driving-power, runs steadily and always the same. The power for the other parts of the machine is communicated from *M* to *N*, to *O* and *P*, and by the spur-gears to the shaft *F*, from which are run the couch-rolls, the press-rolls, the drier, and the calenders, and the speed of all these together may be accelerated or slackened to any desired extent by simply moving the belts on the two pairs of pulleys *M N*, *O P*, while the individual parts are each controlled by the arrangement of cone-pulleys as shown in Figs. 154 to 156.

V V are friction-wheels—one on the main shaft *L*, the other on a counter-shaft, *W*, which extends in the rear of the cone-pulleys to the mitre-gears 8, for driving the felt washers 2 2—which are necessarily driven from the main shaft, to insure the necessary uniformity of speed, which they could not have if driven from the shaft bearing the cone-pulleys. These friction-wheels are controlled by a screw, *X*, engaging in the bearing of the shaft *W*.

Y is a pump used for returning the water used in forming the paper.

Z Z are the screens for straining the pulp. 6 is the pulley that drives the "shake." All these require to be driven by a steady, unvarying power, and are therefore necessarily attached to the main shaft by belts and pulleys.

Drying Cylinders.

Cast-iron cylinders, perfectly true on their surfaces, and having well balanced bodies are used for drying machine-made paper.

It is necessary that the shells of the driers should be of uniform thickness, and that the surfaces should be entirely free from sand holes, it is consequently desirable that these shells should be cast in loam and that they should be turned true on the inside as well as on their surfaces.

Steam is admitted to the drying cylinders through hollow journals. Usually these journals are provided with a valve-seat in which is arranged a valve, which is held to its seat by the pressure of the steam within the drum or cylinder. When a very low pressure of steam is employed in the cylinder, the valve is not held tightly to its seat, and permits foreign substances and sediment to work between the valve and its seat, whereby the parts are caused to wear unevenly and become leaky.

The object of Roach's invention, shown in Fig. 158, is to provide the joint with means whereby the valve is retained in its seat at all times.

Fig. 158 is a vertical section of Roach's improved pipe-joint applied to a hollow journal.

Fig. 158.

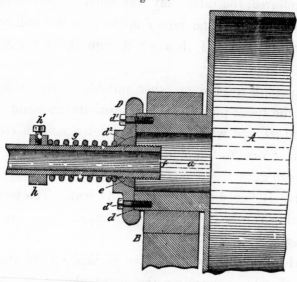

A represents the drum or cylinder, and a the hollow journal upon which the drum rotates, and which is supported in a suitable bearing, B.

D represents a head or cap, which is applied to the end of the hollow journal and provided with a depression or cavity, d, into which the end of the journal is fitted. The head D is secured to the journal by screw-bolts, d', so as to turn therewith ; or, if preferred, the head and journal may be provided, respectively, with screw-threads and the head screwed on the journal. The head D is provided with a valve-seat, d^2, of conical, oval, or other suitable form, in which is seated a valve, e, of similar form. The valve e is secured to a steam-pipe, f, by means of an internal screw-thread, which engages with an external screw-thread formed on the inner end of the pipe f.

If preferred, the valve may be formed in one piece with the steam-pipe f. The pipe f opens into the hollow journal, and may extend through it into the cylinder A, if desired.

g represents a spiral or other suitable spring, which surrounds the steam-pipe f, and bears with its inner end against the head D, and with its opposite end against a collar, h, which is secured to the pipe f by a set-screw, h'. The spring g tends to draw the pipe f and valve e outwardly, whereby the valve is forced tightly against its seat. The tension of the spring can be regulated by means of the collar h and set-screw h'.

By this construction the valve is held on its seat at all times, thereby preventing the accumulation of sediment between the valve and its seat, which results in an unequal wear of the parts, causing leakage, and it also prevents corrosion of the parts when the machine is not in operation.

It has also been a common practice to force the steam for heating directly into the drying-cylinders through the journals, into the ends of which the steam-pipe is packed, the journal, in effect, being made only a continuation of the steam-pipe, but always with a leakage and escape of steam, from the inevitable wear of the end of the stationary steam-pipe being packed into the moving grinding-journal, requiring constant repacking.

The condensed water, which accumulates rapidly in these drying cylinders, has usually been conducted out by means of a pipe attached to the inside of the cylinder-head, extending across its diameter, and curved near the bottom, so as to take up at each revolution a quantity of water which was blown out or discharged through an opening in the cylinder-head as at each revolution of the cylinder it was lifted and turned over. As this water was blown out a considerable amount of steam escaped with it, and continued to do so till the end of the discharge-pipe, in revolving, was again filled with water. The percentage of steam thus escaping and wasted is large enough to very materially affect the heat in the cylinder, and in a battery of four, six, or more, draws very largely on the supply of steam, and adds a heavy item to the operating expenses of the mill.

Fig. 159.

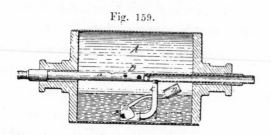

Fig. 159 shows a longitudinal section through the drying cylinder of Jaminson's improvement in steam-traps.

A is the drying-cylinder. *B* is the steam-pipe passing

through the journals of the cylinder and the cylinder longitudinally, having in it perforations, a, for the discharge of the steam to heat the drier, of the same capacity as the pipe itself. In this pipe, midway between the cylinder-heads, is a partition, on one side of which steam enters the cylinder, and on the other the water condensed is discharged. These perforations are in the half of the pipe next the entrance of the steam. On the other side of this partition, in the steam-pipe B, is fastened a pipe, C, which passes almost to the bottom of the cylinder, when it turns and terminates with a valve or gate, E. This gate or valve E has an opening in its side, which is closed by a gate balanced on a pin. On the upper end of this sliding gate is fastened a rod projecting forward some inches. To the end of this rod is secured a float, D, which, rising on the surface as the water increases in the bottom of the cylinder, pushes back the rod attached to it and to the sliding gate, whereby the gate is opened and the water of condensation rushes into the valve-opening, and, by the pressure of the steam, is forced up into the closed half of the steam-pipe and out through it, where, by a connection on its end, it is conveyed into a tank, or back to the boiler, without the loss of heat or the escape of steam, the float falling just as fast as the water is lowered by the discharge, and the gate is closed at the same rate.

It will be seen that in this way, as the water is discharged as the gate is open, and as that is automatically controlled by the amount of water acting on the float, nothing but water can pass out, and not a particle of steam can escape or be wasted.

The steam-pipe passing through the journals of the cylinder is made tight at both ends by a sleeve of brass or other metal, that may be readily renewed, fitting closely into the journal-bearings with ordinary steam-packing, preventing all loss of steam. This of itself is a great advantage over the ordinary way of driving the steam directly through the journal into the cylinder, as it is impossible with the constant wear of the journal against the end of the steam-pipe to make the fitting continue secure and steam tight.

With the device shown in Fig. 159 the water, it is claimed, is kept constantly discharging without the loss of a particle of steam either through the discharge-pipe or by leakage at the journals.

Single Cylinder Machine.

A machine adapted to the forming of thin papers and

Fig. 160.

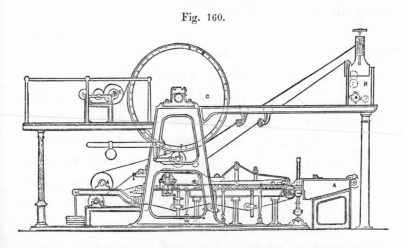

such as are required to be smooth on only one face is illustrated in Fig. 160. It is a modification of the Fourdrinier machine, and consists of an ordinary paper-machine as far as the couch-rolls *A A*. After leaving the couch-rolls the paper is carried backwards on the top of the endless belt *B*, and is brought in contact and adheres at *d* with the large heated cylinder *C*. The press-roll *E* presses the paper against the periphery of the cylinder *C*, and the paper passing around the cylinder is wound up when quite dry on reels at *G*, and is afterward cut or calendered in the usual way.

The felt, as it travels, passes through the box *H*, which is filled with water and acts as a belt washer.

Calendering.

Leading Paper through Calender-Rolls. — Hitherto in the process of calendering the paper web, as it passes continuously from the driers of the machine, has been conducted and guided through the stack of calender-rolls by the fingers of the machine tender, and serious accidents are continually occurring, in which the fingers of the operator get jammed and terribly bruised and the danger multiplied, since the paper web has to be restored every time its continuity is interrupted for any cause whatsoever. Moreover, in the process of "mending up" a large amount of "broken" is produced, because the draft and tension across the paper web are not uniform, and folds or wrinkles are caused, which at once make a crack or break in the paper, and these continue until the tension is properly restored, the paper during this interval being rendered useless for commercial purposes.

Smith's Pneumatic Guide.—To overcome the objections

which have been mentioned, and to render the waste of the paper less and make the effort of mending-up not so laborious to the operative, and reduce the danger to a minimum, various mechanical contrivances have been devised.

Fig. 161 illustrates a pneumatic device for leading paper through calender-rolls ; it is the invention of Mr. Richard Smith, of Sherbrooke, Quebec, Canada.

Fig. 161 is a side elevation of Mr. Smith's invention arranged as a whole and fitted upon the calender-roll stack and drier-frame, the two upper rolls being in section.

Fig. 161.

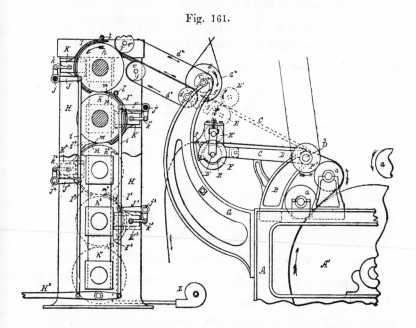

As the inventor proposes to use a blast of air or its equivalent (either suction or pressure) in order to guide the paper web automatically through and around the rolls, it becomes

necessary to partially cover and protect a portion of the
surface of the calender-rolls in order to produce the desired effect, by confining the current of air, and thus oblige
it to assume a certain direction, and carry with it the advance
end of the paper web to be led between the rolls.　There
is therefore shown, as partially closing the exterior peripheries of the rolls, a series of alternately oppositely-disposed
wind-cases, $I\,I'\,I^2$, etc., which are plates bent preferably
concentric with the curvature of the rolls and extending
their entire length.

In revolving bodies centripetal force always exerts a somewhat important function upon an object located upon its
surface ; hence, and more especially in paper-making machines, the paper web has a great tendency to adhere to the
surface of the calender-rolls, and pass continuously around
one roll in lieu of advancing on and around the next adjacent roll.　This fact is especially noticeable in the manufacture of light-weight paper, and it is found necessary to employ in connection with the wind-cases above alluded to, a
device called a " doctor." In this especial instance the
inventor has terminated the upper extremity of the wind-
cases, which enter between two rolls in the shape of a straight
steel bar or doctor, M, which is bevelled to coincide, or
approximately so, with the line of a tangent to the roll at
the point where the doctor touches the roll.　The most
effective position is a point a short distance to one side of
the place of contact between the two rolls where the exterior
surface of the top roll first begins to assume an upward path
of movement in its rotation ; hence, when the paper after a

break is to be mended up, the operator raises the lever and advances the wind-cases towards the rolls and in close proximity thereto, while their respective doctors are brought in close contact with the upwardly-moving surface of each roll, and thus any tendency which the paper web may have as it emerges from one side between two rolls to the other to pass up and wind around the top roll of the pair, is instantly checked by the doctor, which, as the curve of the latter coincides with the interior curve of the wind-case, it guides the paper along in its proper course down around the lower roll of the pair. The same action ensues as the paper web emerges from between the next two rolls and meets the next doctor, and so on down between the rolls composing the stack.

Mounted upon the wind-cases, with which they are suitably connected, there is disposed a series of pipes, l l' l^2, etc. The latter communicate with a blower, L, either exhaust or pressure, and from which the air is obtained as means for guiding and passing the paper web through the rolls.

Operation.—The operation of the apparatus, shown in Fig. 161, is as follows, with the various rolls and pulleys moving in the direction of the arrows, as indicated: The paper web has been interrupted and broken from some cause, and it must be restored; hence the operator drops the table C into a horizontal position and lifts the lever H^2 to bring the doctors to bear against the rolls h h' h^2, etc., and the wind-cases I I I^2 near to the surfaces of the rolls. The paper is then passed around the rolls a a and drier A', thence over the table C between the quick-running rolls

$E\ E'$, which exert sufficient tension upon the web commensurate with its strength, hence adjustable but not enough to break it. Immediately upon the proper restoration of the tension and removal of the slack consequent upon the mending up, the operator swings the table upward, the paper web meantime continuously passing along until the movable cutter g has engaged with and passed the fixed cutter g' upon the arms $G\ G'$. The paper web is then instantly severed, and the draft rolls carry and feed it directly to the tapes $d^2\ d^2\ d^2$, whence it is conveyed to the entrance of the wind-case I in a line directly at right angles to the calender-rolls—an important feature, since it obviates the loss of paper incidental to the ordinary method, where the paper web is introduced at an angle with the axis of rotation of the rolls. Upon the arrival of the advance end of the paper web in front of the wind-case the pressure from the air-current through the pipe l forces it quickly around the roll h, while the interior surface of the case maintains and guides it in a proper direction between the rolls h and h'. Immediately after the web emerges from between the rolls, although in close contact and with a tendency to pass up and wind around the upper one, it at once encounters the doctor, which removes it therefrom and compels it to follow the curve of the wind-case I', when the pressure from the air-current emerging from the pipe l' still further advances it, and so on through the rolls of the stack, when it may either be passed onward and led through in a similar manner by means of a second series of tapes, $d^2\ d^2$, to a second stack, or be led to the reels, upon which it is temporarily stored.

A great advantage accrues by the operation of conducting paper through calender-rolls in the manner just described—that is, the current of air in passing over the surface of the paper exercises a decided influence in cooling the heated continuous paper web as it passes from the driers. Hitherto it has been customary to grind and finish the calender-rolls cold; hence, after being mounted in proper position, and the heated web has passed through them for any length of time, the rolls become hot and expand, and the result is that the faces of the rolls do not coincide, since they were ground to fit when in cool position. Further, it is found in practice that a special cold blast of air applied across the web prior to its entrance between the rolls of the stack and independent of the current inducing the progress of the web through the stack, accomplishes the cooling of the individual rolls to remain cold; hence there is no expansion, and the surfaces coincide exactly with the greatest degree of efficiency.

Cram's Entering Guide.—The device for threading calender-rolls, shown in Figs. 162 to 164, is the invention of Mr. Madison H. Cram, of Pawtucket, R. I., and its object is to provide a simple means for guiding the paper between the rolls with certainty and precision; and it consists in the combination, with the set of calender-rolls, of two narrow endless belts passing together between the rolls, and means for shifting the location of the belt along the rolls.

Fig. 162 is an end elevation of a set of calender-rolls provided with Mr. Cram's improvement. Fig. 163 is a rear elevation of the same. Fig. 164 is a vertical section taken in the line of the inner face of the bearing standard, showing the belts running on the large diameter of the rolls.

Fig. 162. Fig. 163. Fig. 164.

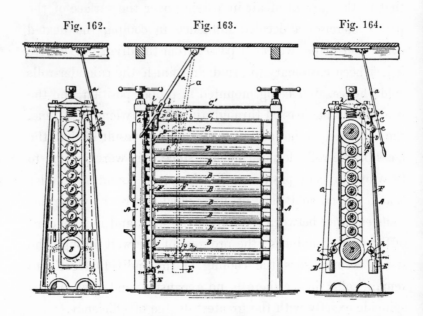

$A\ A$ are the bearing standards, in which the rolls $B\ B\ B$ are placed one above the other with their peripheries touching each other. The upper ends of the standards $A\ A$ are connected by means of the parallel guide-rods $C\ C'$, upon which is placed the sliding belt-carriage D, a sliding movement being imparted to the carriage along the rods by means of the pivoted shipper-handle a, provided with a slot, b, which embraces the smaller outer end of the stud c projecting from the carriage. The carriage D is provided with the loose belt-carrying pulleys d, e, and f, which revolve upon suitable pins or studs; and upon the rods g and h, which extend from one of the bearing-standards to the other, near the base of the same, are placed the loose belt-carrying pulleys i and j, these pulleys being made capable of a sliding movement along the rods; and also upon a rod k, which extends from one of the bearing-standards to the other at the same height and parallel with the guide-rod C', is placed the loose pulley l. The weight E is provided with the upright ears m m, between which is pivoted the loose pulley o, and the opposite weight, E', is in like manner provided with the loose pulley n. The narrow endless guide-belt F, when in its normal position, passes from the carriage-pulley f over the smaller end portion of the top roll B, thence downward from side to side between the smaller end portions of the rolls and under the bottom roll B, thence over the loose pulley j upon the rod h, then under the loose pulley o of the weight E, and upward to the pulley f. The narrow endless belt G, which runs over the rolls B in contact with the belt F, passes from the roll e of the carriage D over the smaller portion of the top roll B in contact with and above the belt F, thence downward with the belt F from side to side between the rolls B and outward over the loose pulley i upon the rod g, thence under the pulley n of the weight E' and upward to the loose pulley l upon the rod k, and thence over the loose carriage-pulley d to the pulley e, the belts F and G moving uniformly in the same direction between the rolls B B.

Operation.—In guiding the paper into the calender-rolls the belts F and G are first moved from the smaller to the larger portion of the rolls by the lateral movement of the carriage D, by means of the shipper a. The paper fed

35

forward between the guiding-belts F and G, and upon the completion of its passage through the rolls of the carriage D, is moved back to its normal position, the guiding-belts being thus carried from the larger to the smaller portion of the rolls B B, the resulting slack of the belts F and G being taken up by the action of the weights E and E'. The calender-rolls are thus claimed to be rapidly threaded without danger to the workmen.

The belt G, instead of running directly upon the belt F, may run side by side with it if preferred.

Moistening the Paper.—When the web of paper leaves the driers it is usually too hard to receive the impression of the calenders as readily as if its surface were slightly humid, and the paper is consequently sometimes moistened with steam just before it passes through the calenders. The paper is dampened by means of a three-eighths inch pipe perforated with small holes about the size of the head of a pin; this pipe is secured to the frames of the calenders a short distance below the sheet where it first enters. Sometimes two pipes are employed, one on each side of the calenders, so as to dampen each side of the paper.

Automatic devices are used for turning off the steam from the pipe in case the paper should break so as to prevent the rolls from getting wet.

Moistening the Calender Rolls. Brewer's Method.—Sometimes it is desirable to moisten the calender-rolls of paper-making machines, the object being to heat the same from the exterior.

The " steam condensing doctor" invented by Mr. Frank

Brewer, of Marseilles, Ill., is shown in Figs. 165, 166, and 167, and it is designed to accomplish the purpose first named.

Fig. 166.   Fig. 165.

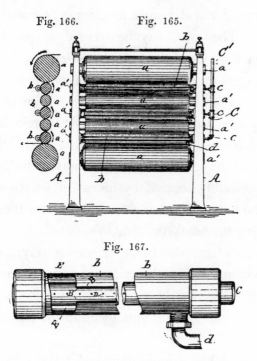

Fig. 167.

The device, besides performing the above-mentioned functions, acts as a doctor to remove any waste from the rolls and prevents them from clogging.

Fig. 165 represents a side elevation of a stand of calender-rolls with Mr. Brewer's device attached. Fig. 166 is a transverse vertical section through the rolls and the device. Fig. 167 is a longitudinal view of one of the pipes of the device partly broken away and partly in section.

A designates the frame, upon which the rolls *a a* are

journaled in proper bearings, the journals being designated by *d' d'*.

b b b are pipes having their ends closed by the screw-caps having central circular openings, for a purpose hereafter explained. The caps may be secured in position, when screwed on the ends of the pipes by set-screws.

Brackets are secured to the standards of the frame in such a position that pipes *b*, near their ends, rest in concave notches of the brackets, and may be firmly held therein by the set-screws which pass through the flanges of the notches. The brackets may be secured in place in any desirable manner, but are preferably secured by means of set-screws passing through the slots and entering the outer sides of the standards, thus making the brackets adjustable.

The pipes *b* cross the frame *A* transversely, and are preferably so situated as to lie adjacent to every second one of the smaller rolls *a*, as shown in Fig. 166, the axis of each roll and its corresponding pipe being in the same horizontal plane.

Each pipe *b* has a longitudinal slot, the edges of which form a close joint against the surface of the roll. The slot does not extend the entire length of the pipe, its end being equally distant from the ends. The axis of the slot lies in the plane with those of the roll and pipe.

C C are small steam-pipes running through the pipes *b*, passing through the openings in the caps and having their ends on one side closed. Their other ends are connected by proper couplings, *c c c*, to the steam-pipe *C'*, Fig. 165, which takes steam from any suitable source of supply.

Each pipe C, Fig. 166, has a longitudinal series of small equidistant openings, D, on the side facing the roll, and lying in the same plane as the axes of the roll and pipe.

d d are discharge-pipes for water of condensation, which pipes depend from the pipes b on the side adjacent to the steam-pipe C'.

Steam being admitted through the pipes C' C passes out of the openings D and fills the spaces E between the pipes b and C, and as the rolls rotate heats and moistens them.

The pipes b, on account of their contact to the rolls, act as doctors to the rolls to remove waste that might clog them.

Newton's Method.—The essential feature of the invention of Mr. Moses Newton, of Holyoke, Mass., illustrated in Figs. 168 and 169, consists in the delivery of steam upon the surfaces of the calender-rolls and in conveying running water to the interior of the rolls and thence out again, causing the condensation of the steam upon the surfaces of the rolls, thereby producing a polished surface on the paper as it passes over and between the calender-rolls.

Fig. 168 shows a front elevation of an ordinary stack or series of calender-rolls provided with water and steam pipes in accordance with Mr. Newton's plan. Fig. 169 represents a vertical cross-section of the same.

A illustrates the main frame of the apparatus; B, the calender-rolls, arranged horizontally in the frame, as usual; C, pipes arranged to deliver cold water into a greater or less number of the calender-rolls at one end; D, pipes which carry off the water from the opposite end of the rolls, and E a steam-pipe having any desired number of perforated arms or branches extending along the outside of the rolls and delivering the steam thereon. Water may be admitted

Fig. 168.   Fig. 169.

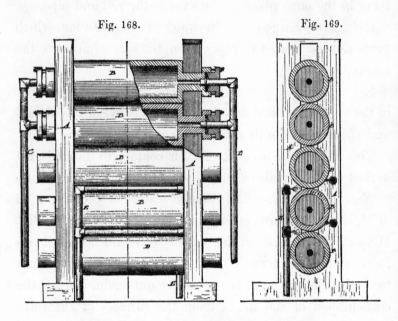

to all or any desired number of the rolls, and the steam may, in like manner, be delivered to a greater or less number of the rolls, as circumstances may require. In most cases the best results are secured by supplying water to the top rolls only and delivering steam upon the bottom rolls only.

The steam condenses upon the surface of the rolls, dampening them slightly, but with perfect uniformity, and the consequence is that the rolls impart to the surface of the paper a much smoother finish and higher polish than can be attained by the ordinary mode of procedure. When the steam and water pipes bear the relation shown in the illustrations the condensation takes place gradually, but mainly upon the upper rolls, with which the steam comes in contact.

Preventing the Burning or Injury by Heating of the Paper or Material of which the Calender-Rolls is composed.—As paper-calender rolls are ordinarily constructed, the great pressure to which they are subjected, together with the speed at which they revolve, operate to heat their bearings, as well as the rolls themselves, and they are often completely ruined from this cause, the heat being so intense as to burn and destroy the paper or material of which the rolls are composed, especially near their ends.

The object of the invention of Mr. H. J. Frink, of Chicopee, Mass., illustrated in Fig. 170, is to provide one or

Fig. 170.

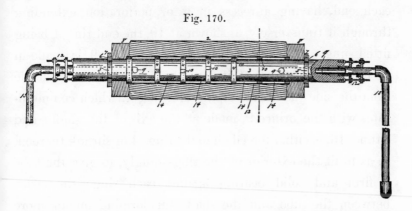

more internal annular chambers in the interior of a paper calender-roll, through which to force water to prevent the paper or material of which the roll is composed from being burned or injured by heating, and to keep the journal-bearings at as uniform temperature as possible.

Fig. 170 is a sectional view of a calender-roll made according to Mr. Frink's invention, at a plane on the line of its axis, with the extreme end of the shaft in section, showing the application of a stuffing-box thereto.

The main shaft should be of suitable size to withstand the great pressure to which it is subjected, in which, near each end, is made an annular groove, 5, into one of which are fitted two half-collars, 6, with preferably a little space between its ends, and another collar, 7, is shrunk on to these two half-collars 6, which keeps them solid and firm in their groove. A retaining-head, 8, is then fitted snugly and firmly on to the main shaft 3 and against the retaining-collar 6, and a series of rings, 4, are then shrunk on to the main shaft 3 at suitable distances from each other, each ring, except that at each end, having a recess in it, or perforation, extending through it transversely, as shown at 10, the end ring 4 being fitted snugly against the head 8. An orifice, 9, is made in each end of the shaft at its axis, and another one is made from the side just inside of the first ring 4, which communicates with the orifice 9, made at the axis of the shaft; and a tube, 13, is either forced on to the rings 4 or shrunk thereon, so as to fit the exterior of the rings snugly, to give the tube a firm and solid bearing against each ring; the space between the tube and the shaft thus forming one or more annular chambers or compartments—if more than one, communicating with each other through the transverse perforation or cavity in the dividing rings 4.

After the tube 13 is fitted properly in place, the paper or other substance, 2, is forced upon the tube 13 and against the fixed head 8 by hydraulic pressure. The other head 8

is then forced into place against the roll 2, and the other
end half-collars 6 are placed in their groove 5, and the collar
7 is shrunk thereon, so that the heads 8, tube 13, and roll 2
are solidly in place, with the annular chambers or compartments 14 inside the tube and between the rings 4. If the
roll is to be made very short for special purposes, there may
be but one of these annular chambers, the orifice 9 at each
end of the shaft opening into the same chamber; but if the
roll is to be longer, there may be several of these annular
chambers or compartments, with the orifice 9 at each end of
the shaft opening into the end chamber or compartment.
The exterior surface of the roll 2 is then turned off and
finished, and a pipe, 15, is connected with the orifice 9 in
each end of the shaft 3 with any ordinary and well-known
packing or stuffing-box attachment, as shown at 12, so that
the roll may revolve and the pipe remain stationary, and
the joint where the two are connected may be water-tight.

The constant flow of water through the annular chambers,
14, between the tube and its shaft, it is claimed, keeps the
whole of the roll 2 perfectly cool, so that it or its bearings
never become heated to any appreciable or injurious extent.
As this introduction of water into the roll operates to keep
its journal-bearings at a more uniform degree of temperature, it follows that the frictional bearing of the journals is
also more uniform, inasmuch as there is little or no expansion of the metal of the journals, and no consequent increase
of friction.

It is evident that instead of securing or shrinking the
rings 4 on to the shaft 3 the latter may be turned down,

leaving a series of annular projections or collars, having substantially the same form as the rings 4.

Method for the easy Removal and Replacement of Calender-Rolls.—Usually calender-rolls of paper, wood, or metal are used in a " stack" as it is called ; or a series placed one above another, have been kept in place within the frame, which is made open from the bottom roll to the top, the opening only large enough to receive the journal-boxes of the rolls, which are perhaps from four to five inches in diameter, while the roll itself is from six to fifteen inches in diameter.

The pillow-block is always solid, with the open-sided frame cast with it, and holds the bearings in which the lower and usually largest cylinder runs, and to which the power is applied, and by the friction of which upon the one next above and resting on it gives motion to the whole composing the stack.

Calender-housings having been constructed in this way, it became necessary, whenever it was required to remove rolls for " turning-off" or for any purpose, to hoist them up and out at the top one by one, even to the last one at the bottom, which is the driving-roll, and very heavy, weighing often several tons, making the removal of any of the lower ones especially a long, tedious, and expensive operation, and not unaccompanied with danger.

The pressure upon the rolls has usually been applied by a screw or weighted lever working upon the upper roll, and it has been proposed to use, instead of this device, a cushion of steam or water or some other fluid, to be forced to any extent needed into a tight cylinder having a piston the rod of which, bearing upon the cap of the upper journal-box when the cylinder is charged, shall ·hold the rolls firmly down, and at the same time allow a slight recoil if anything of an improper thickness passes between the rolls.

In this connection we illustrate in Figs. 171 and 172 the invention of Mr. George E. Marshall, of Turner's Falls, Mass. Fig. 171 is a front view of a "stack" of cylinder-rolls. Fig. 172 is an end view of the same.

Fig. 171.

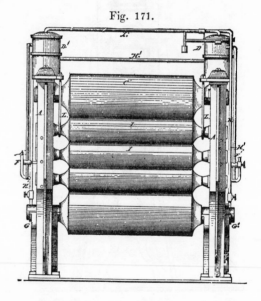

A represents the frame of heavy cast-iron holding the rolls, having between the sides an opening wide enough to allow the withdrawal of the rolls; and these sides, near the bottom, are expanded in their breadth, and correspondingly in the opening, wide enough to allow of the withdrawal of the bottom or driving-roll. This frame has cast on it a base of sufficient width and thickness to sustain solidly the rolls, when it is bolted to the floor.

G shows the boxes, each pair of which holds the journals of the respective rolls, and are of a peculiar construction, made in several parts, so that they can be taken to pieces and removed from the frame without disturbing the roll.

Fig. 172.

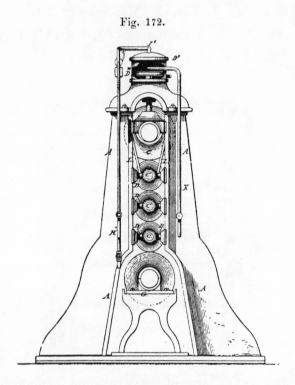

The boxes in which the journal runs have on their sides trunnions *D*, which enter the adjustable sliding sides of the boxes. *C* represents the end of the journal *I*, the roll, having between it and the trunnion-box steel rings or washers.

The trunnion-box can be turned and adjusted in any way or removed with ease by slipping the sliding boxes on the trunnions.

Through the sides of the frame, holes are drilled under each box, to allow a pin to be put in to sustain the boxes when it is necessary to remove any roll below them.

C'' is the upper roll, larger than the others, and upon the boxes of which the pressure is applied from the cylinders and pistons and piston-rods above them, to keep the rolls to a close pressure.

D' is a steam-tight cylinder, one over each end of the rolls into the top of which steam is forced by a pipe I'. Within each cylinder is a piston, D'', both having on their under sides piston-rods, which, when steam is let into the cylinder above the pistons, are forced down upon the boxes of the upper roll, thus giving the desired pressure. The steam is let on through the pipe I', and the pressure is controlled by steam-cocks. Entering the side of each cylinder D' about midway is a discharge-pipe, K, for taking off the water of condensation through a steam-trap.

H' is a pipe, with a trap, for drawing off any water that by leakage may accumulate under the piston.

$L\,L$ are yokes bolted into the caps and passing down under each roll as it is required to be raised.

G' is a pillow-block, which sustains the lower or driving roll in the stack. This is bolted to the base of the frame of the calender-stack, but so that it can be removed by relieving it of the pressure of the rolls.

Stripping Sheets of Paper from off the last Roller of Calendering Machines.—The stripping-fingers for stripping the sheets of paper from off the last roller of calendering machines as usually made are secured to one rod and their ends caused to press against the roller by means of a weight or spring actuating the rod. The fingers are rigid on the rod as regards lateral movement, and consequently bear always on the same parts of the roller, which necessitates frequent refitting of the same by grinding; otherwise these grooves mark the paper being calendered. This objectionable feature in calendering machines may be overcome by makiug the stripping-fingers independent in their action on the last roller by securing them to separate blocks provided with adjustable weights to cause the fingers to bear independently with the proper force against the roller, and the blocks have open bearings, by which they are placed on a fixed rod, thus enabling each finger to be moved laterally, so that its end may bear on different parts of the roller, thus causing an even wear to the surface of the roller. The fingers may be moved laterally, as desired, by the attendant while the machine is in operation, or taken off and placed and arranged on the fixed bar, to suit the sizes of the sheets of paper being calendered.

In this connection we illustrate in Fig. 173 the invention of Mr. John McLaughlin, of Lee, Mass.

Fig. 173 represents an elevation of a calendering-machine with the independent detachable fingers in position to strip the paper from off the last roller.

Only sufficient of a calendering-machine is shown to illustrate the application of the improved stripping-fingers.

$a\ a$ are the side frames, and b, c, and d the rollers. d is the last roller and it is against this that the stripping-fingers $e\ e$ bear to discharge the paper from it into the receiving-box. Three stripping-fingers only are here shown, it of course being understood that the number of them will depend on the size of the machine and quality and size of the paper being calendered.

Fig. 173.

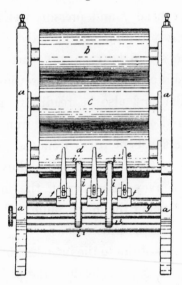

Fig. 173 illustrates the essential features of Mr. McLaughlin's invention. The finger e is secured in an angular position to the block or head f by means of a screw, which passes through a slot in the lower end of the finger to enable the upper end, which is tapered off to a sharp edge, so as to lie close to the roller d, to be set in line with the ends of the other fingers, to strip the paper evenly off the roller d. The head or block f has an open bearing, by which it rests on the fixed rod g, secured in brackets from the side frames $a\,a$; and projecting from its sides in a horizontal position under the free end of the finger e is a rod which may be cast with the block or head f, or be screwed or driven in a hole therein; and on this rod is placed a pressure-weight secured thereto by a screw. The position in which this weight is set on the rod determines the pressure of the end of the finger against the roller d. Each finger, with its carrying block or head f and pressure-weight, being independent, and merely resting on the fixed rod g, it will be observed may be set to bear on any part of the roller d desired, and be shifted about, so as to prevent grooves being worn in the roller, and also be readily removed from and replaced on the rod g, according to the number required to properly strip the sheets of paper from the roller d.

We have shown tapes $i\,i$ on small pulleys $i'\,i^2$, arranged and adapted to convey the paper off the fingers $e\,e$ into a receptacle, placed under the lower pulleys i^2, but the receptacle is not shown in the illustrations.

Plate Calenders.—Prior to the introduction of the method of glazing paper in super-calenders, the paper was first cut into sheets and passed through plate calenders. Plate calenders are still used to a considerable extent in English and Continental mills; but there are probably not two mills in the United States where they are now employed.

With these calenders the paper to be glazed is laid in single sheets between zinc or copper plates until a pile of about twenty-five sheets of paper are so arranged, and this stack of alternate sheets of paper and metallic plates is then passed forward and backward between the rollers, under great pressure, until the polished surface of the plates communicates a gloss to the paper.   If a very highly calendered surface is desired the plates are frequently changed and the paper relaid between them.

Fig. 174 shows a side elevation of a plate calender; it consists of two solid iron press rolls, *B B*, mounted in a framework, *A*.   Pressure is applied to the top roll by means of levers and weights.

Fig. 174.

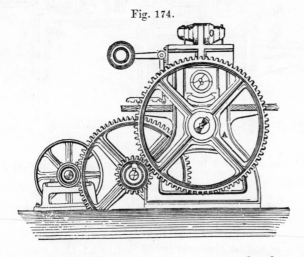

In operating this form of calender the stock of paper and plates is first deposited on the table in front of the rolls and is then pushed forward until the rolls take a " bite" on the stack and carry it through the rolls, the stack then comes in contact with a shifting device which reverses the motion of the rolls and carries the stack backward through the rolls.

36

Cutting and Rolling.—After the paper has been wound up at the end of the drier it is trimmed and slit so as to produce the width of web required.

The usual form of paper cutter employed in British and Continental mills as well as in many of the mills of the United States is shown in Fig. 175, and it cuts from six to eight webs at one time.

Fig. 175.

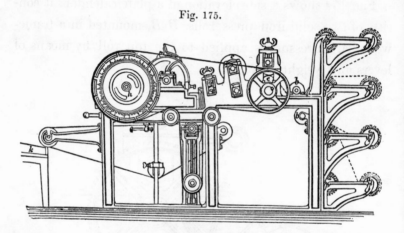

The webs of paper to be operated upon are shown at *a a*, from whence they are led between the leading rolls *b b* through the feeding rolls *c c*, which latter are driven, by means of the change pulley *d*, at such a rate of speed that the paper is fed to the revolving knife at the precise speed requisite to produce the exact length of sheet required.

The paper after passing the feeding rolls *c c* travels on to the slitting-knives *e*, which are circular revolving knives which slit the paper into the required width.

From the slitting-knives *e* the web travels through the drawing rolls *f f* to the revolving knife *g*, which, pressing down with sufficient sheering force against the dead knive g', cuts the web of paper crosswise into the required length of sheet.

The dimensions of the sheet of paper may be increased or diminished by changing the diameter of the expanding pulley h and the change pulley d.

After being cut, the sheets of paper fall upon the endless felt i and are carried forward to the table k, where they are arranged by boys or girls.

Cutting Water-Marked Paper.—In machines for cutting water-marked paper a greater degree of nicety in the adjustment is necessary than in the revolving cutters.

A form of single sheet paper-cutter adapted for cutting water-marked paper is shown in Fig. 176.

Fig. 176.

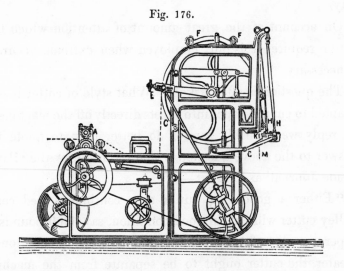

The paper after passing through the slitting-knives A, which are similar to those of the revolving cutter shown in

Fig. 175, it passes over the measuring drum C, which by a crank arrangement, D E, has imparted to it an oscillatory movement and can be adjusted to draw the exact quantity of paper forward for the length of sheet required.

The paper is made to adhere to the drum by means of the gripper rolls F F, arranged so as to rise and fall as the drum oscillates, while the dancing roll B keeps the web at a uniform tension. By means of the knife I, the paper is cut into sheets; this knife is connected with cranks and links, G, and is supported by the link rods H H working horizontally with a swinging motion against the dead knife K. At the same time clamp L holds the web in position. The sheets to be cut may be seen hanging down at the dotted lines M. The sheets are then arranged by boys or girls in the usual way.

On account of the great amount of attention which this cutter requires it is only employed when extreme accuracy is necessary.

The question is often asked, " What style of cutter is best adapted to cut the paper into sheets directly off the machine ?" In reply we probably cannot do better than to quote the answer to the above question which was made by the ' Paper Trade Journal' as follows:—

" Either a good dog cutter or a continuous feed cone-pulley cutter will cut, in good condition, say 6000 pounds of paper per day of twenty-four hours. If the production is greater, the cutter ought to be separate from the machine and the paper should be run on rolls and then carried to the cutter, and the paper from six to eight rolls should be cut

off at one time. When the machine is speeded to make
four to seven tons of paper per day, the machine tender has
not time to attend to the cutter, and, in fact, there is no
cutter made that will do the work as it ought to be done, at
the speed necessary. The best cutters will make from
twenty to twenty-four cuts per minute if they are kept in the
best possible condition, and this is as fast as girls can sort
and lay the paper properly. It is, therefore, better to put in
two cutting machines and run them slow enough to secure
good work and to cut all of the paper in daytime. If the
paper is made up into large rolls, a sufficient quantity is run
on a large reel, which is then removed from the machine
and an empty one substituted, and while it is being filled the
one just removed is run through the winder, which trims the
edges, cuts the paper in proper width, and winds it up on
wooden rolls or spools. The winder is simply the cutting
machine divested of the cross-cut knife and all superfluous
rolls. One or two shafts, as the case may be, are added
with suitable framework; the spools are put on these shafts
and held in place by collars and set screws; the shafts are
driven by friction motion of the same power that drives the
reel on the machine. The spools must be of exactly the
same diameter, since, if they are not, the paper will wind
loose and uneven on the smaller, while that being wound on
the larger, having all the strain to bear, will be likely to
break, and even when the spools are on different shafts, and
each is driven by its own friction belt, it is difficult to make
the rolls uniform. For the purpose of overcoming this
difficulty Messrs. J. and W. Jolly, of Holyoke, have built a

driver, consisting of a combination of pulleys and bevel gears which permit the use of spools of different diameters, and it is likely to prove the best means yet devised for driving the winder. Inasmuch as many of these rolls are made large and heavy, it is necessary that they should be wound solidly in order that they shall be able to withstand the inevitable rough handling which they receive in transportation from the mill to the printing house, and the rewinding they are subjected to for the purpose of wetting down previous to being printed."

Defects in Apparatus in common use for Winding the Cut Web of Paper into Rolls.—In apparatus for winding the cut web of paper into rolls there has been a great drawback, in the form of the impossibility of equalizing the speed of the several rolls in such a manner that rolls of different diameters can be revolved at different speeds corresponding to their respective diameters and still be driven from the same main shaft, so that the paper after it is cut into strips may be wound evenly upon all the rolls regardless of their diameters. If this drawback could be overcome there would be accomplished a great saving in the manufacture of paper, inasmuch as in machinery heretofore used it has been necessary to remove as much good paper from the other rolls as damaged paper from one roll in case the paper on one roll should be damaged by some accident or other. It is also desirable that a new roll may be started while the paper is still being wound upon the other rolls.

Manning's Machine.—By the use of the apparatus illustrated in Fig. 177 the inventor, Mr. John J. Manning, of

Great Barrington, Mass., claims that all the objections which have been named are overcome, and, in short, that the several rolls and their shafts may be revolved at a speed corresponding to their diameter, and that the result is performed automatically.

Fig. 177.

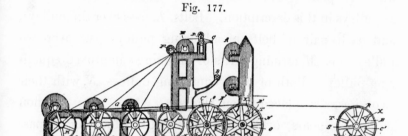

Fig. 177 is a side elevation of as much of a paper-making machine as is necessary to illustrate Mr. Manning's improvement.

A indicates the roll of paper, which is to be cut in suitable widths to be wound upon the rolls, and the web passes from the roll A under a roller or cylinder, B, then upward over a small roller, C, between the revolving shears or cutters D, which cut the web into suitable widths, and between two rollers or cylinders, E and F, from which rollers the several webs or strips of paper pass down to their respective rolls, which are wound each upon a shaft, G, journaled transversely in bearings in the frame of the machine, the bearings being of such a construction that the shafts may be lifted out of their bearings. The shafts, of which there are as many as there are rolls to be wound and webs of paper cut, have small pinions secured upon them inside of the bearings, which pinions mesh with pinions secured upon the ends of shafts which are provided with pulleys, K, or similar gears, such as cog-wheels.

In the illustration the gears K are shown as pulleys having belts passing over them, and we shall refer to the gears as pulleys in this description. Belts, L, pass over the pulleys, and each pair of belts of adjoining pulleys pass over two pulley-rims, M, forming parts of Manning's improved equalizing-pulley. Both of these equalizing pulleys, N, with their two rims, are placed upon shafts, O, which have common pulleys or gears, over which pass belts, which again pass over the two rims of an equalizing-pulley, R, upon the drive or power shaft S.

In the illustration there are shown four rolls, and consequently four shafts, and therefore two equalizing-pulleys, N, are shown, and one equalizing-pulley upon the drive-shaft, and it follows that if more than four shafts are used the number of equalizing-pulleys, N, is increased at the rate of one pulley for each pair of shafts and belts, and the number of equalizing-pulleys, R, is increased at the rate of one for each pair or less of equalizing-pulleys, N, each of which pairs of equalizing-pulleys are again provided with suitable belting, which passes to another equalizing-pulley, the pulleys having separate shafts and belt-pulleys, the number of equalizing-pulleys decreasing gradually by division with two in the number of pulleys until one equalizing-pulley is reached, which is placed upon the drive-shaft. It follows that if an uneven number of rolls and shafts is used, one of the pulleys, *N*, will only have one belt passing over it and will have the two rims revolving together, the manner of connecting the rims being described hereafter. If only two shafts and rolls are used, one equalizing-pulley only is used.

In this manner it will be seen that these pulleys serve as equalizers for the several shafts revolving in concert, automatically regulating the speed according to the tension upon the several pulleys. It follows that, although these pulleys are especially adapted to and intended for paper-cutting apparatus for paper-making machinery, the pulleys may be used in any other machinery, in which it is desirable to automatically regulate the speed of pulleys or cog-wheels according to the tension upon them. By using these pulleys any damaged part of the web winding upon one roll may be removed, and the roll started again without the necessity of removing any paper from the other rolls so as to equalize their thickness and consequently their speed of revolution, or a roll may be removed and a new roll started while the other rolls continue to wind the paper upon the large rolls already formed.

Dangoise's Machine.—There has of late years been invented a large number of machines for trimming, slitting, and rolling paper. Dangoise's machine, which the writer has seen in practical operation in Belgium, is an ingenious contrivance for trimming, slitting, and rolling paper, and we illustrate the machine in Figs. 178 and 179.

Fig. 178 is a side elevation of the machine and Fig. 179 a front view.

To bearings on the opposite side frames of the machine are adapted the journals of a drum, H, which is driven, through the medium of suitable gear-wheels, from the driving-shaft, carrying the smoothing-roller C. On the

Fig. 178.   Fig. 179.

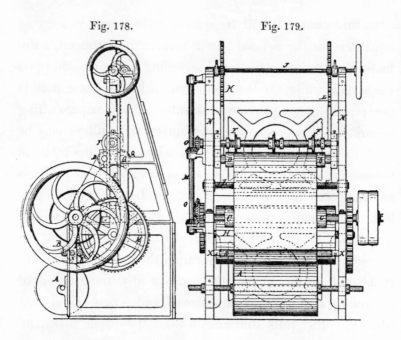

drum H rests the roller E on to which the continuous sheets of paper, after they are cut are wound, the journals of this roller being adapted to bearings arranged in guides P in the opposite side frames of the machine. To these guides are also adapted plates, which form bearings for three rollers, D, G, and F, the two former being arranged to rest on the rolls of paper, which are wound on the roller E, while the roller F carries a series of circular cutting knives, Fig. 179, arranged at points corresponding with the width of paper desired to be cut. The edges of these knives are adapted to corresponding grooves in the roller D, and on the roller G are arranged a series of blades, corresponding in position to the knives on the roller F, for a purpose explained hereafter.

The cutter-roll F is driven from a vertical shaft, M, through bevel-wheels, o o, receiving motion from the driving-shaft through suitable gearing.

The front plates, N, of the guides P are removable, so that the roll E may be withdrawn when it is desired to remove the rolls of paper from the roller E.

The plates which form the bearings for the rollers F D G have secured to them vertical racks, K L, controlled by pinions on a horizontal shaft, J, furnished with retaining-pawls and ratchets, so that the rollers may be raised to any desired height after the required amount of paper has been wound on the roller E.

The roll A of paper to be cut is arranged on the lower part of the frame, as is shown in Fig. 178, and immediately above this roll is the stretcher or tension device X, so secured to the opposite side frames that it can be adjusted to different angles to vary the tension. This stretcher carries two bars, B and B', the paper from the roll A first passing over the bar B and under the bar B', and thence over the roll C, which, being caused to rotate in a direction opposite to that pursued by the paper, smooths the paper and removes all creases therefrom. The paper then passes over the roller D and between the latter and the cutter-roll F, by which the paper is cut into a number of strips of the required width, these strips being wound by means of the drum H on to metal or wooden bobbins arranged on the roller E. The rollers D G, resting on the rolls of paper as they are wound on to the bobbins, give the required tension and pressure to the paper, while the rings or blades Q on the roller G, entering between the strips of paper as they are wound into rolls, guide the strips and insure the formation of even and compact rolls.

If desired, the cutting-knives on the roll F, except those at the ends for trimming the edges of the paper, may be dispensed with, in which case the blades Q on the roller G are so constructed as to cut the paper to the proper widths as it is wound on to the roller E, these blades thus serving the purpose of both cutting and guiding the paper.

Finishing Paper.

After being cut, and, if necessary, calendered, the paper is sorted, or, in other words, it is examined sheet by sheet, and all soiled or torn sheets are thrown out. It is next counted into quires and reams, each quire containing twenty-four sheets, and each ream twenty quires.

The appearance, and consequently the market value, of the paper when finished depends very greatly upon those who have charge of this department.

Carelessness of any kind in handling and finishing the paper should not be tolerated. In putting up the sheets into quires, half-reams, and reams, it should always be remembered that buyers pay a great deal for "fancy" in this world. A neat and attractive package of stationery will catch the eye of a purchaser much more quickly than one that is not so invitingly finished, and although the material in the latter is identically the same as in the former package, the attractively finished paper will be the more valuable for the reason that it commands a quicker sale. American paper-manufacturers usually appreciate this point, but British manufacturers, as a rule, seem to pay little attention to it.

When the paper is sent out in web too much care cannot be given to tightly reeling it and keeping the edges even, thus imparting to it a neat and finished appearance, which makes it more desirable and salable, as there results economy in both the printing and in the cutting.

List of Patents relating to Paper-making Machines, issued by the Government of the United States of America, from 1790 *to* 1885 *inclusive.*

No.	Date.	Inventor.
	Jan. 23, 1833.	W. Cole.
	Feb. 20, 1836.	C. Forbes.
1,059	Dec. 31, 1838.	J. M. Hollingsworth.
1,336	Sept. 25, 1839.	W. and A. L. Knight and E. F. Condit.
5,041	March 27, 1847.	L. W. Wright.
5,671	July 18, 1848.	G. L. Wright.
6,337	April 7, 1849.	J. M. Hollingsworth.
8,698	Jan. 27, 1852.	G. W. Turner.
10,519	Feb. 14, 1854.	S. G. Levis.
12,027 } 12,028 }	Dec. 5, 1854.	O. Marland.
13,913	Dec. 11, 1855.	C. D. Jones.
14,621	April 8, 1856.	P. H. Wait.
15,852	Oct. 7, 1856.	J. Kinsey.
16,430	Jan. 20, 1857.	J. S. Blake.
17,663	June 30, 1857.	E. M. Bingham.
17,817	Aug. 4, 1857.	P. Clark.
19,045	Jan. 5, 1858.	S. Rossman.
21,008	July 27, 1858.	F. Lindsay and W. Geddes.
21,768	Oct. 12, 1858.	J. and R. McMurray.
26,387	Oct. 12, 1858.	T. Vandeventer.
31,215	Jan. 22, 1861.	G. J. Wheeler, G. N. Dunnell, and W. Sharp.

No.	Date.	Inventor.
34,633	March 11, 1862.	J. Harper.
38,684	May 26, 1863.	J. F. Jones.
38,698	May 26, 1863.	G. E. Rutledge.
39,500	Aug. 11, 1863.	J. L. Seaverns.
41,102	Jan. 5, 1864.	G. E. Sellers.
42,854	May 24, 1864.	R. L. Howe.
42,896	May 24, 1864.	J. B. Wortendyke.
Reissues 1,817 1,818	May 24, 1864.	J. B. Wortendyke.
43,280	June 28, 1864.	F. Baker.
43,860	Aug. 16, 1864.	S. Nowlan.
44,059	Sept. 6, 1864.	A. Anderson.
45,149	Nov. 22, 1864.	E. N. Foote.
46,405	Feb. 14, 1865.	J. P. Tice.
48,347	June 20, 1865.	J. Shanlan.
50,323	Oct. 10, 1865.	S. W. Baker.
51,293	Dec. 5, 1865.	H. Chapman.
53,991	April 17, 1866.	C. Lang.
58,051	Sept. 18, 1866.	E. B. Bingham.
59,661	Nov. 13, 1866.	S. G. and G. S. Rogers.
62,958	March 19, 1867.	R. L. Howe.
63,939	April 16, 1867.	E. O. Potter.
70,534	Nov. 5, 1867.	E. Curtis.
71,108	Nov. 19, 1867.	E. Wilmot.
72,564	Dec. 24, 1867.	F. Thing.
79,659	July 7, 1868.	J. Jennings.
82,854	Oct. 6, 1868.	A. B. Lowell.
83,165	Oct. 20, 1868.	A. Howland.
83,616 83,617	Nov. 3, 1868.	E. T. Ford.
84,235	Nov. 17, 1868.	J. Viney.
85,157	Dec. 22, 1868.	J. Wrinkle.
88,035	March 23, 1869.	S. Gwynn.
89,132	April 20, 1869.	A. T. Dennison.
89,766	May 4, 1869.	C. Hofmann.
90,711	June 1, 1869.	J. W. White.
90,898	June 1, 1869.	C. B. Van Walkenburgh.
91,842	June 29, 1869.	R. C. Harris.
92,161	July 6, 1869.	W. Campbell.
92,303	July 6, 1869.	G. F. Goetzie.
92,596	July 15, 1869.	E. F. Ford.
95,153	Sept. 21, 1869.	J. P. Sherwood.

No.	Date.	Inventor.
100,755	March 15, 1870.	W. W. Harding.
101,345	March 29, 1870.	G. S. Barton.
102,265	April 26, 1870.	I. Hoffman.
102,754	May 10, 1870.	W. H. Beasdale.
104,281	June 14, 1870.	L. Dodge.
106,134	Aug. 9, 1870.	L. Dean.
106,179	Aug. 9, 1870.	C. P. Leavitt.
109,552	Nov. 22, 1870.	P. Scanlan.
111,081	Jan. 17, 1871.	C. A. Pease.
111,496	Jan. 31, 1871.	M. and A. Waissnix and C. A. Shecker.
111,751	Feb. 14, 1871.	R. A. Kelly.
112,422	March 7, 1871.	D. Crosby.
118,624	Aug. 29, 1871.	C. McBurney and L. Hollingsworth.
123,573	Feb. 13, 1872.	James F. Marshall.
124,881	March 26, 1872.	J. Burns and J. Campbell.
127,463	June 4, 1872.	L. M. Crane.
128,469	July 2, 1872.	F. Curtis.
131,103	Sept. 3, 1872.	M. J. Kearney.
131,732	Oct. 1, 1872.	C. J. Bradbury.
134,810	Jan. 14, 1873.	M. Lawler.
138,173	April 22, 1873.	W. McLaughlin.
140,418	July 1, 1873.	R. Hutton.
141,358	July 29, 1873.	N. Keely.
144,172	Sept. 11, 1873.	C. Whealan.
143,801	Oct. 21, 1873.	J. Whitehead.
144,902	Nov. 25, 1873.	D. Hamel.
146,520	Nov. 25, 1873.	B. F. Field.
149,381	April 7, 1874.	G. Dunn and R. McAlpine.
150,545	May 5, 1874.	L. A. Duckett.
152,216	June 23, 1874.	C. W. Cronk.
153,277	July 21, 1874.	B. G. Read.
155,027	Sept. 15, 1874.	R. Hutton.
158,204	Dec. 29, 1874.	B. F. Eaton.
158,400	Jan. 5, 1875.	J. Butler.
160,175	Feb. 23, 1875.	J. L. Firm.
164,468	Dec. 1, 1874.	R. McMurray.
164,814	June 22, 1875.	C. L. Crum.
166,122	July 27, 1875.	M. Matthews.
167,574	Sept. 7, 1875.	S. Sellers.
168,746	Oct. 11, 1875.	J. W. Huested.
174,369	March 7, 1876.	A. W. Keeney.
175,724	April 4, 1876.	A. McDermid.
176,344	April 18, 1876.	C. O. Perrine.

No.	Date.	Inventor.
179,161	June 27, 1876.	W. Conquest.
181,921	Sept. 5, 1876.	J. H. DeWitt.
183,112	Oct. 10, 1876.	C. T. Bischoff.
185,536	Dec. 19, 1876.	G. Howland.
194,582	Aug. 28, 1877.	L. Cole.
195,698	Oct. 2, 1877.	W. Buchanan and C. Smith.
195,821	Oct. 2, 1877.	P. W. Hudson.
196,542	Oct. 30, 1877.	D. Scrymgeour.
196,634	Oct. 30, 1877.	M. H. Cornell.
197,004	Nov. 13, 1877.	J. Bacon.
197,502	Nov. 27, 1877.	J. A. Turner and J. T. Stoneham.
199,359	Jan. 22, 1878.	J. Dunbar.
200,209	Feb. 12, 1878.	G. W. Lewthwaite.
200,309	Feb. 12, 1878.	G. F. Jones.
200,337	Feb. 12, 1878.	F. Phillips.
200,367	Feb. 12, 1878.	J. A. White.
200,369	Feb. 12, 1878.	C. Young.
201,757	March 26, 1878.	J. W. Dixon.
206,106 ⎱ 206,107 ⎰	July 16, 1878.	J. Hatch.
207,287	Aug. 20, 1878.	F. A. B. Koons.
208,792	Oct. 8, 1878.	H. Burgess.
209,003	Oct. 15, 1878.	C. Young.
210,097	Nov. 19, 1878.	O. W. Clark.
211,991	Feb. 4, 1879.	J. O. Gregg.
211,362	Feb. 18, 1879.	G. Dunn and F. Hollister.
212,485	Feb. 18, 1879.	J. W. Moose.
215,422	May 13, 1879.	G. Wilson and A. Raymond.
215,946	May 27, 1879.	J. T. F. McDonald.
216,696	June 17, 1879.	J. Peaslee.
216,914	June 24, 1879.	W. S. Tyler.
218,003	July 29, 1879.	J. Dunbar.
222,353	Dec. 9, 1879.	E. B. Hayden.
223,918	Jan. 27, 1880.	N. Kaiser.
Reissue 10,074	April 4, 1882.	
223,381	Jan. 6, 1880.	W. C. Phelps.
225,141	March 2, 1880.	J. Jordan and C. C. Markle.
225,609	March 16, 1880.	J. Jamison.
226,609	April 20, 1880.	H. Hayward.
229,636	July 6, 1880.	S. Pusey.
230,029	July 13, 1880.	A. McDermid.
231,038	Aug. 10, 1880.	J. J. Harris.

No.	Date.	Inventor.
231,169	Aug. 17, 1880.	G. Holloway.
231,579	Aug. 24, 1880.	G. Holloway.
232,031	Sept. 7, 1880.	J. H. Henry.
237,021	Jan. 25, 1881.	R. Hutton.
237,047	Jan. 25, 1881.	C. B. Rice.
239,275	March 22, 1881.	J. M. Shew.
241,522	May 17, 1881.	P. Ambjorn.
242,815	June 14, 1881.	C. W. Cronk.
246,799	Sept. 6, 1881.	C. W. Mace.
247,844	Oct. 4, 1881.	G. H. Moore.
248,001	Oct. 4, 1881.	C. C. Woolworth.
249,992	Nov. 22, 1881.	J. Randall.
252,050	Jan. 10, 1882.	D. McKay.
255,867	April 4, 1882.	R. Hutton.
256,047	April 4, 1882.	J. Randall.
258,710	May 30, 1882.	C. M. Burnett.
258,937	June 6, 1882.	M. A. Martin.
259,391	June 13, 1882.	B. A. Hickox.
260,172	June 27, 1882.	F. Curtis.
260,356	July 4, 1882.	T. P. Barry.
260,988	July 11, 1882.	W. O. Jacobs.
263,012	Aug. 22, 1882.	J. B. Bird.
266,307	Oct 24, 1882.	C. Pareus.
267,704	Nov. 21, 1882.	J. J. Manning.
268,276	Nov. 28, 1882.	R. W. Perkins.
270,718	Jan. 16, 1883.	J. Albey.
274,483	May 27, 1883.	G. Garceau.
275,056	April 3, 1883.	G. E. Marshall.
276,127	April 17, 1883.	C. Young.
280,123	June 26, 1883.	C. Batter.
280,555	July 3, 1883.	H. A. Barber.
280,564	July 3, 1883.	I. Bratton.
281,034	July 10, 1883.	L. Dejonge.
282,096	July 31, 1883.	J. J. Manning.
284,273	Sept. 4, 1883.	G. R. Caldwell.
285,838	Oct. 2, 1883.	S. Pember and S. Bird.
285,954	Oct. 2, 1883.	T. P. Barry.
286,587	Oct. 16, 1883.	H. F. Chase.
288,152	Nov. 6, 1883.	M. Sembritzki.
289,675	Dec. 4, 1883.	H. Marsden and H. Schofield.
291,406	Jan. 1, 1884.	H. Sawyer.
293,471	Feb. 12, 1884.	G. Kaffenberger.
293,785	Feb. 19, 1884.	W. H. and W. S. Ravenscroft.

37

No.	Date.	Inventor.
293,870	Feb. 19, 1884.	J. J. Harris.
296,083	April 1, 1884.	J. V. Stenger.
296,222	April 1, 1884.	B. A. Schubiger.
297,702	April 29, 1884.	G. E. Marshall.
297,775	April 29, 1884.	J. L. Firm.
298,562	May 13, 1884.	F. W. Dunnell.
298,634	May 13, 1884.	T. Stewart.
301,596	July 8, 1884.	R. W. Hopking.
301,732	July 8, 1884.	D. Lockwood.
303,404	Aug. 12, 1884.	C. Smith.
304,091	Aug. 26, 1884.	W. J. Foley.
305,615	Sept. 23, 1884.	J. J. Manning.
305,824	Sept. 30, 1884.	W. D. Kites and E. D. Fillio.
309,658	Dec. 23, 1884.	J. Sinclair.
312,314	Feb. 17, 1885.	C. Young.
313,994	March 17, 1885.	J. Crossley.
316,221	April 21, 1885.	H. A. Barber.
318,378	May 19, 1885.	W. Leishman.
319,567	June 9, 1885.	G. Dunn.
319,615 } 319,616 }	June 9, 1885.	F. C. Plume.
319,969	June 16, 1885.	M. Fitzgibbons.
320,372	June 16, 1885.	J. F. F. MacDonnell.
321,312	June 30, 1885.	W. A. Philpott, Jr.
323,079	July 28, 1885.	J. F. Seiberling.
324,601	Aug. 18, 1885.	R. Smith.
325,165	Aug. 25, 1885.	W. A. Fletcher and W. E. Keightley.
325,973	Sept. 8, 1885.	M. J. Roach.
329,610	Nov. 3, 1885.	C. Smith.

CHAPTER XVI.

THE PREPARATION OF VARIOUS KINDS OF PAPERS.

Asbestos or *amianthus paper* consists usually of two parts of paper pulp and one of amianthus. It is distinguished from ordinary paper by its color, having a yellowish tint. When burned in a flame it leaves a white residue, which, when not violently shaken, retains the form of the paper, and upon which the writing, provided ink containing sulphate of iron has been used, can be traced and deciphered with some trouble by the yellow marks left behind. Experiments in the manufacture of asbestos paper have been made in the United States, where beds of amianthus have been discovered and the price of the material is low.

The asbestos used in the United States is in part mined here, in part imported. In this country the mineral is found in very many localities, but usually in pockets or other small deposits. In most cases of occurrence the amount is not sufficient to warrant the expenditure of the capital necessary for opening the deposits; consequently the number of occurrences is far greater than that of operated mines.

The following are the leading localities at which this mineral is obtained: the towns of Brighton, Sheffield, Pelham, and Winsdor, Massachusetts; Richmond County and elsewhere in New York; near New Brunswick, New Jersey; near Media and Colerain, Pennsylvania; in the western part of Maryland; Hanover and Loudon counties, Virginia; western North Carolina; northwestern South Carolina; Rabun and Fulton counties, Georgia; Butte, Fresno, Los Angeles, Tulare, Mariposa, Placer, and Inyo counties, California. It is reported also from Dakota, Wyoming, Colorada, Utah, and Nevada. This list of occurrences might be increased indefinitely, as the mineral is by no means an uncommon one.

The annual production in 1883 and 1884 was about 1000 short tons. The price in New York ranged from $15 to $40 per ton, the price varying with the quality. The American asbestos is usually characterized by a short fibre, and by being somewhat brittle and harsh. These qualities, while unfitting it to a greater or less extent for such uses as the manufacture of rope, cloth, etc., in which a long fibre is required, do not greatly injure it for the manufacture of paper.

Imported asbestos comes mainly from the province of Quebec, Canada, and is perhaps the best for general uses. The better qualities of the Quebec asbestos bring $75 to $100 per ton in New York, while the price of the poorer grades ranges as low as $40 per ton. For the manufacture of drop-curtains for theatres, etc., Italian asbestos is principally used, as it has a long, silky, tough fibre, well fitted for the purpose. This brings in New York from $100 to $250 per ton.

A demand has lately sprung up for asbestos paper for insulation of electric wires. Sheathing paper is also largely made from asbestos.

Carbolic acid paper is prepared with $3\frac{1}{2}$ ounces of carbolic acid to the square foot. It is used for disinfecting purposes, and also for packing fresh meat. The process of preparing it is as follows: Melt at a moderate heat 5 parts of stearine, 6 of paraffine, and 2 of carbolic acid. Apply the melted mixture to the paper with a brush.

A still more effective paper, and which can be used for a great many purposes, is obtained by the use of a smaller quantity of nitric acid in place of carbolic acid, the remainder of the process being the same.

Improved Cigarette Paper.—Tobacco leaves are ground to an impalpable powder which is sifted in a box upon a moistened sheet of cigarette paper. The sheet thus prepared is covered with another sheet and brought under a press. Other sheets treated in the same manner are placed upon these and the whole finally subjected to strong pressure, whereby the tobacco-powder is intimately united with the moist paper. After remaining in the press for 12 to 24 hours the paper is removed and is ready for use. By a suitable mixture the color, flavor, and smell of the various kinds of tobacco can be successfully imitated. Paper thus prepared burns uniformly, never on one side only, and does not char.

Colored Paper for Tying up Bottles, etc.—The dry aniline colors of all shades are used. Dissolve 15 grains of aniline color in 1 ounce of highly rectified alcohol, dilute the solution with 10 ounces of distilled water, and add 23 grains of tannin dissolved in $\frac{1}{2}$ fluid ounce of alcohol. The object of the addition of tannin is to fix the color permanently upon the fibres of the paper, as without it the color on drying could be easily rubbed off. Now take thin white writing-paper, spread it upon a marble or copper plate, and apply the fluid by means of a sponge. Hang the paper over a cord to dry, and in a few days varnish it with a concentrated solution of sodium water-glass to 100 parts of which have been added 10 parts of glycerine.

Cork paper, patented in America by H. Felt & Co., is prepared by coating one side of a thick, soft, and flexible paper with a preparation of 20 parts of glue, 1 of gelatine, and 3 of molasses, and covering it with fine particles of cork lightly rolled on. The material is used for packing glass, bottles, etc.

Electro-chemical Telegraph Paper, Pouget-Maisonneuve's. —Sufficiently sized paper is treated with a solution of 5 parts of ferrocyanide of potassium and 150 of salammoniac in 100 of water. Telegrams received by means of this paper and Morse's apparatus, before the present system of receiving by sound was introduced, gave very satisfactory results.

Emery Paper.—Fig. 180 represents Edwards's patented apparatus used in the manufacture of emery, sand, glass, and similar papers. *a* is the beam on which the endless paper is rolled. In unrolling it passes over the brush-roller *i*, which takes up the glue from the boiler *h* and applies it to the paper. The boiler containing the glue is constructed of copper or iron, and surrounded with a steam-jacket. The small rollers *k* and *n* act as distributers, both being turned by friction with *b*. As soon as the paper reaches the even plane from *b* to *c* the glue upon it is heated by steam emanating from the apparatus d, and a fine jet of the material, emery, glass, sand, etc., falls from e upon the surface thus heated. The powder penetrates deeply into the soft, sticky mass, and adheres quickly.

Fig. 180.

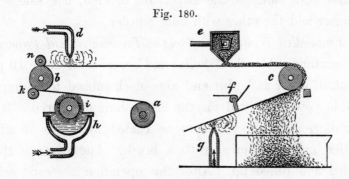

The excess falls off by the paper turning over c, and is collected in a box. The powder in e is heated by a steam-pipe. The fan f sets the paper in motion, whereby all the powder not adhering tightly is shaken off. A jet of steam striking the surface of the paper through g helps to set the powder more securely in the glue.

Water-proof Emery Paper.—The paper is coated on both sides with pulverized emery which is made to adhere to it by means of a water-proof cement, so that moisture can have no injurious effect upon the paper. This flexible water-proof cement is prepared by melting 2 parts of hard, African copal, pouring over this, while yet hot, 3 of boiled linseed oil and adding one part of oil-lacquer, 1 of Venetian turpentine, 1 of Venetian red, $\frac{1}{8}$ of Berlin blue, $\frac{1}{8}$ of litharge, and 1 of dissolved caoutchouc. Mix these ingredients intimately, and should the compound be too thick dilute with some linseed-oil varnish. Then spread it uniformly upon paper, or a suitable cheap fabric, stretched in a frame, and sift finely pulverized emery, or glass, quartz sand, etc., over it; and, when dry, remove the excess of powder. Sometimes both sides of the paper are covered, one side with coarser and the other with finer powder.

Enamelled Writing Surfaces on Pasteboard and Paper.—A mixture of bleached shellac and borax dissolved in 10 per cent. of water and glue and vine-black rubbed to an impalpable powder is used for the first coloring material. It is transferred to the paper to be coated by means of a felt roller, and distributed with a brush. The paper is then dried and rolled up. After this operation a second color consisting of vine-black, pergamentine (water-glass and glycerine) is used, the paper receiving three coats of this. It is then cut into suitable sizes, steamed at a temperature of 248° F., and finally smoothed by calendering. For white tablets Kremnitz white is used in place of vine-black; for colored, ultramarine, etc.

Iridescent Paper.—Boil $4\frac{1}{2}$ ounces of coarsely powdered gall-nuts, $2\frac{3}{4}$ ounces of sulphate of iron, $\frac{1}{2}$ ounce of sulphate of indigo, and 12 grains of gum-Arabic; strain through a cloth, brush the paper with the liquor, and expose it quickly to ammoniacal vapors.

Imitation of Mother-of-Pearl on Paper.—Stout paper with a glossy coating is allowed to float upon a solution of salts of silver, lead, or bismuth. As soon as the paper lies smooth upon the surface of the solution it is slowly lifted and allowed to dry. The dry paper is then placed in a room impregnated with sulphide of hydrogen, and remains here until the surface has assumed a metallic lustre. Diluted collodion is now poured over the paper thus prepared, or it is drawn through a bath of it, when, after drying, the beautiful iridescent colors will appear upon the paper. The most varying effects can be produced by sprinkling reducing substances or salts upon the surface of the paper before submitting it to the action of the sulphide of hydrogen.

This process is not only adapted for paper but can also be employed for finished articles, as boxes, bonbonnières, etc.

Leather Waste—How Prepared for Use in the Manufacture of Paper.—To extract the tannin place the waste for a few hours in a solution of 5 parts of lime, 5 of crystallized soda, and $1\frac{1}{2}$ of salammoniac in 100 of water; then wash first with acidulated and next with pure water. The prepared waste is worked into paper in the ordinary manner, either by itself or mixed with rags.

Photo-lithographic Transfer Paper, and Transfer-color belonging to it.—Paper is treated with a solution of 100 parts of gelatine and 1 of chrome-alum in 2400 of water, and, after drying, with white of egg. It is sensitized in a bath consisting of 1 part of chrome alum, 14 of water, and 4 of alcohol. The addition of the latter prevents the solution of the white of egg. On the places not exposed to the light the white of egg becomes detached, together with the color with which the exposed paper has been coated. The transfer color consists of 20 parts of printing ink, 50 of wax, 40 of tallow, 35 of resin, 210 of oil of turpentine, and 30 of Berlin blue.

Preserving Papers.—Two new varieties of preserving papers have been recently brought into the market. The one is obtained by immersing soft paper in a bath of salicylic acid, and then drying in the air. The bath is prepared by diluting a strong solution of the acid in alcohol with a large volume of water. This paper may then be used for wrapping up apples, etc.

The other paper used as protection against moths and mildew is best prepared from strong Manilla paper by immersing it in the following bath: Seventy parts of tar oil, 5 of crude carbolic acid containing about ½ phenol, 20 of coal-tar at a temperature of 160° F., and 5 of refined petroleum. The paper is then squeezed out, and dried by passing it over hot rollers.

Tar Paper.—Boil 100 pounds of tar for 3 hours, then dissolve in it a quantity of a glue prepared from resin and soap, pour 8 gallons of boiling water upon the mixture, stir carefully, and let the mixture boil. Then stir carefully 100 pounds of potato flour into 60 gallons of water in a vat, mix the dissolved tar with 15 gallons of boiling water, and add this to the potato flour in the vat, stirring constantly. Twenty-four parts of this homogeneous fluid are taken to 20 parts of paper-pulp. From the pulp the tar-paper is manufactured, which can be painted black and varnished to make it water-proof. The prepared tar-solution may also be used to impregnate wood, sail-cloth, etc.

Tracing Paper, Tracing Linen, and Transparent Packing Paper.—The paper is first treated with boiled linseed oil, and the excess of oily particles removed with benzine. The paper is then washed in a chlorine bath.   When dry it is again washed with oxygenated water.

Paper can be made transparent by applying a thin coating of a solution of Canada balsam in turpentine to the paper, then give it a good coating of much thicker varnish on both sides.   Perform the work before a hot fire, to keep the paper warm, and a third or even fourth coating until the paper becomes evenly translucent.   Paper prepared in this manner comes nearer to perfection than any other.

By the following very simple process ordinary drawing paper can be rendered transparent, for the purpose of making tracings, and its transparency removed so as to restore its former appearance when the drawing is completed. Dissolve any quantity of castor oil in one, two, or three volumes of absolute alcohol, according to the thickness of the paper, and apply it by means of a sponge.   The alcohol evaporates in a few minutes, and the tracing paper is dry and ready for immediate use.   The drawing or tracing can be made either with lead-pencil or India ink, and the oil removed from the paper by immersing it in absolute alcohol, thus restoring its original opacity.   The alcohol employed in removing the oil is, of course, preserved for diluting the oil used in preparing the next sheet.

Linen, to prepare it for being used for tracing purposes, is first provided with a coating of starch and then with an application of linseed oil and benzine.   It is finished by being smoothed between polished rollers.

Transfer Paper.—Mix lard to a paste with lampblack, rub this upon the paper, remove the excess with a rag, and dry the paper. A copy of the writing can be transferred on a clean sheet of paper by placing it underneath the prepared paper and writing upon the latter with a lead-pencil or sharp point.

Water-proof paper, transparent and impervious to grease, is obtained by soaking good paper in an aqueous solution of shellac and borax. It resembles parchment paper in some respects. If the aqueous solution be colored with aniline colors very handsome paper, of use for artificial flowers, is prepared.

Peterson's Water-proof Paper.—Dissolve $3\frac{1}{2}$ ounces of tallow soap in water, add sufficient solution of alum until the soap is entirely decomposed, and mix this fluid with a gallon of paper-pulp. The paper is in all other respects prepared in the ordinary manner, and need not to be sized. It is especially suitable for cartridge-shells.

Wrapping Paper for Silver Ware.—The appearance of silver ware is frequently injured by being exposed to air containing sulphuretted hydrogen or sulphurous and other acids. The small quantity of sulphuretted hydrogen contained in illuminating gas and which in burning yields sulphurous acid is frequently sufficient to spoil the appearance of all the articles in a store. To prevent this a prepared paper is recommended. Prepare a solution of 6 parts of caustic soda in water of 20° Baumé, then add 4 of zinc oxide and let the mixture boil for two hours, if possible under a pressure of 5 atmospheres. Dilute the solution, when clear, to 10° Baumé, and it is ready for impregnating the paper.

Writing, Copying, and Drawing Paper which can be Washed.—The paper is made transparent by immersion in benzine and then, before the benzine volatilizes, plunged into a solution of siccative prepared in the following manner: One pound each of lead shavings and oxide of zinc are boiled for 8 hours, together with $8\frac{3}{4}$ ounces of hardened Venetian turpentine in $2\frac{1}{2}$ gallons of purified linseed-oil varnish, and then allowed to stand for a few days to cool and settle. The clear layer is then poured off and to this are added 5 pounds of white West Indian copal and $8\frac{3}{4}$ to 10 ounces of sandarac dissolved in spirit of wine or ether. This paper can be written or drawn upon with pen and ink or water colors; or, by using good copying ink, good copies can be taken from it without a press.

INDEX.

38

TECHNOLOGY AND SOCIETY

An Arno Press Collection

Ardrey, R[obert] L. **American Agricultural Implements.** In two parts. 1894

Arnold, Horace Lucien and Fay Leone Faurote. **Ford Methods and the Ford Shops.** 1915

Baron, Stanley [Wade]. **Brewed in America:** A History of Beer and Ale in the United States. 1962

Bathe, Greville and Dorothy. **Oliver Evans:** A Chronicle of Early American Engineering. 1935

Bendure, Zelma and Gladys Pfeiffer. **America's Fabrics:** Origin and History, Manufacture, Characteristics and Uses. 1946

Bichowsky, F. Russell. **Industrial Research.** 1942

Bigelow, Jacob. **The Useful Arts:** Considered in Connexion with the Applications of Science. 1840. Two volumes in one

Birkmire, William H. **Skeleton Construction in Buildings.** 1894

Boyd, T[homas] A[lvin]. **Professional Amateur:** The Biography of Charles Franklin Kettering. 1957

Bright, Arthur A[aron], Jr. **The Electric-Lamp Industry:** Technological Change and Economic Development from 1800 to 1947. 1949

Bruce, Alfred and Harold Sandbank. **The History of Prefabrication.** 1943

Carr, Charles C[arl]. **Alcoa, An American Enterprise.** 1952

Cooley, Mortimer E. **Scientific Blacksmith.** 1947

Davis, Charles Thomas. **The Manufacture of Paper.** 1886

Deane, Samuel. **The New-England Farmer,** or Georgical Dictionary. 1822

Dyer, Henry. **The Evolution of Industry.** 1895

Epstein, Ralph C. **The Automobile Industry:** Its Economic and Commercial Development. 1928

Ericsson, Henry. **Sixty Years a Builder:** The Autobiography of Henry Ericsson. 1942

Evans, Oliver. **The Young Mill-Wright and Miller's Guide.** 1850

Ewbank, Thomas. **A Descriptive and Historical Account of Hydraulic and Other Machines for Raising Water,** Ancient and Modern. 1842

Field, Henry M. **The Story of the Atlantic Telegraph.** 1893

Fleming, A. P. M. **Industrial Research in the United States of America.** 1917

Van Gelder, Arthur Pine and Hugo Schlatter. **History of the Explosives Industry in America.** 1927

Hall, Courtney Robert. **History of American Industrial Science.** 1954

Hungerford, Edward. **The Story of Public Utilities.** 1928

Hungerford, Edward. **The Story of the Baltimore and Ohio Railroad, 1827-1927.** 1928

Husband, Joseph. **The Story of the Pullman Car.** 1917

Ingels, Margaret. **Willis Haviland Carrier, Father of Air Conditioning.** 1952

Kingsbury, J[ohn] E. **The Telephone and Telephone Exchanges:** Their Invention and Development. 1915

Labatut, Jean and Wheaton J. Lane, eds. **Highways in Our National Life:** A Symposium. 1950

Lathrop, William G[ilbert]. **The Brass Industry in the United States.** 1926

Lesley, Robert W., John B. Lober and George S. Bartlett. **History of the Portland Cement Industry in the United States.** 1924

Marcosson, Isaac F. **Wherever Men Trade:** The Romance of the Cash Register. 1945

Miles, Henry A[dolphus]. **Lowell, As It Was, and As It Is**. 1845

Morison, George S. **The New Epoch:** As Developed by the Manufacture of Power. 1903

Olmsted, Denison. **Memoir of Eli Whitney, Esq.** 1846

Passer, Harold C. **The Electrical Manufacturers, 1875-1900.** 1953

Prescott, George B[artlett]. **Bell's Electric Speaking Telephone.** 1884

Prout, Henry G. **A Life of George Westinghouse.** 1921

Randall, Frank A. **History of the Development of Building Construction in Chicago.** 1949

Riley, John J. **A History of the American Soft Drink Industry:** Bottled Carbonated Beverages, 1807-1957. 1958

Salem, F[rederick] W[illiam]. **Beer, Its History and Its Economic Value as a National Beverage.** 1880

Smith, Edgar F. **Chemistry in America.** 1914

Steinman, D[avid] B[arnard]. **The Builders of the Bridge:** The Story of John Roebling and His Son. 1950

Taylor, F[rank] Sherwood. **A History of Industrial Chemistry.** 1957

Technological Trends and National Policy, Including the Social Implications of New Inventions. Report of the Subcommittee on Technology to the National Resources Committee. 1937

Thompson, John S. **History of Composing Machines.** 1904

Thompson, Robert Luther. **Wiring a Continent:** The History of the Telegraph Industry in the United States, 1832-1866. 1947

Tilley, Nannie May. **The Bright-Tobacco Industry, 1860-1929.** 1948

Tooker, Elva. **Nathan Trotter:** Philadelphia Merchant, 1787-1853. 1955

Turck, J. A. V. **Origin of Modern Calculating Machines.** 1921

Tyler, David Budlong. **Steam Conquers the Atlantic.** 1939

Wheeler, Gervase. **Homes for the People,** In Suburb and Country. 1855